Information Systems Engineering and Management

Volume 24

The book series "Information Systems Engineering and Management" (ISEM) publishes innovative and original works in the various areas of planning, development, implementation, and management of information systems and technologies by enterprises, citizens, and society for the improvement of the socio-economic environment.

The series is multidisciplinary, focusing on technological, organizational, and social domains of information systems engineering and management. Manuscripts published in this book series focus on relevant problems and research in the planning, analysis, design, implementation, exploration, and management of all types of information systems and technologies. The series contains monographs, lecture notes, edited volumes, pedagogical and technical books as well as proceedings volumes.

Some topics/keywords to be considered in the ISEM book series are, but not limited to: Information Systems Planning; Information Systems Development; Exploration of Information Systems; Management of Information Systems; Blockchain Technology; Cloud Computing; Artificial Intelligence (AI) and Machine Learning; Big Data Analytics; Multimedia Systems; Computer Networks, Mobility and Pervasive Systems; IT Security, Ethics and Privacy; Cybersecurity; Digital Platforms and Services; Requirements Engineering; Software Engineering; Process and Knowledge Engineering; Security and Privacy Engineering, Autonomous Robotics; Human-Computer Interaction; Marketing and Information; Tourism and Information; Finance and Value; Decisions and Risk; Innovation and Projects; Strategy and People.

Indexed by Google Scholar. All books published in the series are submitted for consideration in the Web of Science.

For book or proceedings proposals please contact Alvaro Rocha (amrrocha@gmail.com).

Narasimha Rao Vajjhala · Sanjiban Sekhar Roy ·
Burak Taşcı ·
Muhammad Enamul Hoque Chowdhury
Editors

Generative Artificial Intelligence (AI) Approaches for Industrial Applications

Springer

Editors
Narasimha Rao Vajjhala
American University in Bulgaria
Blagoevgrad, Bulgaria

Burak Taşcı
School of Technical Sciences
Firat University
Merkezkoeyler, Elazig, Türkiye

Sanjiban Sekhar Roy
School of Computer Science
and Engineering
Vellore Institute of Technology
Vellore, Tamil Nadu, India

Muhammad Enamul Hoque Chowdhury
Department of Electrical Engineering
Qatar University
Doha, Qatar

ISSN 3004-958X ISSN 3004-9598 (electronic)
Information Systems Engineering and Management
ISBN 978-3-031-76709-8 ISBN 978-3-031-76710-4 (eBook)
https://doi.org/10.1007/978-3-031-76710-4

This Springer imprint is published by the registered company Springer Nature Switzerland AG
The registered company address is: Gewerbestrasse 11, 6330 Cham, Switzerland

Foreword

In the fast-changing world of artificial intelligence, Generative AI and Large Language Models (LLMs) are set to transform our society in significant ways. This book, *Generative Artificial Intelligence (AI) Approaches for Industrial Applications*, explores the impact of generative AI across different sectors. Generative AI, a part of artificial intelligence focused on creating new content, is changing what machines can do. The rise of powerful LLMs like ChatGPT and BARD has highlighted their potential to change communication, creativity, and problem-solving in many areas. The importance of generative AI is in its ability to mimic human creativity, allowing machines to create text, images, music, and even complex designs. This opens new paths for innovation, helping industries improve productivity, customize user experiences, and streamline processes. In health care, generative AI is improving drug discovery, personalized treatment plans, and medical imaging, which leads to better patient outcomes and faster research. In manufacturing, AI-driven predictive maintenance, automated design, and process optimization are boosting efficiency and saving costs.

The creative industries are also changing as generative AI helps artists, musicians, and filmmakers explore new creative possibilities. The ability to generate high-quality content on its own is not only enhancing human creativity but also making creative tools more accessible. However, the power of generative AI also brings ethical, social, and practical challenges. Issues related to data privacy, algorithmic bias, and the ethical use of autonomous decision-making must be addressed to ensure these technologies are used responsibly.

This book offers a detailed look at the uses, challenges, and future of generative AI, making it a valuable resource for academics, professionals, and researchers. Through technical discussions, case studies, and real-world examples, it shows how generative AI can change operations across different fields.

As we approach this technological revolution, it is important to understand the wide-reaching effects of generative AI and LLMs. By using their potential responsibly, we can create a new era of innovation that benefits everyone.

I commend the authors for their effort in creating this important work, and I believe that it will guide those exploring the world of generative AI. Finally, I congratulate the editors for their commendable work in bringing forth this much-needed book.

Prof. Pijush Samui
Dean (P&D)
NIT Patna
Patna, India

Guest professor, University of Science
and Technology Beijing
Beijing, China

Title of Docent
Tampere University
Tampere, Finland

Preface

The emergence of Generative Artificial Intelligence (AI) has significantly transformed various industries, driving innovation, efficiency, and scalability across multiple domains. As industries seek to enhance productivity, streamline operations, and innovate their business models, generative AI offers powerful tools that enable organizations to create new products, optimize processes, and improve decision-making. This book, *Generative Artificial Intelligence (AI) Approaches for Industrial Applications*, explores the diverse applications of generative AI in industrial contexts, showcasing how advanced AI techniques are shaping the future of industries ranging from manufacturing to health care. Each chapter examines a specific area of industrial application, providing a comprehensive understanding of the methodologies, technologies, and potential challenges associated with implementing generative AI in various sectors.

This book is a testament to the power of international collaboration, bringing together the expertise of 36 authors from 13 different countries. These contributors represent diverse academic and professional backgrounds, spanning regions from India, the USA, and Albania to Brunei Darussalam, Kosovo, and beyond. Their collective efforts demonstrate the global commitment to advancing knowledge in the field, promoting cross-cultural dialogue, and promoting innovative solutions to the challenges we face today. This international collaboration underscores the importance of shared insights and cooperation in the pursuit of excellence. The book is structured into 16 chapters, each offering unique insights into generative AI's transformative role across industries. The chapters cover foundational concepts, real-world case studies, and cutting-edge research, making it a valuable resource for academics, professionals, and researchers. The opening chapter introduces the fundamental principles of generative AI and its historical development, laying the groundwork for the subsequent chapters that explore its application in manufacturing, health care, supply chain management, and beyond. The following chapters dive deep into the integration of AI with industrial robotics, the creation of AI-driven predictive models, and the use of generative AI in product design and prototyping. By combining theoretical analysis with practical applications, this book offers readers a comprehensive understanding of how generative AI is shaping industrial landscapes and enabling future advancements.

The first chapter, "Foundations and Emerging Trends in Generative Artificial Intelligence (AI) for Industrial Applications", sets the stage by providing a comprehensive overview of the fundamental concepts, challenges, and future directions of generative AI. This chapter begins with a historical perspective on the evolution of generative AI, starting from the early days of machine learning to the advent of advanced models like Generative Adversarial Networks (GANs), Variational Autoencoders (VAEs), and transformers. The chapter explores the mathematical foundations that underpin these models, such as probability theory, optimization algorithms, and neural networks, giving readers a clear understanding of how generative AI has developed into a critical tool for solving complex problems across industries. The discussion also touches on the technical challenges associated with generative AI, such as training instability and computational costs, offering insight into the trade-offs and limitations of different models. This chapter examines the practical applications of generative AI across industries such as health care, manufacturing, and finance. This chapter highlights how these models are transforming sectors by optimizing product design, enhancing predictive maintenance, and generating synthetic data for more effective decision-making. Additionally, the chapter addresses ethical concerns, including data privacy, security risks, and potential misuse of AI-generated content, making a case for the responsible deployment of AI technologies. Concluding with an exploration of future trends and opportunities, the chapter outlines how advancements in personalized AI systems and scalable models could revolutionize industries, while also emphasizing the need for regulatory frameworks to ensure ethical use and data protection.

The second chapter, titled "Mathematical Foundations and Applications of Generative AI Models", serves as a crucial foundation for understanding the intricate mathematical principles that underpin generative AI. This chapter begins by introducing core concepts in probability theory, such as probability distributions, conditional probability, and Bayes' theorem, which are essential for developing generative models. The chapter then proceeds to explore probabilistic graphical models like Markov random fields and Bayesian networks, which structure the relationships between variables in a dataset. These models lay the groundwork for deep generative models, which use deep neural networks to generate new data samples that closely resemble the input data. By discussing techniques such as loss functions, optimization strategies, and the mathematics behind neural networks, the chapter sets a firm theoretical base for those looking to design innovative generative models in diverse applications. The chapter goes further by explaining how mathematical concepts like linear algebra and calculus are applied to train and refine deep neural networks. The focus shifts to deep generative models, including the application of Generative Adversarial Networks (GANs), which utilize adversarial training techniques to produce highly realistic outputs. Additionally, this chapter covers the utility of generative models in areas such as image synthesis, text generation, and drug discovery. By connecting these mathematical principles to real-world industrial applications, the chapter equips readers with the tools needed to explore the vast potential of generative AI models, whether in product design, optimization, or creative industries.

The third chapter, "Mathematical Frameworks for Generative AI Applications", explores the essential mathematical principles that drive the development and functionality of generative AI systems. This chapter explores the theoretical foundations that support various generative AI models, focusing on key concepts such as probability distributions, optimization algorithms, and deep learning architectures. This chapter introduces discriminant models, probabilistic discriminant models, and probabilistic generative models, illustrating how each model type applies different strategies to solve complex classification and generation tasks. The chapter also highlights the mathematical processes that underpin neural network architectures such as transformers, GANs, and diffusion models, which have transformed industries like health care, computer vision, and natural language processing. In addition to covering foundational concepts, the chapter emphasizes the practical applications of these mathematical models in real-world scenarios. From enhancing image generation in fields such as gaming and filmmaking to improving natural language processing tools like chatbots and content generators, the discussed frameworks provide the tools necessary for the development of innovative solutions across a wide range of industries. The chapter serves as a crucial resource for readers looking to understand the deeper mathematical underpinnings of generative AI and offers insights into how these models contribute to advancing both theoretical research and practical applications in AI-driven industries.

The fourth chapter, titled "Generative Adversarial Networks: A Comprehensive Review", offers an in-depth examination of GANs, which have revolutionized artificial intelligence with their ability to generate synthetic data that closely mimics real-world data. This chapter begins by outlining the fundamental structure of GANs, focusing on the competitive relationship between the generator and discriminator, and explores how this dynamic has enabled advancements in various fields. The review not only covers the basics of GANs but also provides a detailed breakdown of different GAN architectures, including Conditional GANs, Wasserstein GANs, and Progressive GANs, each with their specific use cases such as image generation, data augmentation, and style transfer. Moreover, the chapter addresses the practical applications and challenges associated with using GANs in diverse industries. This chapter highlights the potential of GANs in fields such as health care, finance, and entertainment while also tackling some of the key obstacles, such as training instability, computational demands, and biases in generated data. By examining the advantages and disadvantages of various GAN models, as well as the open research problems in the field, this chapter equips researchers and practitioners with actionable insights to advance GAN research and implementation. This comprehensive review sets the stage for the future development of GANs, guiding researchers toward solving the remaining challenges and expanding the utility of GANs across sectors.

The fifth chapter, "Generative Adversarial Networks: Security, Privacy, and Ethical Considerations", addresses the critical challenges that accompany the rise of GANs, particularly in the areas of data privacy, security, and ethical dilemmas. As GANs become increasingly integrated into various industries, the chapter explores the pressing need to manage their vulnerabilities, such as the creation of deepfakes, synthetic misinformation, and potential misuse in cybersecurity breaches. This chapter begins by reviewing the core principles of GANs and their impact on the generation of synthetic content, highlighting the necessity for responsible use. This chapter discusses various ethical frameworks and philosophical approaches to ensure that GANs are employed ethically across domains like health care, cybersecurity, and media production, where their applications have both positive and negative implications. Additionally, the chapter examines the real-world examples of how GANs are reshaping industries, particularly the risks involved in manipulating data and content. With a focus on privacy concerns, it addresses how GANs might challenge traditional data security measures, especially in the context of biometric data and personal identity theft. Security challenges are explored in detail, with strategies proposed to fortify these AI models against adversarial threats. The chapter concludes with a discussion on the evolving legal and regulatory frameworks needed to protect individuals and organizations from the misuse of GANs, emphasizing the importance of striking a balance between innovation and safeguarding public interests.

The sixth chapter, "Mitigating Hallucinations in LLMs Using Sieve of Fallacies and Truths (SoFT): A Game Theoretic Perspective", explores the increasingly critical issue of hallucinations in LLMs, where these models generate plausible-sounding but factually incorrect or fabricated information. This chapter introduces a novel approach to mitigate such hallucinations by utilizing the Sieve of Fallacies & Truths (SoFT) model, which is based on game theory principles. The authors propose that by integrating SoFT into LLM training processes, it is possible to reinforce the generation of factually accurate content while penalizing false outputs, improving the reliability and accuracy of LLM outputs. This method provides a significant advancement over existing approaches like Retrieval-Augmented Generation (RAG), as it focuses on enhancing the model's intrinsic ability to produce verifiable information without depending heavily on external datasets. Additionally, the chapter discusses the implementation of SoFT within LLMs by operating on their internal word representations and context processing mechanisms. This chapter explains how the game-theoretic framework encourages truthful outputs during the training phase, leading to more reliable performance in real-world applications, particularly in domains like health care, legal contexts, and education, where factual accuracy is paramount. By addressing the problem of hallucinations at a foundational level, the chapter sets the stage for scalable and efficient LLM deployment in knowledge-intensive tasks, reducing the need for continuous oversight or external fact-checking. This approach not only enhances the factual grounding of LLMs but also presents an innovative way to increase their utility across various sectors while minimizing risks associated with misinformation.

The seventh chapter, "Implementing Generative AI in Identity Access Management", explores the significant impact of Generative AI on improving security protocols and enhancing identity verification processes within IAM systems. The chapter begins by introducing Identity Access Management (IAM) as a critical security framework that regulates access to organizational systems and data. This chapter discusses how the rising complexity of digital infrastructures and cyber threats has necessitated the integration of advanced technologies like Generative AI. By simulating user behavior and predicting potential threats, Generative AI enables IAM systems to adapt dynamically to evolving security challenges. Through applications such as fraud detection, anomaly detection, and user behavior simulation, Generative AI models bolster the accuracy of identity verification and contribute to more secure and reliable access management solutions. In addition, the chapter examines specific Generative AI models such as Generative Adversarial Networks (GANs) and Variational Autoencoders (VAEs) that enhance IAM capabilities. These models can simulate complex scenarios, predict identity-related fraud, and detect irregularities in user behavior, all of which contribute to more robust security measures. Through real-world case studies, the chapter illustrates how Generative AI can be successfully implemented within various industries to secure digital assets, improve compliance with regulatory frameworks, and ensure scalable security protocols. The integration of these technologies not only enhances security but also streamlines administrative tasks, paving the way for more efficient and resilient IAM systems in an increasingly digitized world.

The eighth chapter, "Integrating Generative AI in Education: Themes, Challenges, and Future Directions", explores the transformative impact of generative AI on the educational landscape. This chapter offers a systematic analysis of 31 studies, identifying key themes that highlight the integration of generative AI into teaching, learning, and research. It emphasizes personalized learning, curriculum design innovation, and enhanced student engagement as the primary benefits of generative AI in education. Furthermore, it discusses how AI technologies are being used to tailor educational experiences, enabling personalized learning paths and improving overall academic performance. By providing real-time feedback and adaptive learning environments, generative AI is reshaping the way educators and students interact, creating more dynamic and flexible teaching methodologies. However, the chapter also acknowledges the significant challenges that come with integrating AI into higher education. These include ethical concerns, data privacy, technological infrastructure, and the need for faculty and student adaptation. As generative AI becomes more prevalent in educational institutions, ensuring that these technologies are used responsibly and equitably is essential. The chapter concludes with a forward-looking analysis of future research directions, advocating for interdisciplinary collaboration, continuous assessment of AI tools, and the development of robust ethical frameworks. These recommendations aim to help educators, policymakers, and researchers maximize the benefits of generative AI while addressing its inherent challenges in educational settings.

The ninth chapter, titled "The Impact of Generative AI on Healthcare", examines how generative AI is transforming the healthcare industry by enhancing drug discovery, personalized treatment planning, medical imaging, and predictive health analytics. The chapter begins by exploring the revolution in drug discovery, where AI-driven models speed up identifying and optimizing novel drug candidates, significantly reducing time and costs compared to traditional methods. The authors discuss how pharmaceutical companies utilize generative AI to repurpose existing drugs, design innovative treatments, and improve drug safety and efficacy. Through machine learning algorithms, generative AI enables faster molecule generation, which enhances therapeutic effectiveness and shortens the time needed to bring new drugs to market. In addition to drug discovery, this chapter examines the role of generative AI in personalized treatment planning, medical imaging, and diagnosis. Generative AI offers groundbreaking tools for customizing therapeutic strategies based on patient-specific data, enhancing treatment outcomes. This chapter also plays a pivotal role in medical imaging by leveraging technologies such as GANs to improve the accuracy of diagnostic processes and early disease detection. The chapter further explores the application of predictive health analytics, where AI models can anticipate disease progression, assess risks, and enable timely interventions. This comprehensive discussion highlights the potential of generative AI to revolutionize healthcare delivery, paving the way for more efficient, personalized, and preventive healthcare solutions.

The tenth chapter, "Revolutionizing Healthcare with Generative Artificial Intelligence Technologies", provides a comprehensive examination of how generative AI is reshaping the healthcare industry. This chapter examines the key applications of generative AI, such as its role in medical imaging, pandemic prediction, synthetic data generation, clinical administration support, and patient engagement. This chapter emphasizes the potential of AI tools like ChatGPT4 and Google DeepMind Gemini in assisting healthcare professionals by automating tasks, improving diagnostic accuracy, and enhancing patient interaction. These AI models enable healthcare providers to process vast amounts of data efficiently, leading to faster, more accurate medical decisions and optimized patient care, from routine clinical administration to complex diagnostic processes. Furthermore, the chapter explores the challenges associated with integrating generative AI into health care, such as data privacy, security, and ethical concerns. Despite these hurdles, generative AI offers significant promise, particularly in areas like generating synthetic medical data to improve research capabilities and reduce the risk of patient data exposure. The chapter also discusses real-world case studies, providing practical insights into how AI is currently being applied in medical contexts and offering a roadmap for future developments. With its deep analysis and forward-looking perspective, this chapter is essential for understanding how generative AI is poised to revolutionize health care on multiple fronts.

The eleventh chapter, "Identification of Face Age Progression and Rejuvenation Using Generative Adversarial Networks", explores the application of GANs in the complex task of synthesizing human faces across different ages. This chapter introduces the concept of face age synthesis, where the appearance of an individual is altered to predict how their face will look in the future or how it may have looked in the past. With practical applications in areas such as law enforcement and efforts to track victims of human trafficking, the chapter focuses on the use of Conditional Variational Autoencoder Generative Adversarial Networks (CVAE-GANs). The encoder–decoder model generates facial images from late teens to older adults, with the discriminator ensuring the authenticity of these images. The results are evaluated using various performance metrics, demonstrating the effectiveness of this technique in solving age progression and rejuvenation challenges. This chapter also dives into the broader implications of this technology, particularly in law enforcement and missing person cases, where projecting an individual's future appearance can be crucial. The authors compare their CVAE-GAN approach with other GAN-based models, emphasizing improvements in image quality and age-related transformations. The use of the Cross-Age Celebrity Dataset (CACD) for model training further enhances the chapter's credibility, providing readers with insights into how age estimation errors, identity preservation, and image similarity metrics are evaluated. This chapter provides a valuable contribution to the growing body of research in face age synthesis and demonstrates how GANs can be applied to a critical, real-world problem.

The twelfth chapter, titled "Empowering Clinical Decision-Making with Generative AI in Intelligent Decision Support Systems", explores the integration of Generative AI with intelligent decision support systems (IDSS) in the medical field, specifically focusing on how this combination enhances clinical decision-making. The chapter highlights the use of generative AI in analyzing vast amounts of patient data to provide real-time insights, improving diagnostic accuracy and treatment recommendations. The authors discuss how the incorporation of machine learning (ML), the Internet of Things (IoT), and blockchain technology can help create a seamless, secure, and efficient system for monitoring and supporting patients with complex conditions, such as Wilms' tumor. Using IoT devices, patient health data can be collected continuously and analyzed in real time using ML algorithms, while blockchain ensures the security and privacy of this sensitive medical information. In addition to discussing the benefits, the chapter also addresses the challenges associated with implementing such systems in health care, including the need for secure data exchange, privacy concerns, and the high cost of infrastructure development. By integrating blockchain, the chapter demonstrates how sensitive patient information can be protected, while generative AI models enhance predictive capabilities in early disease detection and treatment monitoring. The use of such systems can improve outcomes for patients by enabling healthcare professionals to make informed, timely decisions, ultimately leading to better disease management and higher survival rates. This chapter provides valuable insights into how the combination of AI, IoT, and blockchain can empower healthcare systems to become more responsive and patient-centered.

The thirteenth chapter, "Leveraging Generative AI for Enhanced Predictive Maintenance and Anomaly Detection in Manufacturing", explores the transformative role of generative AI in modernizing maintenance strategies within the manufacturing sector. This chapter highlights how predictive maintenance, powered by AI, enables companies to anticipate equipment failures and identify anomalies, significantly reducing downtime and improving operational efficiency. Through the integration of machine learning algorithms, sensor data, and historical maintenance logs, generative AI can identify critical patterns that inform proactive maintenance decisions. The chapter presents a detailed case study involving a pump-related scenario, where AI-driven models were used to predict failures, optimize maintenance schedules, and diagnose equipment malfunctions more effectively. This proactive approach allows manufacturers to stay ahead of potential issues, ensuring that equipment runs efficiently while minimizing unexpected breakdowns. In addition to predictive maintenance, the chapter examines anomaly detection using advanced AI models like Large Language Models (LLMs) and Retrieval-Augmented Generation (RAG) techniques. By incorporating data from sensor readings, Standard Operating Procedures (SOPs), and maintenance records, these AI tools can generate detailed, context-aware responses to maintenance queries. The chapter demonstrates how these AI models can significantly enhance the accuracy of maintenance strategies by providing timely, actionable insights. This innovative use of generative AI in manufacturing paves the way for a more resilient, data-driven industry, where operational reliability is maximized, and maintenance costs are minimized.

The fourteenth chapter, "The Transformative Role of Big Data Analytics and Generative AI in Redefining FinTech for New Business Models", examines how big data analytics and generative AI are revolutionizing the financial technology (FinTech) sector. The chapter outlines the foundational role of big data, characterized by the four V's—Volume, Velocity, Variety, and Veracity—in enhancing financial decision-making processes. It examines how generative AI leverages big data to predict market trends, optimize investment strategies, and offer personalized financial products. The chapter discusses key applications such as automated financial modeling, predictive analytics, and risk assessment, which empower financial institutions to anticipate customer behavior and detect fraudulent activities. By simulating various market conditions and generating synthetic financial datasets, generative AI facilitates stress testing and scenario analysis, which are essential for resilient financial systems in a rapidly evolving digital economy. Additionally, the chapter highlights the integration of emerging technologies, including blockchain and machine learning, which complement generative AI in transforming FinTech. This chapter explores real-world case studies of FinTech innovations such as virtual currencies, digital wallets, and AI-driven fraud detection systems. The authors also address the challenges of implementing big data analytics in FinTech, including data privacy concerns, ethical issues, and regulatory frameworks like GDPR. By presenting a holistic perspective on how big data analytics and generative AI can redefine financial services, this chapter offers valuable insights into the future of FinTech, emphasizing the need for regulatory evolution and technological infrastructure to support innovation in this dynamic industry.

The fifteenth chapter, "Enhancing Digital Security in Fintech Through Integration of Generative AI in Regulatory Practices", examines the increasingly critical role of generative AI in bolstering cybersecurity and regulatory compliance within the FinTech sector. The chapter begins by exploring the evolution of cyber threats and the growing complexity of financial technology, emphasizing the need for advanced security strategies. This chapter highlights how Generative AI technologies, with their sophisticated learning algorithms, can dynamically enhance security protocols, such as data encryption and anomaly detection, to protect sensitive financial information. The chapter also focuses on the CIA Triad—confidentiality, integrity, and availability—discussing how Generative AI improves each aspect by introducing adaptive security measures, predictive analytics, and real-time threat identification. Moreover, this chapter examines the role of Regulatory Technology (RegTech) in creating more efficient and adaptive regulatory frameworks to mitigate the risks associated with digital financial transactions. By integrating Generative AI into compliance processes, financial institutions can automate regulatory reporting, improve accuracy, and reduce administrative burdens. The chapter also outlines the challenges posed by implementing such technologies, including balancing innovation with cybersecurity risks, and provides case studies demonstrating the successful use of Generative AI in maintaining regulatory compliance. With its forward-looking approach, this chapter serves as a valuable resource for understanding how AI-driven solutions are shaping the future of digital security in the FinTech landscape.

The last chapter, titled "Text Summarization: An Application of Generative AI", examines the use of generative AI for automating the process of condensing large volumes of text into concise, informative summaries. This chapter explores the two primary approaches to text summarization—extractive and abstractive—and how generative AI models like recurrent neural networks (RNNs) and transformer models have revolutionized this field. Extractive summarization selects key sentences directly from the source text, while abstractive summarization generates new sentences that capture the core message of the original text. The chapter compares the effectiveness of these methods in various real-world applications, such as summarizing legal documents, scientific papers, and news articles. The advancements in deep learning and natural language processing (NLP) have enabled the development of sophisticated models that not only preserve the coherence and relevance of the summaries but also enhance the accuracy and fluency of the generated text. Furthermore, the chapter provides practical insights into the implementation of text summarization models, including the challenges and potential future directions for research. It discusses the role of Sequence-to-Sequence (Seq2Seq) models, Long Short-Term Memory (LSTM) networks, and attention mechanisms in improving the quality of abstractive summaries. The chapter also includes performance evaluations, comparing precision, recall, and F-scores of different summarization techniques using metrics like ROUGE and BLEU. By offering a comprehensive analysis of text summarization methods, this chapter highlights the transformative potential of generative AI in improving information processing across industries such as media, education, and research, while also suggesting hybrid approaches to optimize both extractive and abstractive techniques for better results.

In conclusion, *Generative Artificial Intelligence (AI) Approaches for Industrial Applications* provides a comprehensive exploration of the transformative role of generative AI across a wide range of industries. Each chapter explores the theoretical underpinnings, practical applications, and potential challenges of implementing AI-driven models in real-world contexts. From enhancing healthcare decision-making and revolutionizing FinTech security practices to streamlining manufacturing operations and advancing text summarization techniques, this book offers a detailed roadmap for understanding how generative AI can redefine traditional industry practices. By combining cutting-edge research with case studies and practical insights, the book serves as a valuable resource for academics, professionals, and researchers eager to harness the power of AI for innovation and efficiency. As industries continue to embrace the capabilities of generative AI, the need for responsible and ethical implementation becomes increasingly apparent. This volume not only highlights the benefits but also addresses the challenges and ethical considerations associated with AI integration. By fostering a balanced understanding of both the opportunities and risks, *Generative Artificial Intelligence (AI) Approaches for Industrial Applications* equips readers with the knowledge and tools to navigate the evolving landscape of AI technologies and their impact on industrial development. We hope this collection of chapters will inspire future research and encourage thoughtful applications of generative AI, ultimately leading to more sustainable, secure, and innovative industrial ecosystems.

Blagoevgrad, Bulgaria — Narasimha Rao Vajjhala
Vellore, India — Sanjiban Sekhar Roy
Merkezkoeyler, Türkiye — Burak Taşcı
Doha, Qatar — Muhammad Enamul Hoque Chowdhury

Contents

Foundations and Emerging Trends in Generative Artificial Intelligence (AI) for Industrial Applications

Narasimha Rao Vajjhala, Sanjiban Sekhar Roy, Burak Taşcı, and Muhammad Enamul Hoque Chowdhury

Abstract This chapter explores the foundational principles, challenges, and opportunities associated with Generative Artificial Intelligence (AI) across various industries. This chapter begins by examining the mathematical underpinnings of generative AI models, including probability theory, optimization algorithms, and neural network architectures. Key models such as Generative Adversarial Networks (GANs), Variational Autoencoders (VAEs), and transformers are discussed, highlighting their unique capabilities and applications. The chapter then examines the ethical implications of AI deployment, focusing on data privacy, security, and the potential for misuse. Technical challenges such as training instability, bias in data generation, and computational costs are also explored. In parallel, the chapter identifies opportunities for innovation in industries like healthcare, finance, and supply chain management, where generative AI is already driving significant advancements. This chapter outlines potential future advancements in personalized AI experiences, industrial optimization, and scalable AI models, while addressing expected challenges related to regulatory frameworks and data security. Ultimately, this chapter provides a comprehensive overview of how generative AI is shaping the future of industries and the ethical and technical considerations that must be addressed for its responsible deployment.

N. R. Vajjhala (✉)
American University in Bulgaria, Blagoevgrad, Bulgaria
e-mail: nrao@aubg.edu

S. S. Roy
Vellore Institute of Technology, Chennai, India
e-mail: s.roy@vit.ac.in

B. Taşcı
Fırat University, Elazig, Türkiye
e-mail: btasci@firat.edu.tr

M. E. H. Chowdhury
Qatar University, Doha, Qatar
e-mail: mchowdhury@qu.edu.qa

N. R. Vajjhala et al. (eds.), *Generative Artificial Intelligence (AI) Approaches for Industrial Applications*, Information Systems Engineering and Management 24,
https://doi.org/10.1007/978-3-031-76710-4_1

Keywords Generative AI · Generative adversarial networks · Neural networks · Data privacy · AI ethics · Healthcare innovation · Predictive maintenance · Bias in AI · Transformers · Industrial optimization

1 Introduction

1.1 Brief History and Evolution of Generative AI

Generative Artificial Intelligence (AI) has its roots in the broader field of machine learning, which dates to the 1950s [1]. Initially, machine learning models were designed to recognize patterns and classify data. Early models like the Perceptron and linear classifiers focused on supervised learning tasks such as classification and regression [2]. However, the need to generate new data and simulate real-world scenarios drove the development of generative models, which aim to learn underlying patterns and structures in data to create new instances that resemble the original dataset. By the 1980s, advancements in probabilistic models, such as Hidden Markov Models (HMMs) and Bayesian networks, laid the groundwork for generative AI by allowing for the modeling of complex probabilistic dependencies between variables [3].

The evolution of deep learning in the early 2000s marked a significant turning point for generative AI [4]. Neural networks, particularly those with deep architectures, began to outperform traditional machine learning algorithms, and researchers started to explore their potential for generative tasks [5]. One of the landmark breakthroughs came in 2014 with the introduction of Generative Adversarial Networks (GANs) by Ian Goodfellow and his colleagues [6]. GANs introduced a new paradigm in generative modeling, where two networks—the generator and the discriminator—compete in a game-like setting, pushing each other to improve their ability to generate and distinguish between real and synthetic data [7]. This innovation spurred rapid advancements in the field and expanded the potential applications of generative AI across industries.

1.2 Evolution into a Critical Tool for Industry

As the capabilities of generative AI models improved, industries began to recognize their potential for solving complex problems, improving efficiency, and driving innovation. In the early stages, generative AI was primarily used in academic research, particularly for image generation and natural language processing [8]. However, its ability to generate realistic data and simulate scenarios quickly caught the attention of industries such as healthcare, manufacturing, and finance. For instance, in healthcare, generative AI models are now used to create synthetic medical data, enabling the development of new drugs and personalized treatments without compromising patient privacy [9]. Similarly, in manufacturing, generative AI is employed to optimize product designs and predict equipment failures through advanced simulations [10].

The transformative power of generative AI lies in its ability to produce high-quality, realistic data that can be used to augment decision-making, reduce costs, and accelerate processes [11]. For example, in finance, generative AI models can simulate various market conditions and generate synthetic datasets for stress testing and risk analysis, leading to more informed and robust investment strategies [12]. As industries face increasing pressure to innovate and remain competitive, generative AI has emerged as a critical tool, allowing organizations to explore new possibilities in areas such as automation, predictive modeling, and creative design. The flexibility and adaptability of generative AI have made it a cornerstone technology across multiple sectors, enabling industries to harness the power of data in ways that were previously unimaginable.

1.3 Generative AI: Core Concepts

At its core, generative AI refers to a class of machine learning models that can generate new data samples based on the patterns and structures learned from existing data [13]. Unlike discriminative models, which focus on distinguishing between different categories or classes of data, generative models aim to capture the underlying distribution of the data and generate new instances that closely resemble the original dataset [14]. This ability to create realistic data has wide-ranging applications, from image synthesis to text generation and beyond. Generative AI models are particularly valuable in scenarios where labeled data is scarce or difficult to obtain, as they can be used to generate synthetic data for training other models or testing hypotheses [15].

One of the key advantages of generative AI models is their ability to learn unsupervised or semi-supervised, meaning they do not require extensive labeled datasets for training [16]. Instead, they learn from the inherent structure of the data, allowing them to generalize and create new data points that are consistent with the input data. This flexibility makes generative AI models highly versatile and applicable in various fields, including creative industries such as art, music, and design, where they are used to generate novel ideas and solutions [17]. Moreover, the rise of generative AI has led to the development of advanced techniques for improving model performance, such as transfer learning and fine-tuning, further expanding their utility across domains.

1.4 Generative Adversarial Networks (GANs)

GANs are among the most well-known and widely used generative AI models [18]. GANs consist of two neural networks: a generator and a discriminator where the generator's role is to create synthetic data that mimics the real data, while the discriminator's job is to differentiate between real and synthetic data [18]. These two networks are trained simultaneously in a process known as adversarial training, where the generator tries to fool the discriminator by producing increasingly realistic data, and the discriminator improves its ability to distinguish between real and fake data [18]. The competition between these two networks drives both to improve, resulting in high-quality synthetic data generation.

GANs have found applications in various fields, including image generation, video synthesis, and text-to-image conversion [19, 20]. For example, GANs have been used to generate realistic images of faces, even creating entirely new individuals who do not exist in real life [8]. In the creative industries, GANs are employed to generate artwork, music, and even fashion designs [21]. The versatility of GANs extends beyond the creative domain, as they are also used in healthcare for generating synthetic medical images, in finance for data augmentation in fraud detection, and in automotive industries for simulating autonomous driving scenarios. Despite their remarkable potential, GANs face challenges such as training instability and mode collapse, where the generator produces limited or repetitive data [22].

1.5 Variational Autoencoders (VAEs)

Variational Autoencoders (VAEs) are another type of generative model that has gained popularity for their ability to generate data while learning the underlying distribution of the input data [23]. Unlike GANs, which rely on adversarial training, VAEs are based on probabilistic graphical models [24]. A VAE consists of two parts: an encoder and a decoder where the encoder compresses the input data into a latent space, representing the data in a lower-dimensional form, while the decoder reconstructs the data from this latent space [25]. By optimizing the encoder-decoder pair, VAEs can generate new data samples by sampling from the latent space and passing these samples through the decoder to generate realistic data [25].

VAEs are particularly useful for tasks that require generating continuous data, such as image interpolation, data compression, and anomaly detection [23]. In the healthcare sector, VAEs are used to generate synthetic medical data for training machine learning models, thereby addressing privacy concerns and data scarcity issues [23]. In finance, VAEs are employed for risk modeling and portfolio optimization, where they can simulate various market conditions and generate synthetic financial data for analysis [25]. VAEs are also widely used in research for exploring the latent structure of data, providing valuable insights into the relationships between different variables in the dataset.

1.6 Transformers

Transformers have revolutionized the field of natural language processing (NLP) and generative AI with their ability to handle long-range dependencies in sequential data [26]. Introduced in 2017, transformers leverage self-attention mechanisms to capture relationships between different parts of the input data, allowing them to process entire sequences in parallel rather than sequentially, as is the case with recurrent neural networks (RNNs) [27]. This breakthrough has enabled transformers to excel in tasks such as language translation, text generation, and summarization. Models like GPT (Generative Pre-trained Transformer) and BERT (Bidirectional Encoder Representations from Transformers) have set new benchmarks in NLP, demonstrating the immense potential of transformers in generative AI applications [28]. Transformers have found applications beyond NLP, extending into fields such as image generation, where models like Vision Transformers (ViTs) have been developed to handle image data [29]. The self-attention mechanism used in transformers allows them to capture fine-grained details in images, making them well-suited for generative tasks such as image synthesis and style transfer. In addition to their success in language and vision tasks, transformers have also been applied to multimodal tasks, where they integrate different types of data, such as text, images, and audio, to generate comprehensive outputs. Their versatility and scalability have made transformers a dominant model in generative AI research and applications.

1.7 Comparison of GANs, VAEs, and Transformers

While GANs, VAEs, and transformers are all generative models, they differ in their architecture, training methods, and use cases. GANs are highly effective for generating high-quality, realistic images and are widely used in creative industries [20–22, 24]. However, their adversarial training process can be unstable, leading to challenges in training convergence. VAEs, on the other hand, are based on probabilistic methods and are more stable during training [24]. They are well-suited for continuous data generation tasks and are often used for anomaly detection and latent space exploration. However, VAEs typically produce data that is slightly blurrier compared to the sharp outputs of GANs [30]. Transformers, with their self-attention mechanism, are designed to handle sequential data and excel in tasks involving text and language generation. They have also shown promise in vision tasks, especially with the advent of ViTs [28, 29]. Unlike GANs and VAEs, transformers can handle longer-range dependencies in data and are scalable to large datasets [26]. Each of these models has its strengths and weaknesses, and the choice of which to use depends on the specific application and the type of data involved.

2 Key Applications Across Industries

2.1 Overview of Industries Adopting Generative AI

Generative AI is rapidly transforming multiple industries, with manufacturing, healthcare, education, and finance leading the charge in integrating this advanced technology into their operations [9, 10, 15, 31]. Each of these sectors has unique needs and challenges that generative AI addresses, whether through enhanced decision-making, automation, or creating new possibilities for innovation. In manufacturing, for example, generative AI enables more efficient product design, predictive maintenance, and process optimization, reducing waste and improving output quality. Healthcare, on the other hand, benefits from AI's ability to generate synthetic medical data, accelerate drug discovery, and personalize treatment plans, all while maintaining high standards of patient privacy and security [9, 15]. In the education sector, generative AI has made its mark by transforming curriculum design, facilitating personalized learning experiences, and automating administrative tasks such as grading and content creation [8]. Meanwhile, in finance, AI is helping institutions optimize investment strategies, manage risks, and combat fraud by generating predictive models that anticipate market shifts and irregularities [32]. As these industries increasingly rely on data-driven processes, generative AI has proven to be an essential tool for both innovation and efficiency, enabling organizations to adapt to rapidly changing environments and make informed decisions in real time.

2.2 Manufacturing: Product Design and Optimization

In the manufacturing industry, generative AI has revolutionized product design by enabling the creation of optimized designs based on specific constraints and goals [31]. Traditional design processes often rely on iterative methods, where engineers manually adjust parameters to improve functionality. In contrast, generative AI automates this process by generating multiple design alternatives that meet the desired criteria, such as weight reduction, material strength, or cost efficiency [11]. These AI-driven designs are often superior to those created through traditional methods, as they can explore a broader range of possibilities and identify solutions that human designers might overlook. This capability is particularly useful in industries such as automotive, aerospace, and consumer goods, where product design plays a crucial role in performance and competitiveness. For example, in the automotive industry, generative AI is used to optimize car parts for strength and durability while minimizing weight and material usage [33]. This leads to more fuel-efficient vehicles and reduces production costs. In aerospace, AI-driven design is applied to create lighter, more robust components that enhance aircraft performance and safety [34]. By automating the design process, generative AI not only accelerates innovation but also improves the overall quality and sustainability of manufactured products.

2.3 Manufacturing: Predictive Maintenance

Another critical application of generative AI in manufacturing is predictive maintenance [35]. Traditionally, maintenance schedules are based on fixed intervals or reactive responses to equipment failures, both of which can be inefficient and costly [35]. Generative AI transforms this approach by analyzing vast amounts of data from sensors and historical maintenance records to predict when equipment is likely to fail or require servicing [31]. This enables manufacturers to schedule maintenance at the optimal time, preventing costly downtime and extending the lifespan of machinery. Predictive maintenance powered by AI is particularly valuable in industries with heavy reliance on complex machinery, such as automotive manufacturing, heavy equipment, and energy production [36, 37]. For instance, in automotive manufacturing, AI systems monitor the performance of assembly line equipment and predict potential failures before they occur, minimizing disruptions to production [37]. In energy production, generative AI can predict maintenance needs for turbines and generators, ensuring that repairs are made efficiently and without causing outages [38, 39]. By improving maintenance schedules, generative AI enhances operational efficiency, reduces costs, and increases the reliability of critical industrial equipment.

2.4 Healthcare: Drug Discovery and Personalized Treatment

In healthcare, generative AI is driving significant advancements in drug discovery and personalized medicine [24, 40]. Traditionally, the drug discovery process is time-consuming and expensive, often taking years to identify potential drug candidates and bring them to market. Generative AI accelerates this process by generating new molecular structures and predicting their effectiveness as potential drugs [4, 41, 42]. AI models analyze vast datasets of chemical compounds, biological interactions, and patient data to generate novel drug candidates that are more likely to succeed in clinical trials. This has the potential to drastically reduce the time and cost associated with developing new medications. In addition to drug discovery, generative AI is playing a crucial role in personalized medicine [40, 43]. By analyzing patient data, including genetic information, medical history, and lifestyle factors, AI models can generate personalized treatment plans tailored to the individual's unique needs. This approach enhances the effectiveness of treatments and minimizes adverse effects. For example, in cancer treatment, AI models are used to generate personalized drug combinations that target the specific genetic mutations driving the patient's cancer [44, 45]. This level of precision medicine was previously unattainable through traditional methods, highlighting the transformative potential of generative AI in healthcare.

2.5 Healthcare: Medical Imaging and Diagnosis

Generative AI is also making significant strides in medical imaging and diagnosis [46]. AI models can generate highly detailed and accurate medical images, such as MRI scans, by analyzing existing imaging data and predicting the likely structure of new scans [47]. This is particularly valuable in cases where imaging data is incomplete or noisy, as AI can generate clearer and more accurate images, helping doctors make better-informed diagnoses. Additionally, AI can simulate medical conditions in imaging, aiding in training healthcare professionals and improving their diagnostic skills [48]. Moreover, generative AI models are being used to automate the diagnosis of medical conditions by analyzing imaging data for abnormalities, such as tumors or fractures. AI models are often more accurate and faster than human radiologists at detecting subtle changes in images, leading to earlier diagnosis and improved patient outcomes. For instance, in the field of oncology, AI systems can analyze CT and MRI scans to detect early signs of cancer, significantly improving the chances of successful treatment [49]. As AI continues to advance in medical imaging, it has the potential to reduce diagnostic errors and enhance the overall quality of patient care.

2.6 Education: Personalized Learning and Content Generation

In the education sector, generative AI is transforming how students learn and how educators develop instructional materials [8]. One of the most impactful applications of AI in education is personalized learning, where AI systems generate customized learning paths based on a student's abilities, learning style, and progress [50]. By analyzing data on student performance, generative AI models can create tailored lessons, exercises, and assessments that adapt in real-time to the student's needs, providing targeted support and challenges to ensure optimal learning outcomes. In addition to personalized learning, generative AI is being used to automate content generation, such as creating lesson plans, quizzes, and even textbooks [51, 52]. AI can generate instructional materials based on predefined educational goals and standards, freeing up educators' time and allowing them to focus on more interactive and student-centered activities. By automating the creation of educational content, generative AI improves both the quality and accessibility of education, ensuring that learning resources are available to a broader audience [53–55].

2.7 Finance: Risk Management and Fraud Detection

The finance industry has quickly adopted generative AI to improve risk management and fraud detection [56]. Financial institutions face significant challenges in managing risk, as they must constantly monitor market conditions, assess creditworthiness, and detect fraudulent activities. Generative AI models can analyze vast amounts of financial data, including transaction histories, market trends, and customer profiles, to generate predictive models that help institutions anticipate risks and make informed decisions [57]. For example, AI models can predict market downturns, enabling financial institutions to adjust their portfolios and mitigate potential losses [58]. Fraud detection is another critical area where generative AI has made a significant impact [59]. AI systems can generate models of normal transaction behavior, allowing them to detect anomalies that may indicate fraudulent activity. For instance, AI can analyze patterns in credit card transactions to identify unusual spending behavior, flagging potential cases of fraud for further investigation [60]. Generative AI is particularly effective at detecting new and evolving forms of fraud, as it can generate synthetic fraud scenarios based on past cases and use them to train fraud detection systems [61]. By enhancing risk management and fraud detection, generative AI helps financial institutions safeguard their assets and maintain trust with their customers.

2.8 Finance: Investment Strategy and Market Analysis

In addition to risk management, generative AI plays a key role in optimizing investment strategies and market analysis [62]. Financial institutions and investors rely on vast amounts of data to make informed decisions, and generative AI models can generate synthetic datasets that simulate various market conditions, enabling more accurate forecasting and scenario analysis. For example, AI models can generate simulations of economic downturns, interest rate changes, or geopolitical events, allowing investors to test how different strategies would perform under various circumstances [63]. By analyzing historical market data and generating predictions of future market trends, generative AI can also optimize portfolio management and asset allocation. For instance, AI models can generate predictions on stock price movements, helping investors make more informed decisions about buying or selling assets [64]. In the realm of algorithmic trading, generative AI is used to generate and refine trading algorithms that automatically execute trades based on real-time market conditions. These AI-driven strategies can respond more quickly to market changes than traditional human traders, giving institutions a competitive edge in fast-paced financial environments.

2.9 Security: Anomaly Detection and Threat Simulation

Generative AI has also emerged as a powerful tool in enhancing security, particularly in the areas of anomaly detection and threat simulation [65]. In industries such as finance, healthcare, and manufacturing, organizations must constantly monitor their systems for potential security breaches or irregular activity. Generative AI models can generate profiles of normal behavior, allowing them to detect anomalies that may indicate a security threat. For example, in cybersecurity, AI models can analyze network traffic and generate alerts when suspicious activity deviates from normal patterns, enabling faster responses to potential threats [66]. Moreover, generative AI can be used to simulate potential cyber-attacks and test the resilience of security systems. By generating realistic attack scenarios, AI models help organizations identify vulnerabilities and improve their defenses before actual attacks occur. This proactive approach to security is particularly valuable in industries where data breaches or system failures can have serious consequences, such as in finance or healthcare. Through anomaly detection and threat simulation, generative AI enhances the security of digital infrastructures and helps organizations stay ahead of evolving threats.

3 Methodological Foundations

3.1 Basic Mathematical Principles Underpinning Generative AI Models

At the core of generative AI models lie fundamental mathematical principles, particularly those from probability theory, linear algebra, and calculus [67]. Probability theory plays a crucial role in understanding the distributions that generative models seek to learn and replicate [68]. Concepts such as conditional probability, Bayes' theorem, and joint distributions allow models to generate new data that closely mirrors the real-world data they are trained on [69]. For instance, in models like VAEs, probability distributions over latent variables are estimated to generate diverse and realistic outputs [24]. Linear algebra is another critical mathematical foundation for generative AI, especially in defining and operating with data structures like vectors and matrices, which are used extensively in neural networks [70]. Matrix operations, eigenvalues, eigenvectors, and transformations provide the computational backbone of how data moves through the layers of a neural network [71]. Calculus, particularly derivatives and gradients, underpins the optimization processes in these models, enabling them to minimize loss functions and improve their predictive accuracy [72]. Together, these mathematical tools form the basis of the computational frameworks that allow generative models to learn and generalize from data.

3.2 *Optimization Algorithms and Neural Network Architectures*

Optimization algorithms are essential in training generative AI models, allowing them to learn from data by minimizing loss functions, which measure the difference between the generated output and the target data [72]. Gradient descent, and its variants such as stochastic gradient descent (SGD), is one of the most used optimization algorithms [73]. Gradient descent works by iteratively adjusting the model's parameters (weights) in the direction that minimizes the loss. More advanced optimization techniques like Adam (Adaptive Moment Estimation) combine the benefits of momentum and adaptive learning rates to achieve faster and more accurate convergence during training [74]. The architecture of neural networks, particularly deep neural networks (DNNs), also plays a significant role in generative AI models [75]. These architectures typically consist of multiple layers of neurons, including input, hidden, and output layers. Each neuron performs a weighted sum of its inputs, followed by a non-linear activation function, to model complex relationships within the data [75]. Deep learning architectures, such as convolutional neural networks (CNNs) and recurrent neural networks (RNNs), have been especially successful in tasks like image generation and text synthesis, respectively [76].

3.3 *Large Language Models (LLMs)*

Large Language Models (LLMs) are a subset of transformers designed to handle vast amounts of text data and generate coherent, context-aware responses [77]. Models like GPT-3 have demonstrated the ability to generate human-like text, making them valuable in applications ranging from content creation to customer service automation [78]. LLMs are pre-trained on massive corpora of text data, learning to predict the next word in a sentence by understanding the relationships between words and their context [79]. This pre-training enables LLMs to be fine-tuned for specific tasks, such as translation, summarization, or even programming code generation. LLMs like GPT-4, which boast billions of parameters, have also sparked debates around their ethical use, including issues related to bias, misinformation, and privacy [77]. Despite these concerns, LLMs represent a significant leap forward in generative AI, showcasing the immense potential of transformer architectures to generate meaningful and contextually relevant outputs across various domains [79]. Their capacity to handle nuanced language generation and understanding has made LLMs a cornerstone in the advancement of generative AI technologies.

4 Challenges and Opportunities

4.1 Ethical Implications of AI Deployment: Data Privacy, Security, and Potential Misuse

The deployment of generative AI in industries brings significant ethical concerns, particularly around data privacy, security, and the potential for misuse. Generative models rely on vast datasets for training, which often include sensitive information such as personal data, medical records, or financial details. The handling of this data raises concerns about privacy breaches, especially when models generate synthetic data that might inadvertently expose real-world information. For example, AI-generated data in healthcare, if not adequately anonymized, could risk violating patient privacy regulations like the Health Insurance Portability and Accountability Act (HIPAA). Security risks also arise as generative AI becomes more integrated into various industries. The ability to create highly realistic synthetic data, such as deepfake videos or fraudulent documents, poses significant threats to digital security and trust. In finance, for instance, AI-generated synthetic data could be exploited for financial fraud or identity theft. This raises the need for robust cybersecurity measures and ethical AI frameworks to prevent potential misuse. Addressing these ethical implications requires a multi-disciplinary approach, involving not only technical solutions but also regulatory oversight and public policy to safeguard privacy and security.

4.2 Technical Challenges: Training Instability, Bias in Data Generation, and Computational Costs

On the technical side, generative AI faces several challenges that limit its broader adoption and effectiveness. One of the most prominent issues is training instability, particularly in models like GANs [22, 41, 65]. GANs, which rely on an adversarial relationship between two networks, are notoriously difficult to train. Issues such as mode collapse, where the model generates limited variations of output, and oscillating loss functions can prevent the model from converging to a stable solution. These challenges make it difficult to deploy GANs reliably at scale, especially in high-stakes industries like healthcare or finance. Bias in data generation is another critical concern because generative AI models learn from the data they are trained on, and if that data contains biases—such as gender, racial, or socioeconomic biases—the AI will likely replicate and even amplify those biases in its outputs [80]. This poses significant risks, particularly in sectors like criminal justice, hiring, and healthcare, where biased AI systems could lead to unfair or harmful outcomes. Efforts to mitigate bias involve developing more diverse and representative datasets, as well as implementing fairness-aware algorithms that can detect and correct biases during training.

Lastly, the computational costs of training large-scale generative AI models are a significant barrier. Training advanced models, such as GPT-4 or other LLMs, requires massive amounts of computational power and energy, often making them accessible only to organizations with substantial resources [79]. This high cost limits the democratization of AI technology and raises environmental concerns due to the energy consumption involved in training such models. Researchers are working on more efficient training methods, such as model distillation and pruning, to reduce the computational burden without sacrificing performance.

5 Future Research Directions

As AI technology advances, several key areas are poised for significant breakthroughs, particularly in providing personalized experiences, optimizing industrial processes, and making AI models more scalable. Personalized AI-driven experiences are already taking shape in industries such as healthcare, education, and retail, where AI can tailor products, services, or treatments to the needs of individual users. For example, in healthcare, AI could generate personalized treatment plans that consider a patient's genetic profile, lifestyle, and medical history. Similarly, in retail, generative AI could create personalized marketing campaigns or product recommendations that reflect individual consumer preferences and behaviors. Industrial optimization is another area where generative AI is expected to make considerable advancements. In manufacturing, AI models are being developed to optimize supply chains, improve production efficiency, and enhance product design. By simulating complex industrial processes, generative AI can identify inefficiencies and suggest improvements that reduce costs and increase output. Additionally, AI can help industries transition to more sustainable practices by optimizing resource usage and minimizing waste. Scalability is a critical challenge for AI models, particularly for LLMs and deep learning architectures that require substantial computational resources. Future advancements are expected to focus on making AI models more scalable and efficient, enabling wider adoption across industries without the need for massive computational infrastructure. Techniques such as model compression, transfer learning, and distributed training are being explored to make AI more accessible to organizations of all sizes.

As generative AI becomes more prevalent, regulatory frameworks will need to evolve to address the unique challenges posed by this technology. Governments and regulatory bodies will need to develop policies that govern the ethical use of AI, ensuring that AI systems are transparent, fair, and accountable. One of the main challenges is developing frameworks that can keep pace with the rapid advancements in AI technology, while also addressing concerns related to bias, discrimination, and the potential misuse of AI-generated content. Data security will remain a top priority as AI systems become increasingly integrated into industries that handle sensitive information, such as healthcare, finance, and government. Ensuring that generative AI models are secure and do not expose personal or confidential data will require robust cybersecurity measures and secure data handling practices. Additionally, organizations will need to comply with data privacy regulations such as GDPR and HIPAA, which impose strict guidelines on how personal data can be collected, stored, and used in AI systems.

Generative AI holds immense potential to reshape industries and drive unprecedented innovation. From personalized healthcare and optimized manufacturing to enhanced financial systems and more efficient supply chains, AI is enabling industries to become more efficient, adaptive, and responsive to market demands. However, realizing the full potential of generative AI will require addressing the ethical, technical, and regulatory challenges that accompany its deployment. As AI continues to advance, the focus must shift toward developing responsible, scalable, and secure AI solutions that benefit both businesses and society. The transformative power of AI extends beyond technical innovation—it also has the potential to fundamentally change how industries operate and interact with their customers. By embracing AI, organizations can unlock new opportunities for growth, streamline their operations, and create more personalized experiences for consumers. As AI continues to evolve, the future promises a more intelligent and efficient industrial landscape, where generative AI plays a central role in driving progress and shaping the next wave of innovation.

6 Conclusion

This chapter has highlighted the transformative potential of Generative AI across various industries, while addressing the foundational principles, technical challenges, and ethical considerations surrounding its deployment. From the mathematical and architectural foundations of generative models to their widespread application in healthcare, manufacturing, finance, and beyond, AI is becoming an indispensable tool for innovation, efficiency, and optimization. Key models such as GANs, VAEs, and transformers have demonstrated the versatility and power of generative AI in solving complex problems, generating new data, and enabling personalized experiences. However, the chapter also underscored the pressing ethical concerns around data privacy, security, and bias, which must be carefully managed to ensure that AI technologies are used responsibly. Technical challenges such as training instability, computational costs, and the potential for bias in AI-generated content present obstacles that researchers and practitioners must overcome to make AI more reliable and scalable. Looking forward, the future of generative AI holds immense promise, with advancements expected in personalized AI systems, industrial optimization, and more efficient, scalable models. By addressing these challenges and continuing to innovate, generative AI will play a critical role in shaping the future of industries, providing new opportunities for growth and progress while navigating the complexities of an evolving technological landscape.

References

1. Lanzetta, M.: Machine learning, deep learning, and artificial intelligence. In: Artificial Intelligence for Autonomous Networks. Chapman and Hall/CRC, pp. 25–47 (2018)
2. Mohamed, A.E.: Comparative study of four supervised machine learning techniques for classification. Int. J. Appl. Sci. Technol. **7**(2), 1–15 (2017)
3. Yu, D., et al.: Hidden Markov models and the variants. In: Automatic Speech Recognition: A Deep Learning Approach, pp. 23–54 (2015)
4. Gupta, R., et al.: Artificial intelligence to deep learning: machine intelligence approach for drug discovery. Mol. Diversity **25**, 1315–1360 (2021)
5. Liu, W., et al.: A survey of deep neural network architectures and their applications. Neurocomputing **234**, 11–26 (2017)
6. Sabnam, S., Rajagopal, S.: Application of generative adversarial networks in image, face reconstruction and medical imaging: challenges and the current progress. Comput. Methods Biomech. Biomed. Eng. Imag. Visual. **12**(1), 2330524 (2024)
7. Liu, J., et al.: Deep learning for procedural content generation. Neural Comput. Appl. **33**(1), 19–37 (2021)
8. Alqahtani, T., et al.: The emergent role of artificial intelligence, natural learning processing, and large language models in higher education and research. Res. Soc. Adm. Pharm. **19**(8), 1236–1242 (2023)
9. Nova, K.: Generative AI in healthcare: advancements in electronic health records, facilitating medical languages, and personalized patient care. J. Adv. Anal. Healthc. Manag. **7**(1), 115–131 (2023)
10. Elahi, M., et al.: A comprehensive literature review of the applications of AI techniques through the lifecycle of industrial equipment. Discov. Artif. Intell. **3**(1), 43 (2023)
11. Bandi, A., et al.: The power of generative AI: a review of requirements, models, input–output formats, evaluation metrics, and challenges. Future Int. **15**(8), 260 (2023)
12. Gao, G., et al.: Causal data science for financial stress testing. J. Comput. Sci. **26**, 294–304 (2018)
13. Wang, K., et al.: Generative adversarial networks: introduction and outlook. IEEE/CAA J. Automat. Sin. **4**(4), 588–598 (2017)
14. Harshvardhan, G., et al.: A comprehensive survey and analysis of generative models in machine learning. Comput. Sci. Rev. **38**, 100285 (2020)
15. Sai, S., et al.: Generative AI for transformative healthcare: a comprehensive study of emerging models, applications, case studies and limitations. IEEE Access (2024)
16. Van Engelen, J.E., Hoos, H.H.: A survey on semi-supervised learning. Mach. Learn. **109**(2), 373–440 (2020)
17. Takale, D.G., et al.: Advancements and applications of generative artificial intelligence. J. Inf. Technol. Sci. **10**(1), 20–27 (2024)
18. Hughes, R.T., et al.: Generative adversarial networks–enabled human–artificial intelligence collaborative applications for creative and design industries: a systematic review of current approaches and trends. Front. Artif. Intell. **4**, 604234 (2021)
19. Liu, M.-Y., et al.: Generative adversarial networks for image and video synthesis: algorithms and applications. Proc. IEEE **109**(5), 839–862 (2021)
20. Dash, A., et al.: A review of generative adversarial networks (GANs) and its applications in a wide variety of disciplines: from medical to remote sensing. IEEE Access (2023)
21. Shahriar, S.: GAN computers generate arts? A survey on visual arts, music, and literary text generation using generative adversarial network. Displays **73**, 102237 (2022)
22. Park, S.-W., et al.: BEGAN v3: avoiding mode collapse in GANs using variational inference. Electronics **9**(4), 688 (2020)
23. Bao, J., et al.: Variational autoencoder or generative adversarial networks? A comparison of two deep learning methods for flow and transport data assimilation. Math. Geosci. **54**(6), 1017–1042 (2022)

24. Bond-Taylor, S., et al.: Deep generative modelling: a comparative review of VAEs, GANs, normalizing flows, energy-based and autoregressive models. IEEE Trans. Pattern Anal. Mach. Intell. **44**(11), 7327–7347 (2021)
25. Han, K., et al.: Variational autoencoder: an unsupervised model for encoding and decoding fMRI activity in visual cortex. Neuroimage **198**, 125–136 (2019)
26. Bishop, C.M., Bishop, H.: Transformers. In: Deep Learning: Foundations and Concepts, pp. 357–406. Springer (2023)
27. Tay, Y., et al.: Synthesizer: rethinking self-attention for transformer models. In: International Conference on Machine Learning. PMLR (2021)
28. Patwardhan, N., et al.: Transformers in the real world: a survey on NLP applications. Information **14**(4), 242 (2023)
29. Han, K., et al.: A survey on vision transformer. IEEE Trans. Pattern Anal. Mach. Intell. **45**(1), 87–110 (2022)
30. Huang, S., et al.: Evaluating lossy compression rates of deep generative models. In: International Conference on Machine Learning. PMLR (2020)
31. Recio-Román, A., et al.: The future of retail: harnessing generative AI for disruptive innovation and sector transformation. In: Reshaping Marketing Science in Wholesaling and Retailing, pp. 309–333. IGI Global (2024)
32. Ooi, K.-B., et al.: The potential of generative artificial intelligence across disciplines: perspectives and future directions. J. Comput. Inf. Syst. 1–32 (2023)
33. Gazi, M.S.: Implications of generative Al and machine learning on automotive industry development & reduction of carbon footprint: an analysis of the US economy perspective. J. Bus. Manag. Stud. **6**(3), 134–143 (2024)
34. Hassan, K., et al.: Application of artificial intelligence in aerospace engineering and its future directions: a systematic quantitative literature review. Arch. Comput. Methods Eng. 1–56 (2024)
35. Stohr, A., et al.: Generative mechanisms of AI implementation: a critical realist perspective on predictive maintenance. Inf. Organ. **34**(2), 100503 (2024)
36. Theissler, A., et al.: Predictive maintenance enabled by machine learning: use cases and challenges in the automotive industry. Reliab. Eng. Syst. Saf. **215**, 107864 (2021)
37. Çınar, Z.M., et al.: Machine learning in predictive maintenance towards sustainable smart manufacturing in industry 4.0. Sustainability **12**(19), 8211 (2020)
38. Afridi, Y.S., et al.: Artificial intelligence based prognostic maintenance of renewable energy systems: a review of techniques, challenges, and future research directions. Int. J. Energy Res. **46**(15), 21619–21642 (2022)
39. Chatterjee, J., Dethlefs, N.: Scientometric review of artificial intelligence for operations & maintenance of wind turbines: the past, present and future. Renew. Sustain. Energy Rev. **144**, 111051 (2021)
40. Ghebrehiwet, I., et al.: Revolutionizing personalized medicine with generative AI: a systematic review. Artif. Intell. Rev. **57**(5), 1–41 (2024)
41. Gangwal, A., et al.: Generative artificial intelligence in drug discovery: basic framework, recent advances, challenges, and opportunities. Front. Pharmacol. **15**, 1331062 (2024)
42. Visan, A.I., Negut, I.: Integrating artificial intelligence for drug discovery in the context of revolutionizing drug delivery. Life **14**(2), 233 (2024)
43. Zhang, P., Kamel Boulos, M.N.: Generative AI in medicine and healthcare: promises, opportunities and challenges. Future Int. **15**(9), 286 (2023)
44. Tsigelny, I.F.: Artificial intelligence in drug combination therapy. Brief. Bioinform. **20**(4), 1434–1448 (2019)
45. Bhinder, B., et al.: Artificial intelligence in cancer research and precision medicine. Cancer Discov. **11**(4), 900–915 (2021)
46. Kim, K., et al.: Updated primer on generative artificial intelligence and large language models in medical imaging for medical professionals. Korean J. Radiol. **25**(3), 224 (2024)
47. Castiglioni, I., et al.: AI applications to medical images: from machine learning to deep learning. Phys. Med. **83**, 9–24 (2021)

48. Gupta, P., Pandey, M.K.: 2 Role health of diagnosis AI for smart and treatment. In: Smart Medical Imaging for Diagnosis and Treatment Planning, vol. 23
49. Hamamoto, R., et al.: Application of artificial intelligence technology in oncology: towards the establishment of precision medicine. Cancers **12**(12), 3532 (2020)
50. Çela, E., et al.: Foundations of computational thinking and problem solving for diverse academic fields. In: Fonkam, M.M., Vajjhala, N.R. (eds.) Revolutionizing curricula through computational thinking, logic, and problem solving, pp. 1–16. IGI Global, Hershey, PA, USA (2024)
51. Lee, G.-G., Zhai, X.: Using ChatGPT for science learning: a study on pre-service teachers' lesson planning. IEEE Trans. Learn. Technol. (2024)
52. Çela, E., et al.: Risks of AI-assisted learning on student critical thinking: a case study of Albania. Int. J. Risk Contingency Manag. (IJRCM) **12**(1), 1–19 (2024)
53. Ruiz-Rojas, L.I., et al.: Empowering education with generative artificial intelligence tools: approach with an instructional design matrix. Sustainability **15**(15), 11524 (2023)
54. Bozkurt, A.: Generative artificial intelligence (AI) powered conversational educational agents: the inevitable paradigm shift. Asian J. Distance Educ. **18**(1) (2023)
55. Çela, E.: Global agendas in higher education and current educational reforms in Albania. In: Global Agendas and Education Reforms: A Comparative Study, pp. 255–269. Springer Nature Singapore, Singapore (2024)
56. Xu, J., et al.: Predict and optimize financial services risk using AI-driven technology. Acad. J. Sci. Technol. **10**(1), 299–304 (2024)
57. Yusof, S.A.B.M., Roslan, F.A.B.M.: The impact of generative AI in enhancing credit risk modeling and decision-making in banking institutions. Emerg. Trends Mach. Intell. Big Data **15**(10), 40–49 (2023)
58. Javaid, H.A.: AI-driven predictive analytics in finance: transforming risk assessment and decision-making. Adv. Comput. Sci. **7**(1) (2024)
59. Wood, A.: Dark echoes: the exploitative potential of generative AI in online harassment. In: Psybersecurity, pp. 90–129. CRC Press
60. Baratzadeh, F., Hasheminejad, S.M.: Customer behavior analysis to improve detection of fraudulent transactions using deep learning. J. AI Data Min. **10**(1), 87–101 (2022)
61. Fiore, U., et al.: Using generative adversarial networks for improving classification effectiveness in credit card fraud detection. Inf. Sci. **479**, 448–455 (2019)
62. Dahal, S.B.: Utilizing generative AI for real-time financial market analysis opportunities and challenges. Adv. Intell. Inf. Syst. **8**(4), 1–11 (2023)
63. Bai, X., et al.: Leveraging generative artificial intelligence for financial market trading data management and prediction (2024)
64. Weng, B., et al.: Predicting short-term stock prices using ensemble methods and online data sources. Expert Syst. Appl. **112**, 258–273 (2018)
65. Lim, W., et al.: Future of generative adversarial networks (GAN) for anomaly detection in network security: a review. Comput. Secur. 103733 (2024)
66. Bécue, A., et al.: Artificial intelligence, cyber-threats and Industry 4.0: challenges and opportunities. Artif. Intell. Rev. **54**(5), 3849–3886 (2021)
67. Michelucci, U.: Fundamental Mathematical Concepts for Machine Learning in Science. Springer (2024)
68. Testolin, A., Zorzi, M.: Probabilistic models and generative neural networks: towards an unified framework for modeling normal and impaired neurocognitive functions. Front. Comput. Neurosci. **10**, 73 (2016)
69. Zhang, B., et al.: Toward the third generation artificial intelligence. Sci. China Inf. Sci. **66**(2), 121101 (2023)
70. Szegedy, C.: A promising path towards autoformalization and general artificial intelligence. In: Intelligent Computer Mathematics: 13th International Conference, CICM 2020, Bertinoro, Italy, 26–31 July 2020, Proceedings, vol. 13. Springer (2020)
71. Xanthidis, D., et al.: Introduction to neural networks and deep learning. In: Handbook of Computer Programming with Python, pp. 449–484. Chapman and Hall/CRC (2022)

72. Huang, K., et al.: Foundations of generative AI. In: Generative AI Security: Theories and Practices, pp. 3–30. Springer (2024)
73. Dogo, E.M., et al.: A comparative analysis of gradient descent-based optimization algorithms on convolutional neural networks. In: 2018 International Conference on Computational Techniques, Electronics and Mechanical Systems (CTEMS). IEEE (2018)
74. Yang, J., Long, Q.: A modification of adaptive moment estimation (ADAM) for machine learning. J. Ind. Manag. Optimiz. **20**(7), 2516–2540 (2024)
75. Smys, S., et al.: Survey on neural network architectures with deep learning. J. Soft Comput. Paradigm (JSCP) **2**(03), 186–194 (2020)
76. Dhruv, P., Naskar, S.: Image classification using convolutional neural network (CNN) and recurrent neural network (RNN): a review. Mach. Learn. Inf. Process. Proc. ICMLIP **2020**, 367–381 (2019)
77. Raiaan, M.A.K., et al.: A review on large language models: architectures, applications, taxonomies, open issues and challenges. IEEE Access (2024)
78. Imamguluyev, R.: The rise of GPT-3: implications for natural language processing and beyond. Int. J. Res. Pub. Rev. **2582**, 7421 (2023)
79. Patil, R., Gudivada, V.: A review of current trends, techniques, and challenges in large language models (LLMS). Appl. Sci. **14**(5), 2074 (2024)
80. Choi, K., et al.: Fair generative modeling via weak supervision. In: International Conference on Machine Learning. PMLR (2020)

Mathematical Foundations and Applications of Generative AI Models

Naru Venkata Pavan Saish, J. Jayashree, and J. Vijayashree

Abstract This chapter examines the mathematical foundations of generative models, key components of modern machine learning (ML) and artificial intelligence (AI). Generative models are algorithms designed to replicate complex patterns in data by generating new samples that resemble the original dataset. This chapter reviews the essential concepts in probability theory, such as conditional probability, probability distributions, and Bayes' theorem, which are fundamental to understanding generative modeling. This chapter also explores core techniques in generative modeling, including probabilistic graphical models like Markov random fields and Bayesian networks, which provide structured frameworks for representing relationships between variables. The chapter also examines deep generative models, which utilize deep neural networks to learn hierarchical data representations, enabling the creation of highly accurate examples. Further, this chapter also discusses the importance of loss functions in guiding the learning process and evaluating the quality of generated data. The chapter also covers the mathematical principles behind neural networks, using linear algebra and calculus to explain how these networks learn and make predictions. Ultimately, this chapter equips readers with a solid understanding of the mathematical underpinnings of generative models, enabling them to develop innovative algorithms for applications like text generation, drug discovery, and image synthesis.

Keywords Probability distributions · Deep learning · Loss functions · GANs · Optimization · Generative models · Probability theory · Bayesian networks · Deep neural networks · Machine learning

N. Venkata Pavan Saish · J. Jayashree (✉) · J. Vijayashree
School of Computer Science and Engineering, Vellore Institute of Technology, Vellore, India
e-mail: jayashree.j@vit.ac.in

J. Vijayashree
e-mail: vijayashree.j@vit.ac.in

N. R. Vajjhala et al. (eds.), *Generative Artificial Intelligence (AI) Approaches for Industrial Applications*, Information Systems Engineering and Management 24,
https://doi.org/10.1007/978-3-031-76710-4_2

1 Introduction

AI is the result of human desire to create machines that can imitate the cognitive functions that are usually related to the human mind, like learning and problem-solving. This interdisciplinary field has many subfields, such as machine learning, natural language processing, computer vision, robotics, and so on. The main goal of AI is to create systems that can reason, understand, and adapt to complex environments, and in many cases, surpass human performance in certain tasks [1–4]. The origin of AI can be dated to ancient times when myths and stories were told about artificial beings with human-like intelligence. Nevertheless, the formal start of AI as a scientific discipline was in the 1950s. The term "artificial intelligence" was created by John McCarthy, who was the organizer of the famous Dartmouth Conference in 1956, which was the starting point of the AI as a field of study [5, 6]. Initially, the AI researchers believed human intelligence could be duplicated with the help of symbolic reasoning and rule-based systems.

The machine learning techniques, especially deep learning, pushed AI to a new level, thus, the remarkable achievements in image recognition, natural language understanding, and game-playing algorithms were accomplished [7]. Generative AI or known as generative artificial intelligence is that category of AI that is revolutionizing the creative processes. While traditional AI is geared towards reading and comprehending available data, generative AI takes it to another level by incorporating the past knowledge it's acquired into creating novel and impressive content [8]. The capability of these AI models to explore and process tremendous amounts of data makes them smart enough to discover relationships and patterns in that data [9]. This covers, for example, the format of a sentence or how objects are depicted in an image or the rhythm and melody of a musical composition. The computer learns to understand and replicates the patterns later when it produces fresh content that keeps with the same principle. The application area of generative AI is experiencing a very fast-paced development. One of the most impressive features is the amount of reality that these algorithms can mirror. Thanks to generative AI, it is now possible to create images that are so persuasive and real that one can hardly tell the difference between them and actual photos [10]. This subsequently arouses interest in using the technology for application in product design and marketing.

This chapter is largely centered on the developments of AI and the explanation of such progression in thematically related mathematics. In this chapter we explore the history of AI while focusing on the mathematical foundations of today's deep neural networks [11]. These frameworks include Linear Algebra which introduces the tools for manipulating data in high dimensional space and Calculus which is all that is needed to understand the gradient based optimization which forms the basis of learning in Neural Networks. In the following sections where we present the models, we explain them with rigor using mathematical notations and formulations [12]. The intended objective of this chapter is to definitively establish the mathematics required for generative modeling where the readers are provided with comprehensive knowledge of the various concepts allowing them to design new algorithms for real life use. From using generative models for generating realistic images, synthesizing text, or discovering new drugs, the insight to be obtained from this paper will enable the researchers and practitioners to unlock the full potential of generative models.

2 Review of Literature

The twenty-first century was characterized by an unparalleled speed in AI development, resulting from the progress in computing power, big data, and algorithmic innovation. Another aspect of AI that one cannot ignore is the growing availability of generative AI tools. With the accessibility of these tools on the rise, generative AI will be in the hands of more people compared to ever before. The democratization of AI also represents the wave of creativity and innovation, across various industries. Another significant advancement is in natural language processing (NLP), where models like OpenAI's GPT (Generative Pre-trained Transformer) series have gained widespread attention [2].

2.1 Deep Learning

Deep learning is the focused branch of the machine learning algorithm in the field of artificial intelligence that has changed the world of data and technology. It is defined by its ability to handle data with massive dimensional representations using multi-layered neural networks referred to as 'deep'. The most important consequence of deep learning is Automatic feature extraction—Deep learning presents deep architectures for automatic feature extraction from raw data devoid of dependence on handcrafted features which are often time-consuming and expertise hungry. This presents an important breakthrough for the existing disk space optimization work in Generative AI (GenAI), which seeks to develop systems that can generate new works that exhibit certain characteristics or attributes of human creativity and cognition. The application of artificial neural networks (ANNs), which are modelled after the architecture and operations of the human brain, is a fundamental component of deep learning. These networks are made up of layers of connected nodes, or neurons. After processing inputs, each neuron generates an output that is transferred to the layer above [3]. The term "deep" refers to the number of layers in a neural network, and deep learning usually incorporates networks with several layers. From low-level characteristics like image edges to high-level features like shapes and objects, these layers can capture hierarchical representations of data.

In GenAI, deep learning has been crucial to the development of various applications in areas such as NLP, CV, and audio. For instance, deep learning algorithms such as Transformers have revolutionized Natural Language Processing (NLP) tasks by significantly improving the way machines interpret and produce human language [13]. One of the key contributions of the paper "Attention is All You Need" is the invention of a new machine translation algorithm known as Transformer, which relies on self-attention to compute weights for different words of a sentence, to capture longer-range context than was previously possible. This architecture forms the foundation of effective language models such as GPT-3 and its iterations that can produce human-like text, translate between languages, summarize documents, and in some cases, generate poetry and code. In the same way, deep learning enhanced the capability of machines to process visual information in computer vision. For tasks like image classification object detection and segmentation CNNs have been the go-to method. These models can find, detect, and outline objects in an image, detect facial landmarks, and even create images from nothing. For example, Generative Adversarial Networks (GANs) are a particular kind of deep learning model, where the model is two neural networks, known as a generator and a discriminator, that are trained in a 'competitive' fashion [7]. The generator generates fake images that appear more realistic with each generation while the discriminator aims to identify images that are real or fake. This process is adversarial and leads to images that can be realistically used for artistic, creative, or even data augmentation purposes for training other models. Deep learning is also heavily involved in audio processing and makes it possible to teach machines to recognize speech and even produce high-quality human voice imitations [14]. RNNs and its modern successors like LSTM networks and GRUs are particularly suitable for analyzing sequential data like audio. These models have been employed to design advanced speech recognition technologies, text-to-speech converter, and music synthesis systems. For instance, WaveNet, a model created by DeepMind, simulates audio by mimicking the waveform of sound and explains details that earlier models failed to [15].

When coming to sentiment analysis, The IMDB Movie Reviews dataset is one of the most popular datasets mainly employed in the field of sentiment analysis, which includes 50,000 extremely positive and negative reviews. In the healthcare industry, these technologies help in the generation of artificial datasets that are used to build diagnostic algorithms, to test drug reactions, and to generate individualized treatment plans. In marketing, GenAI can create personalized content for adverts, social media, and customer service, which will boost the engagement and conversion rates. In addition, the potential of deep learning models to be portable and applicable to GenAI projects makes them suitable [16]. Such models can be trained on huge datasets and then further optimized for a certain task, which can then be optimized again for a new dataset and a new task. This is important in GenAI where there is a need for bringing something new and original into the world. For instance, deep learning models can be trained on various datasets ranging from text, images and audio to generate cross-media content that encompasses multiple types of media and presents a more engaging experience to the user [17].

Perceptron

The Perceptron model proposed by Frank Rosenblatt in 1957 is one of the basic models of artificial neural networks. It depicts the basic configuration of a neural network that can carry out binary classifications. It is helpful for one to understand the Perceptron algorithm and its mathematical formulation to get the big picture of the workings of machine learning and neural networks. The Perceptron is a type of linear classifier which determines its decision based on the result of a weighted sum of the input features and a comparison to a given threshold. The model has input features, weights associated with features, a bias term, and an activation function [18]. The inputs are features in the dataset and every input is associated with weight value; this denotes the significance of feature in decision making process. The bias term helps in shifting the decision boundary away from the origin and hence it helps in improving the flexibility of the model. The perceptron updates the weights and bias based on the error between the true label and the predicted label, aiming to reduce this error over time.

The Perceptron's prediction is based on a linear combination of the input features x, the corresponding weights w, and the bias term b. The output y is determined by applying an activation function to this linear combination. G is the Activation function which introduces Non-Linearity and Y is the output of the perceptron [19]. The mathematical formula for the Perceptron can be expressed as follows:

$$Y = G\left(\sum_{i=1}^{n} w_i x_i + b\right)$$

Dense Neural Network

A neural network is a dense neural network or fully connected neural network is the basic architecture in deep learning. It comprises of numerous connected neurons wherein each neuron is linked with all the neurons in the succeeding and prior layers. This interconnected feature enables DNNs to learn complex representations of data as a function of combinations of inputs. A DNN comprises three layers: the input layer, the hidden layer, and the output layer. The first layer is the input layer which takes the input data this could be in terms of pixel values in an image or features in a dataset [20]. One neuron in this layer corresponds to one attribute of the data. The output layers give the final prediction or classification. The hidden layers are positioned between these two layers; they are the actual site of processing and learning. The functionality of a DNN can be mathematically described through the concept of a weighted sum. Each neuron in a layer receives inputs from all neurons in the previous layer. These inputs are multiplied by corresponding weights, summed up, and then passed through an activation function. The formula for the output of a neuron j in layer l can be expressed as:

$$z_j^l = \sum_{i=1}^{n_{l-1}} w_{ij}^l a_i^{l-1} + b_j^l$$

The weighted sum z is then passed through an activation function to introduce non-linearity into the model, enabling it to learn complex patterns. Common activation functions include the sigmoid function, hyperbolic tangent (tanh), and rectified linear unit (ReLU). For example, the ReLU activation function is defined as:

$$a_j^l = \mathrm{ReLU}\left(z_j^l\right) = \max\left(0, z_j^l\right)$$

The DNN will modify the weights and biases parameters during training to achieve the least error between the predicted values and target values. This is usually achieved using a method called backpropagation in conjunction with an optimization method e.g. stochastic gradient descent (SGD). The adjustments are made based on the loss function which represents the difference between the actual output and the predicted one. A common loss function for classification tasks is the cross-entropy loss: A common loss function L for classification tasks is the cross-entropy loss:

$$L = -\sum_{i=1}^{m}\left[y_i \log(a_i) + (1 - y_i)\log(1 - a_i)\right]$$

Backpropagation calculates the gradient of the loss function with respect to each weight by applying the chain rule of calculus. This gradient indicates how much the loss would change with a small change in the weight [21]. The weights are then updated in the opposite direction of the gradient to reduce the loss. The weight update rule can be expressed as:

$$w_{ij}^l \leftarrow w_{ij}^l - \eta \frac{\partial L}{\partial w_{ij}^l}$$

where η is the learning rate, a hyperparameter that controls the size of the weight updates. DNNs have been used in numerous tasks such as images and speech, natural language, and games. Their unique characteristics of automatically extracting features from the given raw data make them a useful tool in the domain of machine learning. The major drawback of deep networks is their increased computational cost and the need for hyperparameter optimization to obtain the best results [22].

RNN

RNNs are a special type of artificial neural networks that are well suited for processing serial data. Unlike the traditional feedforward neural networks where there are no directed cycles, the RNNs have directed cycles which retain information about the previous inputs in terms of a hidden state. This characteristic makes RNNs suitable for applications in which the input is sequential, like time-series analysis, speech recognition, and natural language processing. The key distinction between RNNs and feedforward neural networks is the presence of recursive connections that enable RNNs to maintain information from prior times. This is done by reinserting the output of the neurons back into the network at the next time step along with the input [23].

An RNN processes an input sequence x = (x_1, x_2, x_3, …, x_T) over T time steps. At each tine step t, the RNN updates its hidden state h_t based on the current input x_t and the previous hidden state h_(t − 1). The hidden state h_t serves as memory that captures information about the sequence up to time t. Mathematically, the hidden state update can be expressed as:

$$\mathbf{h}_t = f(\mathbf{W}_{xh}x_t + \mathbf{W}_{hh}\mathbf{h}_{t-1} + \mathbf{b}_\mathrm{h})$$

where

h_t is the hidden state.
x_t is the input at time t.
$\boldsymbol{W}_{xh}$ is the weight matrix for the input hidden state.
$\boldsymbol{W}_{hh}$ is the weight matrix for the hidden state to hidden state.
$\boldsymbol{b}_h$ is the bias vector for the hidden state.
f is an activation function, typically the hyperbolic tangent (tanh) or rectified linear unit (ReLU).

The output y_t at each time step t is computed using the current hidden state h_t

$$y_t = g\left(\boldsymbol{W}_{hy}h_t + b_y\right)$$

where

$\boldsymbol{W}_{hy}$ is the weight matrix for the hidden state to output.
b_y is the bias vector for the output.
g is an activation function, often a SoftMax function for classification tasks.

Training RNNs involves adjusting the weights and biases to minimize a loss function over the entire sequence. This is typically done using backpropagation through time (BPTT), an extension of the standard backpropagation algorithm for sequences. BPTT computes the gradients of the loss with respect to the weights by unrolling the RNN over time and applying the chain rule of calculus. The loss function L for a sequence can be defined as the sum of the losses at each time step:

$$L = \sum_{t=1}^{T} L_t(y_t, \hat{y}_t)$$

One major challenge with training RNNs is the problem of vanishing and exploding gradients. During backpropagation through time, the gradients can become very small (vanish) or very large (explode), making it difficult to learn long-term dependencies [24]. This problem is especially pronounced for long sequences. To mitigate these issues, variants of RNNs such as Long Short-Term Memory (LSTM) networks and Gated Recurrent Units (GRUs) have been developed.

LSTM

LSTMs are a type of RNN specifically designed to address the vanishing gradient problem. They introduce a more complex structure called a memory cell, which can maintain information over long periods. Each LSTM cell contains gates that regulate the flow of information:

- **Forget Gate**: Decides what information to discard from the cell state.

- **Input Gate**: Decides what new information to store in the cell state.
- **Output Gate**: Decides what information to output based on the cell state.

The equations for an LSTM cell are:

$$\begin{aligned}
f_t &= \sigma\left(\boldsymbol{W}_f.\left[\boldsymbol{h}_{t-1}, x_t\right] + \boldsymbol{b}_f\right)\\
i_t &= \sigma\left(\mathbf{W}_\mathrm{i}.\left[\mathbf{h}_{\mathrm{t}-1}, \mathrm{x}_\mathrm{t}\right] + \mathbf{b}_\mathrm{i}\right)\\
\tilde{C}_t &= \tanh\left(\mathbf{W}_C.\left[\mathbf{h}_{\mathrm{t}-1}, \mathrm{x}_\mathrm{t}\right] + \mathbf{b}_\mathrm{C}\right)\\
C_t &= f_t * C_{t-1} + i_t * \tilde{C}_t\\
\mathrm{o}_\mathrm{t} &= \sigma\left(\mathbf{W}_\mathrm{o}.\left[\mathrm{h}_{\mathrm{t}-1}, \mathrm{x}_\mathrm{t}\right] + \mathbf{b}_\mathrm{o}\right)\\
\mathbf{h}_t &= \mathrm{o}_t * \tanh(C_t)
\end{aligned}$$

RNNs and their variants are widely used in various applications due to their ability to process sequential data. Some notable applications include:

- Natural Language Processing (NLP): RNNs are used for tasks like language modeling, machine translation, text generation, and sentiment analysis. They can capture the temporal dependencies in text data, making them suitable for understanding and generating human language [25].
- Speech Recognition: RNNs are employed to convert spoken language into text. They can model the temporal structure of audio signals, allowing for accurate transcription of speech.
- Time Series Prediction: RNNs are used for predicting future values in a sequence based on past observations. This is useful in fields like finance, weather forecasting, and stock market analysis.
- Music Generation: RNNs can be trained to compose music by learning patterns in sequences of musical notes.
- Video Analysis: RNNs can analyze video data by processing sequences of frames. This is useful for tasks like action recognition and video captioning [26].

RNNs are one of the most important and prominent architectures in the field of deep learning and have seen huge advances in many aspects of sequential data analysis including time-series, speech, and video applications. Their capabilities to keep context and to address temporal relations rendered them useful in the fields mentioned above. One of the most important applications of RNNs is in the field of NLP where they have helped to achieve progress in the field of language modelling, machine translation, text summarization, and sentiment analysis. The following paragraph will provide further insights into how RNNs are used in NLP and the influence they have had in this field.

2.2 Natural Language Processing

Natural Language Processing (NLP) is a subtopic of computer science and artificial intelligence that is concerned with human language. It refers to the field of study that focuses on computers that can interpret and manipulate human language. NLP involves a broad spectrum of applications ranging from basic operations such as spell checking and text categorization to advanced applications, for example machine translation and natural language understanding. The recent development in terms of powerful machine learning algorithms, especially Recurrent Neural Networks, has led to significant strides in NLP that has revolutionized the applications and industries [27]. While different, all NLP systems share several tasks that are essential to the field and have their specific methods and difficulties. Some of the primary tasks include:

- Tokenization: This is the process of parting a text into individual words, phrases, symbols, or any other significant unit of the text called tokens. Tokenization is an initial phase in several NLP processes and the most critical method for text preprocessing.
- Sentiment Analysis: This task aims to classify the text's attitude or emotional feeling in a document, for example, is the feeling positive, negative, or neutral. Social sentiment analysis is a widely used technique for social media monitoring or customer feedback analysis.
- Machine Translation: This involves a process where text is converted from a specified language to another language. It is rather one of the most hard and beneficial NLP applications.
- Text Summarization: This task involves using a summary writing format to write an abstract of a body of work in a compact form while maintaining the content. It is helpful in the sense that it allows for the grasping of the overall picture of large document.

RNNs have been at the forefront of NLP. While ordinary neural networks do not have feedback connections, RNNs do, enabling them to remember their previous state in so-called hidden layers. This makes RNNs particularly ideal for data in the form of sequences, where the order of elements is important, such as in text data [28]. Language modeling is the problem where we are given a sequence of words and asked to compute the probability of some future word conditioned on the past words. It is one of the central concepts in NLP and is used in text and translation generation. RNNs, especially their deeper versions like LSTM and GRU, are particularly useful for language modeling as they facilitate learning dependencies in text. One of the most amazing ways through which NLP has been put into use is machine translation. First and second generations of machine translation technology were based on rule-based approaches and statistical machine translation respectively; both of which were computationally expensive and time consuming; they also called for large parallel corpora [29]. But with the help of RNNs and especially more recently with the rise of Transformer models, machine translation has seen progress. Among the models that have been used, RNN-based sequence-to-sequence models with attention mechanisms have been particularly successful. The use of encoder-decoder recurrent neural networks (RNNs) where the encoder RNN receives the input sentence and the decoder RNN produces the output sentence as the final translation has become a standard way of performing the task. This is further strengthened by the attention mechanism which helps the model to choose parts of the input sentence during the decoding stage with the aim of producing a more accurate translation [30].

Text generation is the task of producing text that reads naturally and is relevant after seeing some input. RNN has been implemented for text generation in different styles and domains such as poetry or stories generation and technical articles generation. RNNs are trained in large amounts of text, where they learn the structure and patterns that exist in a human language to generate human-like text. For example, creating poetry or writing original music lyrics means teaching RNNs enormous amounts of Shakespeare's writing or text collections with lyrics. The RNN can then go ahead to create new text that has similar structure to the original content. Another important usage of NLP is sentiment analysis, which is applied often in market research, customer feedback analysis, and social media monitoring. When RNNs consider the word order and context, they may accurately represent the sentiment conveyed in text. RNNs learn to predict the sentiment of new texts by training on labeled datasets where texts are tagged with sentiment labels.

For instance, it is possible to categorize movie reviews as favorable or bad using a basic RNN model. As the model analyzes each word in the review, the hidden states of the RNN record the changing sentiment and produce an overall sentiment forecast.

Text Processing

A key component of NLP is text processing, which gives computers the ability to manage, interpret, and evaluate textual data. This is a set of procedures and methods that turn unstructured text into a format that is suitable for a range of natural language processing (NLP) applications, including text summarization, sentiment analysis, and machine translation [31]. Every stage of the text processing process is essential to getting the text ready for more in-depth modeling and analysis. The first and most important stage in Natural Language Processing (NLP) is text preprocessing. To guarantee that the text data is in an appropriate format for additional analysis, it must be cleaned and prepared. Text preparation involves the following steps:

- Tokenization: Tokenization is the process of dividing a document into discrete words or units of speech. For languages with rich morphology, tokenization may entail more complicated rules than just dividing text into spaces. The initial stage in comprehending the text structure is tokenization.
- Lowercasing: Making all the text's characters lowercase guarantees consistency and lessens the complexity brought on by case sensitivity. "Python" and "python" are two examples of tokens that would be interpreted similarly.
- Removing Punctuation: In most NLP jobs, punctuation is eliminated from text because it adds nothing to the meaning. This cuts down on the number of tokens and simplifies the language.

- Removing Stop Words: Common words that appear repeatedly but have little meaning, including "is," "and" and "the," are considered stop words. Eliminating stop words makes it easier to concentrate on the words that provide sense to the text.
- Stemming and Lemmatization: Lemmatization considers the morphological analysis of words and returns the base form (lemma) that exists in the language (e.g., "better" to "good"). Stemming reduces words to their base or root form (e.g., "running" to "run"). Lemmatization is favored for more precise text processing.

Feature Extraction

A crucial step in machine learning and natural language processing (NLP) is feature extraction, which entails converting unprocessed data into a format that models can use efficiently. Feature extraction in NLP seeks to translate textual data into numerical representations while maintaining the text's semantics and organization. Since most machine learning algorithms need numerical input to function, this transformation is crucial [32]. In NLP, feature extraction techniques span a wide range of approaches, from straightforward ones like Bag of Words (BoW) to more complex ones like word embeddings and deep learning-based algorithms. This material examines several NLP feature extraction techniques.

Bag of Words (BoW)

One of the simplest and most used feature extraction methods in NLP is the Bag of Words (BoW) model. BoW treats a text as an unordered collection of words, retaining diversity while ignoring word order and grammar. With this method, each document is represented by a vector that counts the occurrences of each word, and a vocabulary of all the unique terms in the text corpus is created. Every word's frequency in these phrases would be tallied by the BoW representation. BoW is straightforward and simple to use, but it has drawbacks. For example, it ignores context and word order, which are important factors in meaning interpretation.

Term Frequency-Inverse Document Frequency (TF-IDF)

An improvement over the BoW model is Term Frequency-Inverse Document Frequency (TF-IDF). TF-IDF takes into consideration both term frequency, which measures how frequently a word appears in a document, and inverse document frequency, which measures how common or unique a word is across all texts in the corpus. This contributes to the weight of uncommon and important words being increased and common words like "is" and "the" being decreased. The number of times a term appears in a document divided by the total number of terms in the document yields the term frequency (TF). The logarithm of the total number of documents divided by the number of documents containing the phrase yields the inverse document frequency, or IDF. For applications like document categorization and information retrieval, TF-IDF is a more effective feature extraction method since it helps to emphasize essential words and de-emphasize common terms. While word embeddings capture the meaning of individual words, sentence embeddings capture the meaning of entire sentences or paragraphs. These embeddings are useful for tasks that require understanding the context and relationships between words in a sentence [33].

Word Embeddings

A major development in feature extraction for natural language processing is word embeddings. Word embeddings depict words as dense vectors in a continuous vector space, in contrast to BoW and TF-IDF, which represent words as sparse vectors. Words with comparable meanings are grouped together in the vector space to represent the semantic links between words.

- Word2Vec: Created by Google, Word2Vec learns word associations from big text corpora using neural networks. It uses two primary methods to build embeddings: Skip-gram and Continuous Bag of Words (CBOW). Whereas Skip-gram predicts the context based on a given word, CBOW predicts a word based on its context.
- GloVe (Global Vectors for Word Representation): a count-based approach developed at Stanford that aggregates global word-word co-occurrence statistics from a corpus to produce word embeddings. Both local and global statistical information about words are captured by the ensuing embeddings.
- FastText: Facebook created FastText, an extension of Word2Vec that takes subword information into account. It can create embeddings for words that were not observed during training (out-of-vocabulary words) by piecing together the character n-grams that represent each word.

One essential part of NLP is feature extraction, which converts unprocessed textual data into numerical representations that may be used with machine learning models. Every approach, from cutting-edge strategies like word embeddings and deep learning-based models to more conventional ones like Bag of Words and TF-IDF, has specific benefits for certain natural language processing applications. The ability of models to comprehend, interpret, and produce human language has been greatly improved by the ongoing development of feature extraction techniques, which has accelerated the development of numerous NLP applications [34]. Feature extraction will always be essential to creating increasingly complex and precise language processing systems as NLP advances.

2.3 Generative AI

Within AI, generative AI is a quickly developing topic that focuses on building models that can produce fresh, original content, including text, photos, music, and even films. In contrast to conventional AI models, which are mainly intended for tasks such as classification, prediction, or optimization, generative AI models acquire the ability to recognize patterns and structures in data, which allows them to generate new outputs that bear similarities to the training set. This innovative technology has the power to completely change several sectors, including healthcare, design, entertainment, and more. Deep learning models, especially those based on neural networks, are the foundation of generative artificial intelligence. Introduced by Ian Goodfellow and colleagues in 2014, Generative Adversarial Networks (GANs) represent one of the most notable advances in generative AI. Generator-discriminator networks (GANs) are made up of two neural networks that are trained concurrently using adversarial learning. The discriminator compares the newly generated data samples to actual data and provides input to the generator so that it can enhance its outputs. The generator generates new data samples. Until the generator generates data that cannot be distinguished from actual samples, this iterative procedure is repeated [35].

NLP has been transformed by the Transformer architecture, another groundbreaking generative AI paradigm. The Transformer model, presented by Vaswani et al. in 2017, makes efficient use of self-attention mechanisms to process and produce text sequences. Later models such as BERT (Bidirectional Encoder Representations from Transformers) and GPT (Generative Pre-trained Transformer) were built on top of this design. One prominent example of a generative AI model that can generate text that is both coherent and contextually relevant on a broad range of topics is OpenAI's GPT-3 model. Its capacity to comprehend and produce text that resembles that of a person has created new opportunities for chatbots, content production, and even code development. Generative AI has considerably more uses than just text. GANs have been applied to the visual arts to produce realistic visuals, improve photo resolution, and even produce artwork in the vein of well-known painters. For example, users can turn common images into works of art that emulate the techniques of renowned artists like Picasso, Van Gogh, and others using DeepArt and other AI-driven technologies [36]. This capacity to produce high-quality visual content has important ramifications for sectors where visual appeal is crucial, such media, gaming, and advertising.

Furthermore, training big generative models requires a significant amount of computer resources. Large datasets and a lot of processing power are needed for models like GPT-3, which raises questions regarding accessibility and environmental sustainability. As the area progresses, it will be imperative to build more efficient algorithms and utilize approaches such as transfer learning to mitigate these issues. With generative AI providing new means to improve our lives and broaden our creative horizons, its influence on society is only going to increase as researchers and practitioners continue to investigate and improve this technology [37].

GAN (Generative Adversarial Networks)

A novel family of machine learning frameworks called Generative Adversarial Networks (GANs) is created to produce fresh, synthetic data that closely resembles a specified distribution of real data as shown in Fig. 1. Since their introduction by Ian Goodfellow and associates in 2014, generative architecture networks (GANs) have transformed the area of generative modeling by providing innovative solutions across a range of fields, such as drug discovery, image production, and text synthesis. This material explores the design, operation, uses, and difficulties of GANs, offering a thorough understanding of this revolutionary technology.

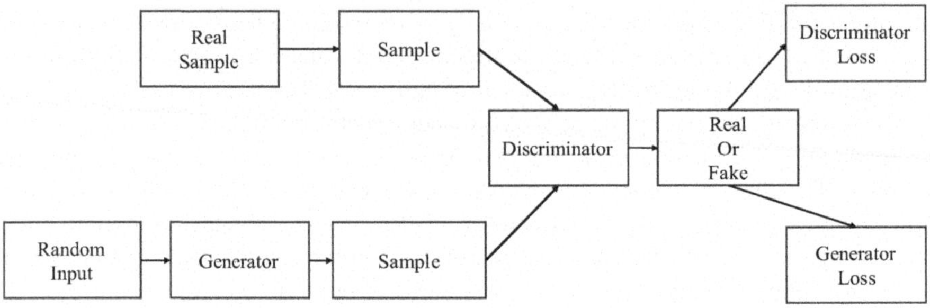

Fig. 1 GAN structure

Architecture and Functioning of GANs

- Generator: The generator network creates synthetic data samples by converting random noise as input. To create data that is indistinguishable from actual data is its aim. It keeps learning from the discriminator's feedback to make its generated samples more realistic.
- Discriminator: In contrast, the discriminator network is a binary classifier that determines if input data is created or real by using real-world examples. Its main goal is to accurately differentiate the bogus data produced by the generator from the real data samples [38]. The discriminator seeks to decrease the likelihood of misclassification during training, whereas the generator seeks to maximize the likelihood that the discriminator would classify its output as real. One way to formalize this adversarial process is as follows:

$$\min_{G} \max_{\mathrm{D}} \mathbb{E}_{x \sim p_{data}(x)}\left[\log D(x)\right] + \mathbb{E}_{z \sim p_z(z)}[\log(1 - D(G(z))]$$

where

- G represents the generator
- D represents the discriminator
- x is a sample from the real data distribution $P_{data}(x)$ $P_{data}(x)$
- z is a sample form the noise distribution $P_z(z)$.

The training phase is iteratively updating the generator and discriminator using backpropagation, guided by the respective loss functions derived from the equation above. Over time, the generator improves at providing realistic data, while the discriminator improves at detecting fake data, until an equilibrium is established in which the discriminator can no longer confidently discriminate between genuine and fake data. These two networks are trained simultaneously in a process known as adversarial training, where the generator improves its data generation capabilities based on feedback from the discriminator, and the discriminator enhances its ability to distinguish real data from fake data [39].

Applications of GANs

GANs have found applications in a variety of fields due to their capacity to provide high-quality, realistic data. Here are a few notable applications:

- Image Generation: Image synthesis is one of the most common applications for GANs. GANs may generate realistic images from noise, allowing for the development of new artworks, photorealistic human faces, and even fictitious items. For example, GANs have been used to create images of non-existent celebrities, high-resolution photos from low-resolution inputs (super-resolution), and art inspired by great painters.
- Data Augmentation: In machine learning, a vast and diverse dataset is essential for developing robust models. GANs can improve current datasets by creating new synthetic samples that closely resemble the original data. This is especially important in industries like medical imaging, where getting tagged data can be costly and time-consuming [40].
- Text-to-Image Synthesis: GANs can convert text descriptions into related images. This application has important implications for content production and design, as it enables the development of visual material from textual inputs. For example, given a description such as "a small bird with a red belly and blue wings," a GAN can generate an image that matches the description.
- Video Generation and Prediction: GANs are used to generate and predict video sequences. This includes producing totally new videos from a set of initial frames, as well as anticipating future frames in a video sequence. These capabilities are useful in applications such as autonomous driving, which requires predicting the movement of objects in video frames.
- Multi-modal Generation: Using GANs to generate multi-modal data, such as text, images, and audio, can result in more sophisticated and versatile generative models. This can improve applications for virtual reality, augmented reality, and cross-modal content development.
- Adversarial Training approaches: Advancements in adversarial training approaches, such as adversarial domain adaptation and adversarial feature learning, show promise for increasing the resilience and generalization of GANs in a variety of applications [41].

While GANs have already had a considerable influence in domains such as image synthesis, data augmentation, and creative content generation, they are still evolving, with continuing research tackling present issues and opening new possibilities. As GAN technology evolves, it is critical to manage its ethical implications appropriately, ensuring that its transformational potential is used for the betterment of society. In summary, GANs represent a significant advancement in artificial intelligence, enabling the generation of realistic and diverse data.

Transformers

Transformers have transformed the fields of natural language processing (NLP) and machine learning since their debut by Vaswani et al., in 2017. Their capacity to analyze data sequences in parallel and understand complicated connections has made them the foundation for numerous cutting-edge models, including BERT, GPT, and T5. Transformers were developed to overcome the constraints of recurrent neural networks (RNNs) and convolutional neural networks (CNNs) in handling sequential input. Transformers, unlike RNNs, can process complete sequences at the same time, allowing for faster training and greater capture of long-range relationships. Transformers developed out of an idea to overcome the constraints of existing NLP techniques, which frequently relied on RNNs to analyze sequential input such as text. RNNs, while effective, were plagued by difficulties such as vanishing gradients, which hampered their ability to learn long-range dependencies within words. This ground-breaking technique established the concept of self-attention, a process that enables the model to focus on significant sections of the input sequence regardless of their position. Transformers, unlike RNNs, may examine all sections of an input sentence at the same time, allowing them to capture complicated word associations. This transition to a parallel processing architecture provided various benefits. Transformers excel at tasks such as machine translation, which requires understanding the context of the full sentence to generate appropriate translations. The success of Transformers has spurred additional study and innovation in NLP [42]. Variants of the fundamental architecture have been created to handle distinct tasks and domains. For example, pre-trained language models like BERT and GPT-3 use Transformer-based architectures to deliver cutting-edge performance in tasks such as text generation and question answering. The architecture of transformers is shown in Fig. 2.

Attention Mechanism

The attention mechanism is a critical component of the Transformer model, allowing it to properly interpret input sequences by focusing on essential areas of the input when encoding or decoding tokens. The most important innovation in the Transformer architecture is the self-attention mechanism, which enables the model to dynamically balance the value of distinct tokens in a sequence. This section digs into the specifics of the attention mechanism and its mathematical representation. Transformers' attention mechanism assists the model in determining the relative importance of each token in the input sequence. This approach allows the model to capture relationships between tokens regardless of their position, making it especially useful for activities with long-range dependencies [43].

The attention mechanism is a fundamental component of the Transformer model, enabling it to process input sequences effectively by focusing on relevant parts of the input when encoding or decoding tokens. The most critical innovation in the Transformer architecture is the self-attention mechanism, which allows the model to weigh the importance of different tokens in a sequence dynamically. The mathematical principles underlying Transformers, including self-attention, multi-head attention, and positional encoding, are crucial to their success. As research continues, Transformers are expected to evolve further, finding new applications and improving performance across various domains. Now we delve into the details of the attention mechanism and its mathematical formulation. The attention mechanism in Transformers helps the model determine the importance of each token in the input sequence relative to

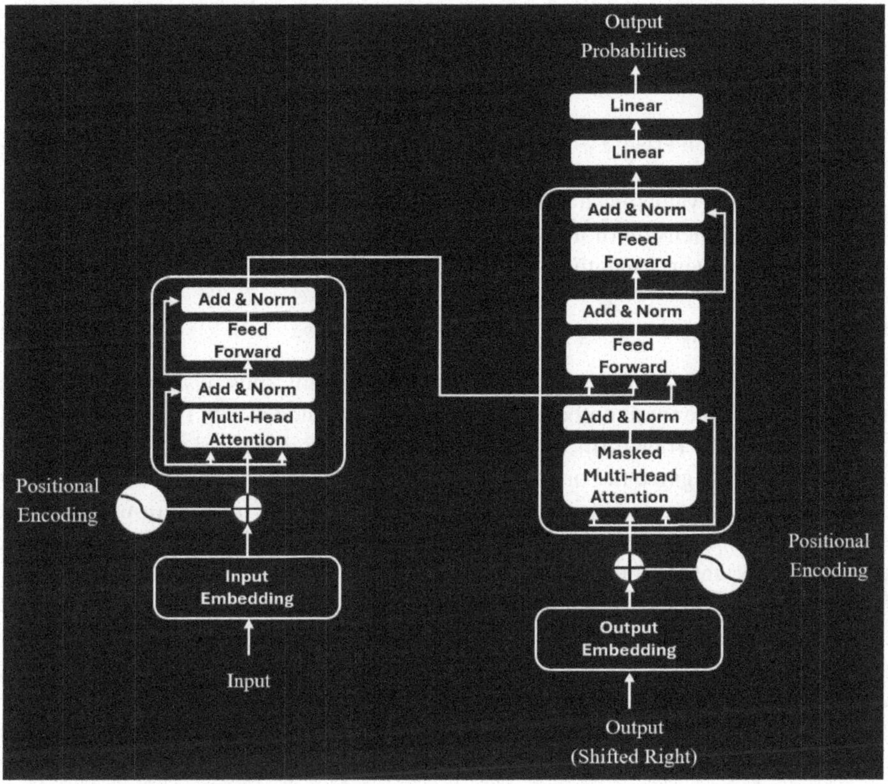

Fig. 2 Transformer architecture

others. This mechanism enables the model to capture dependencies between tokens irrespective of their positions, making it particularly effective for tasks involving long-range dependencies [44]. The primary type of attention used in Transformers is the scaled dot-product attention. It involves the following steps:

- **Input Vectors**: For each token in the sequence, the model generates three vectors: Query (Q), Key (K), and Value (V). These vectors are computed by multiplying the input embeddings with learned weight matrices.
- **Attention Scores**: The attention score for a pair of tokens is computed by taking the dot product of their query and key vectors. These scores are then scaled by the square root of the dimension of the key vectors (d_k) to ensure stable gradients during training.
- **Softmax Function**: The scaled scores are passed through a softmax function to obtain the attention weights, which are essentially probabilities indicating the importance of each token in the context of the current token.
- **Weighted Sum**: Finally, the attention weights are used to compute a weighted sum of the value vectors, producing the output for each token.

$$\textbf{Attention}(Q, K, V) = \text{softmax}\left(\frac{Q.K^T}{\sqrt{d_k}}\right)V$$

$Q \in \mathbb{R}^{n \times d_k}$ (queries) is a matrix of queries derived from the input embeddings.
$K \in \mathbb{R}^{n \times d_k}$ (keys) is a matrix of keys derived from the input embeddings.
$V \in \mathbb{R}^{n \times d_k}$ (values) is a matrix of values derived from the input embeddings.
d_k is the dimension of the key vectors.
n is the sequence length.

Positional Encoding

Transformers do not comprehend token order by default, thus positional encodings are appended to the input embeddings to provide the model with information about token placements. Transformers, unlike recurrent neural networks (RNNs), do not automatically recognize the order of tokens in a sequence. To overcome this, positional encodings are added to the input embeddings, which offer information about each token's position in the sequence. This allows the model to include word order, which is important for interpreting context and meaning in natural language. Transformers process tokens in parallel, hence they lack a built-in mechanism to recognize the sequential character of the data. Positional encodings allow the model to incorporate tokens' relative or absolute positions. This is critical for jobs such as translation, where the meaning of a word varies based on its place in a phrase.

Positional encoding allows us to inject information about the positions of tokens into the model. The formulation employs sine and cosine functions of varying frequency. The sine and cosine functions were chosen because they indicate positions in a continuous space that the model can easily learn. Using distinct frequencies allows the model to capture both short-term and long-term dependence. Specifically, the positional encoding for a token at position *pos* is given by:

$$PE_{(pos,2i)} = \sin\left(\frac{pos}{10{,}000^{2i/d}}\right)$$

$$PE_{(pos,2i+1)} = \cos\left(\frac{pos}{10{,}000^{2i/d}}\right)$$

where

pos is the position of the token in the sequence.
i is the dimension of the positional encoding.
d is the dimension of the model (same as the dimension of the input embeddings).

Using sine and cosine functions, the model can distinguish between different positions and calculate the relative distance between tokens. The use of $10{,}000^{2i/d}$ as a denominator ensures that the sine and cosine functions fluctuate smoothly across dimensions.

Multi-head Attention

Multiple attention heads are used to capture different aspects of the input sequence:

$$\mathbf{head_i} = \text{Attention}\left(\text{Q}W_i^Q, \text{K}W_i^K, \text{V}W_i^V\right)$$

where $\text{Q}W_i^Q$, $\text{K}W_i^K$, $\text{V}W_i^V$ are learned projection matrices.

The outputs of the attention heads are concatenated and linearly projected to form the final output:

$$\mathbf{Multihead}_{(\boldsymbol{Q},\boldsymbol{K},\boldsymbol{V})} = \text{Concat}(head_1, head_2, \ldots, head_h)W_O$$

Feed Forward Neural Network

Each layer in BERT contains a feedforward neural network that applies non-linear transformations to the representations:

$$FFN(x) = ReLU(xW_1 + b_1)W_2 + b_2$$

Here W_1 and W_2 are weight matrices, and b1 and b2 are bias vectors.

BERT has set new benchmarks for different NLP tasks, inspiring multiple adaptations, and enhancements. The success of BERT has prompted the creation of models such as GPT (Generative Pre-trained Transformer), which focuses on generative tasks and unidirectional contexts. Future research attempts to make BERT more efficient, lowering computational requirements while maintaining or improving performance. To attain these objectives, techniques such as model pruning, quantification, and knowledge distillation are being intensively investigated [45]. Furthermore, continuing work aims to expand BERT's capabilities to include new languages and specialized topics, making it even more versatile and powerful.

Encoder

The encoder serves as both an information gatherer and interpretation. It takes an input sequence, such as a sentence or paragraph, and processes it to extract meaning and context. This processing occurs through numerous stages, each with two main building blocks:

- Self-attention: This method enables the encoder to grasp the relationships between various words in the input sequence. It assigns a score to each word pair, indicating how important one word is to comprehend another. By paying attention to these scores, the encoder may concentrate on the most essential connections in the input, just like humans do with key parts of a sentence.
- Feed-Forward Network: This is a standard neural network layer that refines the encoded representation. It takes the self-attention mechanism's output and applies nonlinear transformations to extract deeper information and capture complicated word associations.

The encoder stacks these layers several times, allowing it to gradually develop a more detailed and sophisticated knowledge of the input [46]. The encoder's ultimate output is a condensed representation, also known as a context vector, that captures the important meaning of the full input sequence. The encoder processes the input sequence and generates a contextualized representation for each token. It consists of a stack of identical layers, each containing two main components:

- Self-attention Mechanism: Allows the model to focus on different parts of the input sequence when encoding a token.
- Feedforward Neural Network: Applies a non-linear transformation to the encoded representation.

Each encoder layer can be summarized as follows:

$$\textbf{Encoder Layer}(\boldsymbol{x}) = FFN(SelfAttention(x))$$

where x is the input sequence.

Decoder

The decoder takes up the baton from the encoder. Its duty is to generate the output sequence, one element at a time, using the encoder's information [47]. The decoder, like the encoder, has a stacked layer architecture, but with one crucial difference:

- Self-attention: Like the encoder, the decoder uses self-attention to grasp the relationships between items in the generated output sequence thus far. This provides coherence and uniformity across the output text.
- Cross-attention: This is the decoder's secret weapon. It allows the decoder to focus on the context vector generated by the encoder. By focusing on relevant bits of the encoded representation, the decoder can guarantee that the output sequence corresponds to the meaning of the input.

The decoder builds the output element by element. At each step, it uses self-attention to consider previously created items and cross-attention to address the encoder's context vector. This combined information allows the decoder to estimate the next most likely element in the output sequence [48]. The decoder repeats the operation until the whole output sequence is created. The decoder generates the output sequence, using the encoded input representations and previously generated tokens. Like the encoder, the decoder consists of a stack of identical layers, each containing three main components:

- Masked Self-attention Mechanism: Prevents the decoder from attending to future tokens.
- Encoder-Decoder Attention: Allows the decoder to focus on relevant parts of the encoded input sequence.
- Feedforward Neural Network: Applies a non-linear transformation to the decoded representation.

Each decoder layer can be summarized as:

$$\textbf{Decoder Layer}(\boldsymbol{y}, \boldsymbol{e}) \\ = FFN(EncoderDecoderAttention(MaskedSelfAttention(y), e))$$

where y is the target sequence and e are the encoded input sequence.

The essential brilliance of transformers lies in the interaction of the encoder and decoder. The encoder thoroughly analyzes the input, capturing its essence. The decoder then uses the encoded knowledge to create a meaningful output, seamlessly integrating the information from the input [49]. This collaboration enables transformers to do difficult tasks such as machine translation, which requires the decoder to consider the complete source sentence (encoded by the encoder) to provide an accurate translation in the destination language [50]. Transformer encoders and decoders have transformed how machines process and generate text by combining the power of self-attention, feed-forward networks, and innovative architectural decisions. As research advances, these components are expected to play an even bigger role in realizing the full potential of language and human–machine communications [51].

2.4 Metrics

Testing of generative AI models involves a strong set of standards that can adequately measure the quality, the degree of variation and the similarity between the model's outcomes and real data. Herein, the following are the definitions of measures often employed in estimating the generative AI models' performance, examples of each, and their real-world applications.

Perceptual Path Length (PPL): The PPL is a measure that quantifies how continuous the generative model's latent space is and how well transitions between points in the latent space can be made. It quantifies the difference in perception between interpolated images produced from latent vectors, offering information on the continuity of the model's output transition.

Mathematical Foundation

Interpolation in Latent Space: Given two latent vectors z_1 and z_2. Interpolate between them using a linear interpolation parameter t:

$$z(t) = (1 - t)z_1 + tz_2$$

Generated Images: Generated Images I(t) from the interpolated latent vectors z(t) using the generative model G:

$$\mathrm{I(t) = G(z(t))}$$

Perceptual Distance: Calculate the perceptual distance between consecutive images I(t) and $I(t + \delta t)$ using a pre-trained model, typically a deep convolutional network such as VGG. The perceptual distance d_p can be computed as:

$$d_p(I(t), I(t + \delta t)) = \phi(I(t)) - \phi(I(t + \delta t))_2$$

where ϕ is the feature representation from the pre-trained network

Perceptual Path Length: Average the perceptual distances along the path of interpolation:

$$PPL = \frac{1}{L} \sum_{t=0}^{L-1} d_p(I(t), I(t + \delta t))$$

1. Log-Likelihood and Perplexity: Log-likelihood and Perplexity are metrics commonly used for evaluating the performance of generative models, especially in natural language processing. Log-likelihood measures the probability of the test data under the model, while Perplexity provides a normalized measure of how well the model predicts the test data.

Mathematical Foundation

Log-Likelihood: For a given generative model M and the dataset D with N samples {x1, x2, ..., xn} the log-likelihood is calculated as:

$$\log_e L(M) = \sum_{i=1}^{N} \log p(x_i|M)$$

Perplexity: Perplexity is the exponentiated average negative log-likelihood, providing a measure of how well the model predicts a sample:

$$Perplexity = \exp\left(-\frac{1}{N} \sum_{i=1}^{N} \log p(x_i|M)\right)$$

Perplexity can be understood as the geometric mean of the inverse probabilities assigned by the model to each word in the sequence. Lower perplexity indicates better model performance. In a language model, if the average perplexity is low, it means the model is assigning higher probabilities to the correct next words in a sequence, indicating better predictive performance. In text generation tasks like chatbots or predictive text, lower perplexity ensures that the generated text is more coherent and contextually appropriate, enhancing user interaction.

Table 1 Comparative analysis of GenAI architectures

Architecture	Strengths	Weaknesses	Applications
Generative adversarial networks (GANs)	High-quality image generation Versatile for various tasks	Training instability Prone to mode collapse Resource-intensive	Image synthesis, data augmentation, art creation, video generation
Transformers	Handles long-range dependencies Highly parallelizable during training State-of-the-art performance in NLP tasks	Computationally expensive Large memory requirements Can be slower at inference time for very long sequences	Text generation, machine translation, summarization, language modeling
Diffusion models	High quality and diversity of generated images Stable and reliable training dynamics	Slow sampling process Requires substantial computational resources	Image and video generation, data augmentation, molecular design

3 Comparative Analysis of GenAI Architectures

A comparative analysis of GenAI Architectures is presented in Table 1.

4 Challenges and Future Research Directions

Generative models have several well-known and severe technical problems restricting their progress and usage. First, there is the problem of mode collapse, which is when the model only produces a small range of outputs, that is, it only uses a part of the data set. This leads to low diversity in generated samples thus limiting the model's use especially in applications that may demand for diverse output such as in generation of creative content or data augmentation. Such instability may lead to failure of the model to converge, and one is forced to adjust the hyperparameters and employ regularization methods such as gradient penalty or spectral normalization to get the model converge and generalize well.

Despite the availability of large amounts of data, generative models that rely on many parameters for the generation process can place substantial amounts of computational load and may even need the help of dedicated hardware like GPUs or TPUs. These requirements can be disadvantageous, since they can hinder persons with disabilities and small organizations or individuals from trying out and using generative AI solutions. Solving these issues is important for the progress of science and expanding the sphere of usage of generative AI models in different fields. While there is not much a question mark over the capabilities of generative AI, there are multiple, and rather unique, ethical issues that this creation brings into focus: the generation of deepfakes, for example, may be used for propagating fake news, influencing the populace and ruining celebrities' reputations. Another issue is around generating AI that will create fake but very realistic articles including news or posts on social networks that can split society and lead to social unrest. Some of the concerns arising from the ease of creating a lot of content include concerns to do with copyright especially because of the large amount of content that can be produced quickly, and the problem of the proliferation of fake or inferior goods. Efficient hardware configurations, such as multi-GPU setups or specialized hardware like TPUs, are recommended to manage these demands. Additionally, optimizing the software stack with libraries like TensorFlow, PyTorch, and efficient data pipelines can significantly enhance performance.

Moving forward, the following are some of the future paths for generative AI research: one of the trends is the acceleration of model architectures themselves, including high-probability improvements such as the occurrence of sparse attention in transformers or better ways for regularization of GANs. As the approaches in unsupervised or self-supervised learning continue improving the generative models' efficiency in learning from sparse data, they are likely to be applied widely in areas like healthcare that are characteristically data-poor. A more promising future enables integration of generative models with reinforcement learning to build more interactive generation systems. The resultant evolutions are bound to produce better experiences in creative computing as well as scientific research and auto-catalytic systems, where the strides of generative AI are currently converging, it is imperative for these domains to advance both the hardware and the software, in unison, to cope with changing demands.

5 Conclusion

Deep learning powers generative AI models. This chapter examined the mathematical foundations that underpin these models, including essential building elements such as perceptron's and artificial neural networks. This chapter demonstrated how these networks learn by altering connections between processing units, like the way the human brain recognizes patterns and predicts outcomes. Deep learning, a subset of machine learning, is based on neural network mathematical concepts and is inspired by the structure and function of the human brain. Deep learning is fundamentally about training multi-layered neural networks to recognize patterns and predict outcomes from vast datasets. The FNN employs backpropagation, a gradient descent optimization process, to reduce the error between anticipated and actual outputs by iteratively modifying the weights. This procedure uses the chain rule of calculus to compute gradients, which are then utilized to update the weights in the direction that minimizes the error. FNNs are crucial in a variety of applications, including classification, regression, and pattern recognition, and serve as the foundation for more complicated deep learning systems. The chapter then examined deeper into various generative AI models, including GANs and Transformers.

The introduction of Transformers and the attention mechanism has further transformed generative AI. This competition drives the generator to deliver more realistic results. This chapter examined Transformers, a groundbreaking architecture that uses attention mechanisms to analyze complicated relationships in text. Transformers excel in machine translation and text summarization because they focus on relevant bits of the input. Transformers, particularly models such as GPT, rely on attention to manage relationships and context in extended sequences of input, allowing for coherent text production. The attention mechanism computes weighted averages of input vectors, allowing the model to concentrate on important segments of the input sequence. This approach has been expanded beyond NLP to domains such as computer vision, proving the adaptability and efficacy of attention-driven models in generative AI.

References

1. Zhang, B.J., Liu, S., Li, W., Katsoulakis, M.A., Osher, S.J.: Wasserstein proximal operators describe score-based generative models and resolve memorization. arXiv preprint: arXiv:2402.06162 (2024)
2. Rane, N.: Enhancing mathematical capabilities through ChatGPT and similar generative artificial intelligence: roles and challenges in solving mathematical problems. Available at SSRN 4603237 (2023)
3. Yılmaz, B., Korn, R.: Understanding the mathematical background of generative adversarial networks (GANs). Math. Modell. Numer. Simul. Appl. **3**(3), 234–255 (2023)
4. Foster, D.: Generative deep learning. O'Reilly Media, Inc. (2022)
5. Ramezanian-Panahi, M., Abrevaya, G., Gagnon-Audet, J.C., Voleti, V., Rish, I., Dumas, G.: Generative models of brain dynamics. Front. Artif. Intell. **5**, 807406 (2022)
6. Megahed, F.M., Chen, Y.J., Ferris, J.A., Knoth, S., Jones-Farmer, L.A.: How generative AI models such as ChatGPT can be (mis)used in SPC practice, education, and research? An exploratory study. Qual. Eng. **36**(2), 287–315 (2024)
7. Relmasira, S.C., Lai, Y.C., Donaldson, J.P.: Fostering AI literacy in elementary science, technology, engineering, art, and mathematics (STEAM) education in the age of generative AI. Sustainability **15**(18), 13595 (2023)
8. Regenwetter, L., Nobari, A.H., Ahmed, F.: Deep generative models in engineering design: a review. J. Mech. Des. **144**(7), 071704 (2022)
9. Feuerriegel, S., Hartmann, J., Janiesch, C., Zschech, P.: Generative Ai. Bus. Inf. Syst. Eng. **66**(1), 111–126 (2024)
10. Bandi, A., Adapa, P.V.S.R., Kuchi, Y.E.V.P.K.: The power of generative AI: a review of requirements, models, input–output formats, evaluation metrics, and challenges. Future Internet **15**(8), 260 (2023)
11. Zhao, Z., Ye, J.C., Bresler, Y.: Generative models for inverse imaging problems: from mathematical foundations to physics-driven applications. IEEE Signal Process. Mag. **40**(1), 148–163 (2023)
12. Lencastre, P., Gjersdal, M., Gorjão, L.R., Yazidi, A., Lind, P.G.: Modern AI versus century-old mathematical models: how far can we go with generative adversarial networks to reproduce stochastic processes? Physica D **453**, 133831 (2023)
13. Krishna, E.A., Sha, A., Anvesh, K., Reddy, N.A., Raj, B.S., Nisha, K.S.: (2023) Generative AI-driven approach to converting numerical code into mathematical functions. In: 2023 2nd International Conference on Automation, Computing and Renewable Systems (ICACRS), pp. 661–666. IEEE

14. Lee, M.: A mathematical interpretation of autoregressive generative pre-trained transformer and self-supervised learning. Mathematics **11**(11), 2451 (2023)
15. Bengesi, S., El-Sayed, H., Sarker, M.K., Houkpati, Y., Irungu, J., Oladunni, T.: Advancements in generative AI: a comprehensive review of GANs, GPT, autoencoders, diffusion model, and transformers. IEEE Access (2024)
16. Testolin, A., Hou, K., Zorzi, M.: Large-scale generative AI models lack visual number sense. arXiv preprint arXiv:2402.03328 (2024)
17. Feng, G., Zhang, B., Gu, Y., Ye, H., He, D., Wang, L.: Towards revealing the mystery behind chain of thought: a theoretical perspective. Adv. Neural Inf. Process. Syst. **36** (2024)
18. Thomas, C.K., Chaccour, C., Saad, W., Debbah, M., Hong, C.S.: Causal reasoning: charting a revolutionary course for next-generation ai-native wireless networks. IEEE Veh. Technol. Mag. (2024)
19. Chinthapatla, S.: Unleashing the future: a deep dive into AI-enhanced productivity for developers. Int. J. Sci. Technol. Eng. Math. **13**(03). Homepage: http://www.ijmra.us (2024)
20. Esmaeili Nezhad, A., Samimi, M.H.: A review of the applications of machine learning in the condition monitoring of transformers. Energy Syst. **15**(1), 463–493 (2024)
21. Sanford, C., Hsu, D.J., Telgarsky, M.: Representational strengths and limitations of transformers. Adv. Neural Inf. Process. Syst. **36** (2024)
22. Yadav, B.: Generative AI in the era of transformers: revolutionizing natural language processing with LLMs (2024)
23. Han, X., Zhiqin, W., Dexin, L., Wenqiang, T., Xiaofeng, L., Wendong, L., Ning, Y., et al.: AI enlightens wireless communication: a transformer backbone for CSI feedback. China Commun. (2024)
24. Nicola, G., Jackson, S., Queen, Z.: Implementation of ChatGPT artificial intelligence in learning. Blockchain Front. Technol. **3**(2), 138–143 (2024)
25. Jiang, P., Obi, T., Nakajima, Y.: Integrating prior knowledge to build transformer models. Int. J. Inf. Technol. 1–14 (2024)
26. Takale, D.G., Mahalle, P.N., Sule, B.: Advancements and applications of generative artificial intelligence. J. Inf. Technol. Sci. **10**(1), 20–27 (2024)
27. Beura, C.P., Wolters, J., Tenbohlen, S.: Application of pathfinding algorithms in partial discharge localization in power transformers. Sensors **24**(2), 685 (2024)
28. Zhang, Y., Liu, C., Liu, M., Liu, T., Lin, H., Huang, C.B., Ning, L.: Attention is all you need: utilizing attention in AI-enabled drug discovery. Brief. Bioinform. **25**(1), bbad467 (2024)
29. Vaswani, A., Shazeer, N., Parmar, N., Uszkoreit, J., Jones, L., Gomez, A. N., Polosukhin, I., et al.: Attention is all you need. In: Advances in Neural Information Processing Systems, vol. 30 (2017)
30. Wu, C., Wu, F., Qi, T., Huang, Y., Xie, X.: Fastformer: additive attention can be all you need. arXiv preprint arXiv:2108.09084 (2021)
31. Dong, Y., Cordonnier, J.B., Loukas, A.: Attention is not all you need: pure attention loses rank doubly exponentially with depth. In: International Conference on Machine Learning, pp. 2793–2803. PMLR (2021)
32. Harrer, S.: Attention is not all you need: the complicated case of ethically using large language models in healthcare and medicine. EBioMedicine **90** (2023)
33. Shen, Z., Zhang, M., Zhao, H., Yi, S., Li, H.: Efficient attention: attention with linear complexities. In: Proceedings of the IEEE/CVF Winter Conference on Applications of Computer Vision, pp. 3531–3539 (2021)
34. Chang, B., Wang, Y., Zhao, X., Li, G., Yuan, P.: A general-purpose edge-feature guidance module to enhance vision transformers for plant disease identification. Expert Syst. Appl. **237**, 121638 (2024)
35. Wang, C., Pan, J., Lin, W., Dong, J., Wang, W., Wu, X.M.: Selfpromer: self-prompt dehazing transformers with depth-consistency. Proc. AAAI Conf. Artif. Intell. **38**(6), 5327–5335 (2024)
36. Kong, S.C., Yang, Y.: A human-centred learning and teaching framework using generative artificial intelligence for self-regulated learning development through domain knowledge learning in K–12 settings. IEEE Trans. Learn. Technol. (2024)

37. Sahu, A.: Gated transformer-based architecture for automatic modulation classification. Doctoral dissertation. Virginia Tech. (2024)
38. Wang, J., Gao, Y., Wang, F., Zeng, S., Li, J., Miao, H., Zhou, Y., et al.: Accurate estimation of biological age and its application in disease prediction using a multimodal image transformer system. Proc. Natl. Acad. Sci. **121**(3), e2308812120 (2024)
39. Aksamit, N., Hou, J., Li, Y., Ombuki-Berman, B.: Integrating transformers and many-objective optimization for cancer drug design (2024)
40. Gunal, A., Lin, B., Bouneffouf, D.: Conversational topic recommendation in counseling and psychotherapy with decision transformer and large language models. arXiv preprint arXiv: 2405.05060 (2024)
41. Radanliev, P.: Artificial intelligence: reflecting on the past and looking towards the next paradigm shift. J. Exp. Theoret. Artif. Intell. 1–18 (2024)
42. Saish, N.V.P., Vijayashree, J.: Image classification of lung X-ray images using deep learning. In: 2022 5th International Conference on Contemporary Computing and Informatics (IC3I), pp. 1970–1976. IEEE (2022)
43. Gupta, R., Nair, K., Mishra, M., Ibrahim, B., Bhardwaj, S.: Adoption and impacts of generative artificial intelligence: theoretical underpinnings and research agenda. Int. J. Inf. Manag. Data Insights **4**(1), 100232 (2024)
44. Putra, F.R., Ciptaningrum, D.S.: Understanding the role of generative pre-trained transformer (GPT) in improving learning quality and practices. QALAMUNA J. Pendidikan Sosial dan Agama **16**(1), 91–100
45. Jayagopal, A., Xue, H., He, Z., Walsh, R.J., Hariprasannan, K.K., Tan, D.S.P., Rajan, V.: Personalised drug identifier for cancer treatment with transformers using auxiliary information. arXiv preprint arXiv:2402.10551 (2024)
46. Suwinski, P., Liesch, A., Liu, B., Schnitzer, F., Kohlsmann, T., Janschek, K.: Image based landing site detection on planetary surfaces by vision transformers and nested convolutional neural networks. In: AIAA SCITECH 2024 Forum, p. 1745 (2024)
47. Song, B., Dharma Raj, K.C., Yang, R.Y., Li, S., Zhang, C., Liang, R.: Classification of mobile-based oral cancer images using the vision transformer and the Swin transformer. Cancers **16**(5), 987 (2024)
48. Kumar, P., Gupta, V., Grover, M.: Dual attention and channel transformer based generative adversarial network for restoration of the damaged artwork. Eng. Appl. Artif. Intell. **128**, 107457 (2024)
49. Pan, Y., Yuan, Y., Yin, Y., Shi, J., Xu, Z., Zhang, M., Liu, Q.: Preparing lessons for progressive training on language models. arXiv preprint arXiv:2401.09192 (2024)
50. Kumar, A.: Long-term, multi-variate production forecasting using non-stationary transformer. In: International Petroleum Technology Conference, p. D021S084R008. IPTC (2024)
51. Braşoveanu, A.M., Andonie, R.: Visualizing transformers for NLP: a brief survey. In: 2020 24th International Conference Information Visualisation (IV), pp. 270–279. IEEE (2020)

Mathematical Frameworks for Generative AI Applications

Yasir Fahim and Suman Kumar Maji

Abstract Generative AI creates new material, including text and images, by identifying patterns and data from large datasets. This chapter uses mathematical models like deep neural networks, which leverage probabilistic frameworks and intricate architectures. These models simulate the creative processes observed in human cognition. Key concepts covered in this chapter include probability distributions, optimization algorithms, and complex neural network architectures. These elements enable AI systems to generate original content rather than merely repeating learned examples. This chapter provides a comprehensive insight into the mathematical models of generative AI and the role of deep neural networks. This chapter also discusses various neural network architectures essential in generative AI, such as transformers, generative adversarial networks (GANs), and diffusion models.

Keywords Generative AI · Mathematical models · Deep neural networks · Probabilistic frameworks · Generative adversarial networks (GANs) · Transformers · Diffusion models

1 Introduction

Generative Artificial Intelligence (AI) signifies a monumental advancement in AI technology, shifting its focus from traditional data processing and analysis to the active creation of content. Numerous industries, including computer vision, audio creation, natural language processing (NLP), healthcare, and others will be greatly impacted by this shift [1]. By harnessing vast datasets, generative AI systems can produce new outputs that closely resemble authentic data, creating life-like images,

Y. Fahim · S. K. Maji (✉)
Department of Computer Science and Engineering, Indian Institute of Technology Patna, Patna 801106, India
e-mail: smaji@iitp.ac.in

Y. Fahim
e-mail: yasir_2421cs04@iitp.ac.in

N. R. Vajjhala et al. (eds.), *Generative Artificial Intelligence (AI) Approaches for Industrial Applications*, Information Systems Engineering and Management 24,
https://doi.org/10.1007/978-3-031-76710-4_3

articulate text, original music compositions, and simulated environments [2]. These capabilities are revolutionizing industries and fundamentally transforming human interaction with digital content.

This chapter aims to provide a comprehensive overview of the various applications and impacts of generative AI across different industries. We will explore the following key areas:

1. **Computer Vision**: We discuss advancements in technologies like variational autoencoders (VAEs) and generative adversarial networks (GANs), highlighting their applications in gaming, filmmaking, real estate, and the automotive industry [3, 4].
2. **Natural Language Processing (NLP)**: We examine how generative AI is enhancing automated conversational agents, content generation tools, and real-time translation technologies [5, 6].
3. **Audio Creation**: We look into the synthesis of audio and music composition, and its implications for entertainment, advertising, and assistive technologies [7].
4. **Healthcare**: We analyze the contributions of generative AI to drug discovery, medical imaging, diagnostic processes, and medical training [8, 9].
5. **Fashion and Design**: We cover how generative AI is automating design processes, analyzing trends, and customizing consumer goods [10].

Despite the numerous benefits of generative AI, its deployment raises significant ethical concerns. We will address issues such as the creation of deceptive content and data privacy [11]. We will discuss the importance of establishing frameworks and guidelines to ensure responsible use of generative AI, promoting innovation while safeguarding societal interests.

Generative AI represents a transformative leap in artificial intelligence, offering unprecedented capabilities in content creation across various industries. From enhancing visual experiences in computer vision and generating coherent text in natural language processing to revolutionizing audio production and advancing healthcare, the impact of generative AI is profound. To make sure that these technologies serve society as a whole, it is imperative to manage the ethical issues raised by their application (Table 1).

2 Mathematical Models to Generative AI

One of the main objectives in the vast fields of statistical analysis and machine learning is classification, which is the prediction of categorical outcomes from given input factors, such as class labels. These input variables can be continuous or discrete. Generative AI derives its foundation from the field of statistical classification. Among the many methodologies developed to tackle classification problems, three fundamental approaches are particularly noteworthy: discriminant models, probabilistic discriminant models, and probabilistic generative models. Each of these models employs a unique strategy and leverages different statistical techniques to

Table 1 Impact of generative artificial intelligence in different industrial sectors

Industrial sectors	Impacts
Computer vision	Generating images, enhancing photographs, and creating detailed 3D models used in gaming, filmmaking, real estate, and automotive industries
Natural language processing (NLP)	Transforming natural language processing by enabling machines to generate coherent, context-aware text
Audio synthesis and music creation	Synthesize audio and compose music is changing the entertainment and advertising industries, as well as enhancing assistive technologies
Healthcare and pharmaceutical development	Accelerating drug discovery and enhancing medical imaging
Fashion and product design	Automation of design processes and customization of consumer products
Security and ethical concern	Creation of deceptive content, data privacy issues, and the impact on employment

achieve accurate predictions. We will go into great detail on each model type in this discussion, emphasizing the operational and mathematical underpinnings of each.

2.1 Discriminant Models

Generative AI derives its foundation from the field of statistical classification, which focuses on predicting categorical outcomes from given input factors. Among the many methodologies developed to tackle classification problems, three fundamental approaches are particularly noteworthy: discriminant models, probabilistic discriminant models, and probabilistic generative models. Each of these models employs a unique strategy and leverages different statistical techniques to achieve accurate predictions. This section will delve into the operational and mathematical underpinnings of each model type, emphasizing their unique strategies and practical applications.

Mathematical Foundation

Finding a projection that optimizes the distance between several classes is the aim. Let us consider a shared covariance matrix Σ and two classes with means μ_1 and μ_2.

Key Algorithms

The discriminant function can be written as:

$$y(x) = w^T x + w_0 \tag{1}$$

where

$$w = \Sigma^{-1}(\mu_1 - \mu_2) \tag{2}$$

and w_0 is a constant calculated to satisfy some condition (like equal class priors).

In more detail, let assumes x be a feature vector. The data for each class i are normally distributed with mean μ_i and shared covariance matrix Σ. The prior probabilities of the classes are π_1 and π_2. The discriminant function for class i is:

$$\delta_i(x) = \log p(x|y = i) + \log \pi_i \tag{3}$$

For class 1:

$$\delta_1(x) = -\frac{1}{2}(x - \mu_1)^T\Sigma^{-1}(x - \mu_1) + \log \pi_1 \tag{4}$$

For class 2:

$$\delta_2(x) = -\frac{1}{2}(x - \mu_2)^T\Sigma^{-1}(x - \mu_2) + \log \pi_2 \tag{5}$$

The decision rule is to assign x to class 1 if $\delta_1(x) > \delta_2(x)$, and to class 2 otherwise. The decision boundary is where $\delta_1(x) = \delta_2(x)$. Equalizing Eqs. (4) and (5), we get:

$$-\frac{1}{2}(x - \mu_1)^T\Sigma^{-1}(x - \mu_1) + \log \pi_1 = -\frac{1}{2}(x - \mu_2)^T\Sigma^{-1}(x - \mu_2) + \log \pi_2 \tag{6}$$

Expanding and simplifying the quadratic terms, we get:

$$\left(x^T\Sigma^{-1}\mu_1 - \frac{1}{2}\mu_1^T\Sigma^{-1}\mu_1 + \log \pi_1\right) = \left(x^T\Sigma^{-1}\mu_2 - \frac{1}{2}\mu_2^T\Sigma^{-1}\mu_2 + \log \pi_2\right) \tag{7}$$

After re-arranging the terms, we get:

$$x^T\Sigma^{-1}(\mu_1 - \mu_2) = \frac{1}{2}(\mu_1^T\Sigma^{-1}\mu_1 - \mu_2^T\Sigma^{-1}\mu_2) + \log\left(\frac{\pi_1}{\pi_2}\right) \tag{8}$$

where, $w = \Sigma^{-1}(\mu_1 - \mu_2)$ and $w_0 = \frac{1}{2}\left(\mu_1^T\Sigma^{-1}\mu_1 - \mu_2^T\Sigma^{-1}\mu_2\right) + \log\left(\frac{\pi_1}{\pi_2}\right)$. Equation (8) forms linear discriminant function as given in Eq. (1) which acts like the linear decision boundary that separates the two classes [12, 13].

Practical Applications

LDA is widely used in fields such as bioinformatics, finance, and marketing for tasks like image generation, photo enhancement, and 3D model creation.

2.2 *Probabilistic Discriminant Models:*

Probabilistic discriminant models, such as logistic regression, focus on modeling the probability of class membership directly from the input features. Unlike discriminant models, these probabilistic models do not firmly depend on the input data's underlying distribution. Rather, their objective is to calculate the likelihood that a specific input is a member of a specific class.

Mathematical Foundation

In a binary classification context, logistic regression precisely predicts the likelihood that an input, X, belongs to a certain class, Y = 1 as:

$$P(Y = 1|X = x) = \frac{1}{1 + e^{-(w^T x + b)}} \tag{9}$$

where the input feature vector is denoted by x, the weight vector by w, and the bias by b.

Key Algorithms

By maximizing the likelihood of the observed data, which frequently entails the use of numerical optimization techniques like gradient descent, the parameters w and b are determined from the data.

Given a dataset with n independent observations (x_i, y_i) where $y_i \in \{0, 1\}$, the likelihood func tion L(β) is the product of the individual probabilities for the observed outcomes:

$$L(\beta) = \prod_{i=1}^{n} P(y_i|x_i) \tag{10}$$

Using the logistic regression model, this becomes:

$$L(\beta) = \prod_{i=1}^{n} \left(\frac{1}{1 + e^{-(x_i^T \beta)}}\right)^{y_i} \left(1 - \frac{1}{1 + e^{-(x_i^T \beta)}}\right)^{1-y_i} \tag{11}$$

Equation (11) is the log-likelihood equation. The equation can work more conveniently in the form:

$$l(\beta) = L(\beta) = \sum_{i=1}^{n} \left[y_i \log \sigma\left(x_i^T \beta\right) + \left(1 - y_i\right) \log\left(1 - \sigma\left(x_i^T \beta\right)\right)\right] \tag{12}$$

To maximize the log-likelihood function, we take the gradient with respect to β:

$$\nabla l(\beta) = \sum_{i=1}^{n} [y_i - \sigma(x_i^T \beta)] x_i \tag{13}$$

The parameters β are typically estimated using iterative optimization algorithms such as gradient descent. For gradient descent, the update rule is:

$$\beta^{(t+1)} = \beta^{(t)} - \eta \nabla l(\beta^{(t)}) \tag{14}$$

where η is the learning rate [14, 15].

Practical Applications

Logistic regression is extensively used in medical diagnosis, credit scoring, and marketing for predicting probabilities of class membership.

2.3 Probabilistic Generative Models

Probabilistic generative models, on the other hand, use the Bayes theorem to infer the posterior distribution P (Y | X) from the joint probability distribution P (X, Y). These models consist of two basic steps: estimating the class-conditional distributions of the features and the prior probability of the classes. Regarding how the input data for each class is distributed, they make clear assumptions.

Mathematical Foundation

For example, the likelihood P (X | Y = k) is represented as a multivariate normal distribution in Gaussian Naive Bayes:

$$P(X = x|Y = k) = \frac{1}{(2\pi)^{d/2} |\Sigma_k|^{1/2}} \exp\left(-\frac{1}{2}(x - \mu_k)^T \Sigma_k^{-1} (x - \mu_k)\right) \tag{15}$$

where μ_k and Σ_k represent the class k feature's mean and covariance, respectively.

Key Algorithms

In training data, the relative frequencies of each class are often used to estimate the class priors P (Y = k). Then, using the Bayes theorem, the posterior P (Y | X) is determined:

$$P(Y = k|X = x) = \frac{P(X = x|Y = k)P(Y = k)}{P(X = x)} \tag{16}$$

where the entire probability of x, as determined by the law of total probability, is denoted by P (X = x). The class label for a new observation x is determined by maximizing the posterior probability [16]:

$$\hat{y} = \arg\max_{k} p(y = k|x) \tag{17}$$

Practical Applications

Probabilistic generative models are used in fields such as text classification, spam filtering, and document classification.

By comparing the models, we found that, discriminant models prioritize maximizing class separability directly in the feature space and often have linear decision boundaries, whereas probabilistic discriminant models explicitly model the conditional probability of the target given the features and typically assume a logistic form for binary classification.

Apart from both, when the assumptions of normality and independence are met, probabilistic generative models can better handle complex classification scenarios because they capture the underlying distribution of each class and feature dependencies by modeling the entire process of data generation. Understanding these models involves appreciating their assumptions, complexity, and the scenarios in which they are most appropriately applied.

3 Role of Deep Learning

Deep learning has substantially contributed to advancements in statistical modeling, particularly in discriminant analysis and probabilistic modeling. These models are essential for tasks like classification, where the objective is to classify inputs according to their features into predetermined categories. Here, we explore the ways in which discriminant models, probabilistic discriminant models, and probabilistic generative models, combined with deep learning, help comprehend and forecast events.

Discriminant models, particularly in the realm of deep learning, are often implemented as classification networks that directly map input features to specific categories. The most popular type of discriminant model in deep learning is the feed-forward neural network, widely employed in applications like speech recognition and image recognition. These models learn to create boundaries between categories based on the labeled examples provided during training. Deep learning enhances traditional discriminant analysis by automatically learning complex, nonlinear boundaries between classes. For instance, deep convolutional neural networks (CNNs) can learn hierarchical representations without the need for manual feature extraction, which is particularly effective in image classification tasks [17]. The architecture of a CNN involves layers of convolutional filters that detect various features, followed by pooling layers that downsample the spatial dimensions. This hierarchical learning enables CNNs to capture intricate patterns and details in images. CNNs are used to analyze medical images, such as MRI scans, to detect tumors and other anomalies. For instance, Google's DeepMind developed a CNN-based model to diagnose eye diseases from retinal scans with high accuracy. In autonomous driving, discriminant models help classify objects on the road, such as pedestrians, vehicles, and traffic signs. Tesla's Autopilot system uses deep learning models to process camera inputs and make real-time driving decisions.

Probabilistic discriminant models, such as logistic regression and softmax regression, extend discriminant models by not only predicting categorical outcomes but also providing probabilities associated with each category. In deep learning, this is often seen in the output layer of classifiers, where softmax functions are used to estimate these probabilities, offering a measure of uncertainty or confidence in predictions. Deep learning networks improve probabilistic discriminant models by learning feature representations that are more effective at distinguishing between classes under complex, real-world conditions. Neural network architectures such as deep Boltzmann machines and deep belief networks have shown their ability to recognize intricate patterns in the data [18]. Logistic regression can be extended in neural networks with multiple hidden layers, where each layer captures different levels of abstraction. The softmax function at the output layer calculates the probability distribution over classes. Probabilistic models are used in credit scoring to predict the likelihood of loan defaults. Companies like FICO use deep learning models to analyze credit data and improve the accuracy of their scoring systems. Businesses use probabilistic models to predict customer churn by analyzing customer behavior data. For instance, telecom companies use these models to identify customers likely to switch to competitors and target them with retention offers.

Probabilistic generative models attempt to model how the data is generated. By understanding the underlying data distribution, these models can generate new data instances. Examples include Gaussian Mixture Models (GMMs) and more complex deep learning-based systems such as Generative Adversarial Networks (GANs) and Variational Autoencoders (VAEs). Generative models benefit greatly from deep learning, which offers a foundation for creating more intricate models that can successfully imitate data distributions. VAEs, for example, use neural networks to produce a latent space representation by compressing the important properties of the input data [19]. GANs utilize a competitive framework between two neural networks to produce high-fidelity outputs [1]. The encoder network maps input data to a latent space, while the decoder network reconstructs data from this latent space. This framework enables the generation of new, similar data instances. Consist of a generator that creates fake data and a discriminator that evaluates the authenticity of the data. The generator improves based on feedback from the discriminator, producing increasingly realistic data. GANs are used in the film industry to create special effects and digital characters. For example, GANs were employed to de-age actors in movies like "The Irishman." VAEs are used to design new clothing items by generating novel patterns based on existing designs. Companies like Zalando use generative models to create unique fashion items tailored to customer preferences. In drug discovery, VAEs help generate new molecular structures with desired properties, accelerating the development of new medications. Insilico Medicine uses generative models to design potential drug candidates, significantly reducing the time and cost of drug development.

The integration of deep learning with these models has led to notable progress in several domains. Deep learning-based discriminative models enhance diagnostic accuracy in the medical field by categorizing images. In autonomous driving, probabilistic models help in understanding and reacting to dynamic environments. Deep learning's role in these models primarily lies in its ability to automatically and efficiently learn complex representations from large datasets, which traditional statistical techniques struggle with. The probabilistic nature of these models also introduces robustness into deep learning applications, accounting for uncertainty and variability in real-world data.

Apart from it, there are various neural network architectures, which help in implementing Generative AI in practical applications. Transformers, Generative Adversarial Networks (GANs), and Diffusion models are a few examples of these neural network architectures. A brief description about these three are given below:

3.1 Transformers

Vaswani et al. presented Transformers, a revolutionary kind of deep learning architecture, in their landmark work "Attention is All You Need" in 2017 [20]. This creative architecture has various uses outside of natural language processing (NLP), yet it has been a wonderful asset to the area. The key strength of transformers lies in their ability to efficiently handle sequential data through a novel mechanism called self-attention.

Before transformers were invented, the most widely used designs for processing sequential input were recurrent neural networks (RNNs) and their variants, such as gated re-current units (GRUs) and long short-term memory (LSTM) networks. However, these architectures struggled with capturing long-range dependencies due to their sequential nature, which made training time-consuming and prone to issues like vanishing gradients.

Transformers overcome these drawbacks by completely doing away with convolutions and recurrence in favor of a self-attention mechanism. This enables the modeling of dependencies among all elements in a sequence, regardless of how far off they are from one another. This parallelizable approach drastically improves training efficiency and enables the handling of much longer sequences than was previously possible.

Key Components

- *Attention Mechanism*: Natural language processing and other sequential data tasks will be revolutionized by the Transformer design, which is centered around the attention mechanism. The attention mechanism of transformers allows the model to focus on relevant parts of the input sequence while producing an output. Transformers process input simultaneously, as opposed to standard sequential models like Recurrent Neural Networks (RNNs), which process input sequentially. This allows for more effective and efficient information processing. The attention mechanism comprises several key components:

- **Query, Key, and Value**: Three vectors are linked to each input token: a query vector, a key vector, and a value vector. The content and position of the token in the sequence are encoded by these vectors, which are learned during training.
- **Attention Weights**: The importance of each input character for producing the output is determined by the attention mechanism's computation of attention weights. The similarity between the query and key vectors is used to calculate these weights.
- **Weighted Sum**: The value vectors are added together and weighted, with the attention weights acting as the sum's weights, to produce the final output.

• *Encoder-Decoder Architecture*: Machine translation, text summarization, and question answering are examples of sequence-to-sequence tasks in natural language processing (NLP) that have been altered by the Encoder-Decoder architecture under the Transformer model framework. Sequence-to-sequence activities, in which an input sequence is converted into an output sequence, are handled by the encoder-decoder architecture. This architecture comprises two distinct modules: the Encoder and the Decoder, each responsible for different aspects of sequence processing.

  - **Encoder**: The encoder module obtains the contextual representations from the input sequence after processing it. It is made up of several identical layers, each with two sublayers:

    Self-attention Layer: In order to capture dependencies and relationships within the sequence, this layer computes attention scores between each place in the input sequence.
    Feed-Forward Neural Network: The model is able to capture complex nonlinear interactions because the representation of each position is supplied through a feed-forward neural network subsequent to the self-attention layer.

    A series of contextualized representations, with each position encoding details about the context in which it is situated, are the encoder's output.
  - **Decoder**: The contextual representations that the Encoder produces are used by the Decoder module to create the output sequence. Like the Encoder, it is made up of several identical layers, each with three sub-layers:

    Self-attention Layer: The Decoder's self-attention layer calculates attention scores within the output sequence, just like the Encoder does. This makes it possible for the model to focus on relevant data when decoding.
    Encoder-Decoder Attention Layer: By monitoring the output representations of the encoder, this layer makes it possible for the decoder to use information from the input sequence throughout the decoding process.
    Feed-Forward Neural Network: After the attention layers, the representation of each position is supplied through a feed-forward neural network,

which facilitates the modeling of complex relationships in the output sequence [21].

- *Multi-head Attention*: An essential element of the Transformer architecture is Multi-Head Attention, which is responsible for identifying connections between various segments of a sequence. The foundation of Multi-Head Attention is the notion of attention mechanisms, which allow models to concentrate on pertinent segments of the input sequence during prediction. The first introduction of attention mechanisms occurred in the area of machine translation tasks using sequence-to-sequence models.
  - **Self-attention Mechanism**: The fundamental component of Multi-head Attention is the self-attention mechanism, which enables each element in a sequence to pay attention to its neighboring elements. Self-attention allows for global interactions between all sequence parts, in contrast to conventional recurrent or convolutional architectures, which successfully captures long-range relationships.
  - **Multi-head Attention**: Multi-head Attention extends the self-attention mechanism by using many attention heads in simultaneously. As a result of each attention head learning to concentrate on different segments of the input sequence, the model is able to identify a wide range of patterns and correlations. The outputs from the several attention heads are then concatenated and linearly transformed to produce the final output.
  - **Architectural Design**: Multi-head Attention's architecture can be characterized as follows:

    Input Transformation: The input sequence is transformed linearly to produce the query, key, and value vectors. By using learnt weight matrices to parameterize these modifications, the model is able to extract information from the input in an adaptable manner.
    Attention Computation: The query, key, and value vectors are utilized to calculate attention scores for each attention head. The importance of each element in the sequence in relation to the current element is determined by these scores. A compatibility function, like the dot product or scaled dot product, is used to calculate the attention scores.
    Weighted Summation: Utilizing the attention scores as a starting point for the weight calculations, a weighted sum of the value vectors is computed. The attended representation for every element in the sequence is represented by this weighted sum.
    Multi-head Fusion: The ultimate output of multi-head attention is created by concatenating and linearly transforming the outputs from each attention head. The model is able to capture various patterns and correlations from various attention heads because to this fusion step [22].

Applications

- *Machine Translation*: Machine translation is one of the earliest and most prominent applications of transformers. Traditionally, statistical and rule-based approaches were prevalent in machine translation systems. However, transformers, with their self-attention mechanism, have enabled significant improvements in translation quality. In machine translation, transformers process input sequences in one language and generate corresponding sequences in another language. The transformer model consists of an encoder-decoder architecture, where the encoder analyses the input sentence and the decoder generates the output sentence. Transformers have substantially improved translation quality, especially for long and context-rich sentences. They excel at capturing complex syntactic and semantic structures across languages, leading to more accurate and fluent translations. Additionally, transformers can handle multiple languages simultaneously, making them versatile for multilingual translation tasks. Machine translation technology has advanced significantly as a result of the widespread use of transformer-based models, such as Google's Transformer and OpenAI's GPT (Generative Pre-trained Transformer) series. Major translation services and platforms have been enabled by these models, enabling smooth cross-language communication on a worldwide scale source [6, 20].
- *Text Generation*: The process of producing logical and contextually appropriate text in response to an input or prompt is known as text generation. Transformers are now cutting edge text creation models that can generate human-like language in a variety of fields. Transformers forecast the subsequent token in a sequence by examining the tokens that came before it in text production, utilizing their autoregressive features. Sampling from the probability distribution of the subsequent token, conditioned on the input sequence, is done iteratively in this procedure. Transformers have pushed the boundaries of text genera tion, producing high-quality outputs that closely mimic human-written text. They excel at generating diverse and contextually relevant content across domains, including language modeling, storytelling, and creative writing. Transformers can also be modified for particular activities or domains, which enables customized and customized text production. Transformer-based text generation models are extensively employed in many different fields, including content creation, chatbots, and virtual assistants. Examples of these models are OpenAI's GPT series and BERT (Bidirectional Encoder Representations from Transformers). These models have fueled advancements in natural language generation technology, enabling more engaging and interactive human-machine interactions [23, 24].
- *Language Understanding*: Language understanding involves extracting meaning and knowledge from textual data, enabling machines to comprehend and process natural language input effectively. In tasks involving language understanding, including as text classification, sentiment analysis, and question answering, Transformers have shown impressive competence. Transformers have led to state-of-the-art results on benchmark datasets for a range of tasks by improving the performance of language comprehension systems. They excel at capturing subtle linguistic nuances, contextual cues, and domain-specific knowledge, leading

to more accurate and robust language understanding capabilities. Transformer-based models, such as BERT, XLNet, and RoBERTa (Robustly optimized BERT approach), have become indispensable tools for natural language understanding applications. These models have been deployed in various contexts, including search engines, social media platforms, and recommendation systems, enhancing user experience and facilitating more efficient information retrieval and processing [24–26].

3.2 Generative Adversarial Networks (GANs)

A new class of generative models known as Generative Adversarial Networks, or GANs, was suggested by Ian Goodfellow and colleagues in 2014 [1]. These models have brought significant advances to the disciplines of artificial intelligence and machine learning, particularly in areas such as image generation, data augmentation, and synthetic data creation. A technique called adversarial training is used to train two competing neural networks, the generator and discriminator, concurrently. This is the foundation of GANs.

Key Components

- *Generator*: It is the generator's responsibility to produce artificial data that is as authentic as feasible. It converts an input random noise vector, z, into a data sample, G(z). In order to fool the discriminator into believing the created data to be real, the generator aims to produce data that is exact replicas of real data. The generator can be represented mathematically as:

$$G(z; \theta_G) \tag{18}$$

  where the generator network's parameters are represented by θ_G.
- *Discriminator:* In contrast, the discriminator serves as a classifier that assesses the authenticity of the data samples. It takes an input in the form of a produced sample or a sample of actual data, and it returns a probability that indicates how legitimate the input is to be. The discriminator's task is to distinguish between synthetic and genuine data with accuracy. It can be expressed as:

$$D(x; \theta_D) \tag{19}$$

  where the discriminator network's parameters are represented by θ_D, whereas x is the sample of input data.
- *Adversarial Training*: A two-player minimax game is used in the adversarial process of GAN training. The discriminator and generator are simultaneously optimized with conflicting objectives. While the discriminator seeks to maximize this probability, the generator seeks to decrease the likelihood that it will accurately detect fake samples. A formalizable objective function exists for GANs as:

$$\min_{G} \max_{D} V(D, G) = E_{x \sim p_{data}(x)}[\log D(x)] + E_{z \sim p_z(x)}[\log(1 - D(G(z)))] \quad (20)$$

In Eq. (20), the distribution of the input noise vector is represented by $p_z(z)$, while the distribution of the real data is shown by $p_{data}(x)$. Whereas, the discriminator seeks to maximize both terms, the generator seeks to decrease the second term.

Applications

- *Image Generation*: The ability of GANs to produce diversified, high-quality, and photorealistic images has transformed the field of image generation. In a GAN, the discriminator network assesses the veracity of created images, while the generator network learns to map random noise vectors to realistic images. The generator is encouraged to produce images that are identical to genuine ones through this adversarial training process. GAN applications for image generation include:
  - **Art Generation**: New possibilities for artistic expression are opened up by the ability of GANs to produce artwork in a variety of genres, from classical paintings to abstract compositions.
  - **Content Creation**: GANs are employed in industries such as advertising, marketing, and entertainment for generating visual content, including product images, advertisements, and virtual scenes [1, 27, 28].
- *Image-to-Image Translation*: GANs perform remarkably well when transferring images from one domain to another while preserving semantic content. Whether paired or unpaired datasets are used, GANs can learn complex mappings across multiple domains and generate precise translations. Notable applications include:
  - **Style Transfer**: GANs can create artistic effects like superimposing an image's style over another, turning a snapshot into a painting that resembles a work by a well-known artist.
  - **Domain Adaptation**: Domain adaption tasks like translating daytime images to nighttime images, which are taken in one lighting condition and made to appear as though they were taken in a different lighting situation, are made easier by GANs [29, 30].
- *Data Augmentation*: One of the main elements of data augmentation is the use of GANs. The generalization and durability of machine learning models are enhanced by this technique, which also enhances the quantity and diversity of training datasets. GANs can address difficulties with imbalance and scarcity of data by creating synthetic data samples that closely mimic real data. Applications include:
  - **Medical Imaging**: GANs produce artificial medical images to supplement small datasets and help train diagnostic and segmentation algorithms for applications like tumor identification and organ segmentation.
  - **Autonomous Driving**: GANs create synthetic scenes and scenarios to augment real-world driving datasets, enabling the training of more robust autonomous vehicle systems under various environmental conditions [31–33].

- *Super-Resolution*: GANs have been employed for super-resolution tasks, where low-resolution images are enhanced to higher resolutions with improved clarity and detail. By learning complex mappings between low-resolution and high-resolution image spaces, GANs can generate visually pleasing and realistic high-resolution images. Applications include:
  - **Medical Imaging**: GANs increase the clarity of medical pictures, facilitating the identification and examination of irregularities and minute structures. Examples of these images include MRI scans and microscope images.
  - **Surveillance and Security**: GANs improve the resolution of low-quality surveil lance footage, enabling better identification and tracking of objects and individuals in security applications [33–35].

3.3 *Diffusion Models*

Diffusion models are a relatively new class of generative models that have garnered a lot of attention because of their exceptional image quality and iterative denoising procedure. In their seminal study "Diffusion Models Beat GANs on Image Synthesis," Dhillon et al. presented these models as a strong substitute for Generative Adversarial Networks (GANs) in image production challenges [36].

Key Components

- *Diffusion Process*: Diffusion models are a subclass of generative models that use an iterative denoising approach to produce high-quality data via the process of diffusion. This process is fundamentally rooted in the concepts of stochastic processes and probabilistic modeling. Here, the diffusion process will be thoroughly discussed, with special attention paid to the forward and reverse diffusion processes and the nuances of their application.
  - **Forward Diffusion Process**: When noise is gradually introduced to real data through a sequence of time steps, it eventually corrupts the data, marking the beginning of the forward diffusion process. This procedure is comparable to modeling a physics-based diffusion process, in which particles disperse over time as a result of random motion.
  - **Reverse Denoising Process**: The generative magic of diffusion models occurs in the reverse diffusion phase. This process involves transforming the noisy data x_T, which resembles Gaussian noise, back into a realistic data sample. The forward phase introduces noise, which is gradually removed in the reverse process, which is effectively a denoising sequence.

- *Diffusion Probabilistic Model*: Offering a probabilistic viewpoint on data creation, the Diffusion Probabilistic Model is a potent paradigm within diffusion models. The Diffusion Probabilistic Model is grounded in stochastic processes and Bayesian inference principles. Fundamentally, it views the process of creating data as a probabilistic diffusion process in which noise is progressively introduced into the data and the opposite process repeatedly seeks to remove the noise. The Diffusion Probabilistic Model is based on the Markovian property, which claims that a system's future state depends only on its current state and is independent of its previous states. This property enables us to model the diffusion process as a series of conditional distributions, where each step is conditioned only on the previous state. In the Diffusion Probabilistic Model, the noise added at each step follows a Gaussian distribution, allowing for a simple yet effective representation of uncertainty. The approach addresses the inherent uncertainty in the data generating process by mod eling the noise as a probabilistic variable, offering deeper insights into the underlying data distribution.

- *Invertible Neural Networks*: Diffusion models use Invertible Neural Networks (INNs) as the central component of the reverse denoising process. At the heart of INNs lies the concept of invertibility, which ensures that every input corresponds to a unique output and vice versa. This characteristic is essential for the diffusion models' reverse denoising procedure, which aims to accurately recover the original data from the noisy observations. INNs achieve invertibility through the use of reversible transformations, which enable bidirectional map ping between the input and output spaces. These transformations are carefully designed to ensure that no information is lost during the forward and backward passes, allowing for precise reconstruction of the data. One of the key principles guiding the design of INNs is the preservation of information. This entails ensuring that all relevant information from the input data is retained throughout the transformation process, enabling accurate reconstruction at the output.

Applications

- *Image Generation*: Diffusion models excel at generating high-quality, realistic images by iteratively denoising noise-corrupted data. High-detail images are produced as a result of the diffusion process, which progressively changes a basic noise distribution into a complicated data distribution. This application has broad implications across various fields, including art generation, content creation, and computer graphics.
  - **Art Generation**: Diffusion models can create visually stunning artworks by synthesizing images that exhibit intricate textures, patterns, and structures. Artists and designers can leverage diffusion models to generate unique and captivating visual con tent, pushing the boundaries of creativity and artistic expression.
  - **Content Creation**: In industries such as advertising, media, and entertainment, diffusion models enable the automated generation of compelling visual content for advertisements, promotional materials, and digital media. By producing photorealistic images, diffusion models streamline the content creation process, reducing costs and accelerating production timelines.
  - **Computer Graphics**: Diffusion models have applications in computer graphics for generating synthetic images used in virtual environments, video

games, and augmented reality applications. These models can produce realistic scenes, objects, and characters, enhancing the immersive experience of digital simulations and interactive media [28, 37, 38].

- *Image Editing and Manipulation*: Beyond image generation, diffusion models facilitate advanced image editing and manipulation techniques, allowing users to modify existing images in sophisticated ways. By leveraging the learned representations of image data, diffusion models enable precise control over various aspects of an image, including color, texture, and composition.
  - **Image Restoration**: Diffusion models can restore degraded or damaged images by removing noise, artifacts, and imperfections. This application is valuable in fields such as forensics, medical imaging, and historical preservation, where the quality of images may deteriorate over time or due to external factors.
  - **Style Transfer**: Style transfer methods are supported by diffusion models, which allow images to be altered to resemble reference images' artistic style. This method can be used to provide artistic interpretations of photos, produce visual effects, and improve the visual attractiveness of digital content.
  - **Object Manipulation**: Diffusion models allow users to manipulate objects within images by selectively modifying their appearance, position, or orientation. For operations like object insertion, object removal, and scene building, image editing software with this feature is helpful [39, 40].
- *Anomaly Detection*: The identification of strange or anomalous patterns in data can reveal potential risks, fraud, or anomalies, making anomaly detection a crucial task in a variety of industries, including cybersecurity, banking, and healthcare. Diffusion models offer a novel approach to anomaly detection by leveraging their ability to model complex data distributions and detect deviations from normal behavior.
  - **Cybersecurity**: Diffusion models are used in cybersecurity to examine user behavior, system logs, and network traffic in order to identify unusual activity that may point to malware infestations, cyber threats, or illegal access attempts. By identifying deviations from baseline behavior, diffusion models help organizations strengthen their cybersecurity defenses and mitigate risks.
  - **Financial Fraud Detection**: Diffusion models are employed in financial systems to detect fraudulent transactions, money laundering activities, and suspicious behavior in banking and payment networks. By analyzing transaction data and identifying irregular patterns, diffusion models assist financial institutions in preventing financial fraud and safeguarding the integrity of their operations.
  - **Medical Diagnosis**: In healthcare, diffusion models contribute to anomaly detection in medical imaging, patient monitoring, and disease diagnosis. These models analyze medical data, such as MRI scans, ECG signals, and patient records, to identify anomalies indicative of diseases, abnormalities, or adverse health conditions. By flagging unusual findings, diffusion models support

healthcare professionals in early detection and intervention, improving patient outcomes and healthcare delivery [21, 41, 42].

4 Conclusion

The advent of generative AI, which departs from the conventional approach of processing and repeating learnt instances, is one significant paradigm shift in the field of artificial intelligence (AI) [1]. Instead, generative AI ventures into the realm of creativity, striving to produce new and original content that transcends the boundaries of traditional machine learning. At its core, this transformative capability is powered by sophisticated mathematical models, prominently deep neural networks, meticulously crafted to not only comprehend but also replicate the intricate patterns and underlying structures inherent in vast and diverse datasets. The essence of generative AI lies in its ability to harness probabilistic frameworks, which afford the flexibility to model uncertainty and variability in the process of content generation [43]. These frameworks, coupled with advanced optimization algorithms, empower AI systems to delve deep into the nuances of data distribution, extracting complex representations and simulating creative processes akin to human cognition. Through the lens of probability distributions and optimization techniques, generative AI explores the intricate interplay of randomness, optimization objectives, and model parameters, orchestrating a symphony of mathematical principles to bring forth original content.

To truly grasp the essence of generative AI, one must embark on a journey into the depths of neural network architectures and computational frameworks that underpin its operation. Every architecture, from the revolutionary Generative Adversarial Networks (GANs) to the more recent developments in Diffusion models and Transformers, represents a unique approach to content generation, with pros and cons of its own [1, 20, 37]. Understanding these architectures is akin to deciphering the language of creativity encoded within the neural networks, unraveling the mechanisms through which AI systems learn, adapt, and generate content across diverse domains. The chapter's noble endeavor to provide comprehensive insights into the mathematical models of generative AI, with a keen focus on deep neural networks, serves as a beacon of knowledge in the vast sea of AI research [44]. By delving into the intricacies of model architectures, optimization algorithms, and probabilistic frameworks, the chapter lays the groundwork for a deeper understanding of how AI systems can transcend mere replication and venture into the realm of creative expression.

Furthermore, the exploration of key deep neural network architectures such as Transformers, GANs, and Diffusion models serves as a testament to the rich tapestry of approaches within generative AI [1, 37, 45]. Each architecture brings its own unique perspective to the table, offering novel solutions to the age-old problem of content generation. Future generations of researchers and practitioners will be motivated to push the bounds of creativity and invention by this chapter, which sheds light on these many approaches and encourages a deeper understanding for the breadth and complexity of generative AI research. In conclusion, a critical turning point in the development of generative AI has been reached with the fusion of mathematical concepts with state-of-the-art deep learning architectures [43]. Through the fusion of mathematical rigor, computational prowess, and creative ingenuity, generative AI stands poised to revolutionize various industries and reshape the landscape of artificial intelligence as we know it. As we continue to unlock the mysteries of creativity encoded within neural networks, the journey towards AI-driven innovation and artistic expression promises to be as exhilarating as it is transformative.

References

1. Goodfellow, I., et al.: Generative adversarial nets. In: Advances in Neural Information Processing Systems, vol. 27 (2014)
2. van den Oord, A., et al.: Wavenet: a generative model for raw audio. arXiv preprint arXiv:1609.03499 (2016)
3. Karras, T., Laine, S., Aila, T.: StyleGAN2: improved styleGAN for realistic image synthesis. In: Proceedings of the IEEE/CVF International Conference on Computer Vision (ICCV) (2019)
4. Brock, A., Donahue, J., Simonyan, K.: Large scale GAN training for high fidelity natural image synthesis. In: Proceedings of the IEEE/CVF International Conference on Computer Vision (ICCV) (2018)
5. Radford, A., et al.: Language models are unsupervised multitask learners. https://www.openai.com/blog/better-language-models/. OpenAI Blog (2019)
6. Brown, T.B., et al.: Language models are few-shot learners. In: Advances in Neural Information Processing Systems, vol. 33, pp. 1877–1901 (2020)
7. Engel, J., et al.: Neural audio synthesis of musical notes with WaveNet Autoencoders. arXiv preprint arXiv:1704.01279. https://arxiv.org/abs/1704.01279 (2017)
8. Esteva, A., et al.: A guide to deep learning in healthcare. Nat. Med. **25**(1), 24–29 (2019)
9. Aetesam, H., Maji, S.K.: Deep variational magnetic resonance image denoising via network conditioning. Biomed. Sig. Process. Control **95**(Part A), 106452 (2024). ISSN 1746-8094. https://doi.org/10.1016/j.bspc.2024.106452
10. Alpaydin, E.: Introduction to Machine Learning, 4th edn. MIT Press, Cambridge, MA (2020)
11. Vamplew, P., et al.: An ethical framework for robot control. Robot. Auton. Syst. **112**, 12–21 (2019)
12. Fisher, R.A.: The use of multiple measurements in taxonomic problems. Ann. Eugen. **7**(2), 179–188 (1936)
13. McLachlan, G.J.: Discriminant analysis and statistical pattern recognition. In: Wiley Series in Probability and Statistics. Wiley, New York (1992)
14. Cox, D.R.: The regression analysis of binary sequences. J. Roy. Stat. Soc. Ser. B (Methodol.) **20**(2), 215–232 (1958)
15. Hosmer, D.W., Lemeshow, S.: Applied logistic regression. In: Wiley Series in Probability and Statistics, 2nd edn. Wiley, New York (2000)
16. Duda, R.O., Hart, P.E., Stork, D.G.: Pattern Classification, 2nd edn. Wiley, New York (2000)
17. LeCun, Y., et al.: Gradient-based learning applied to document recognition. Proc IEEE **86**(11), 2278–2324 (1998)
18. Hinton, G.E., Osindero, S., The, Y.-W.: A fast learning algorithm for deep belief nets. Neural Comput. **18**(7), 1527–1554 (2006)
19. Kingma, D.P., Welling, M.: Auto-encoding variational bayes. arXiv preprint arXiv:1312.6114 (2013)
20. Ashish, V.: Attention is all you need. In: Advances in Neural Information Processing Systems, vol. 30, p. I (2017)

21. Aetesam, H., Maji, S.K.: Attention-based noise prior network for magnetic resonance image denoising. In: 2022 IEEE 19th International Symposium on Biomedical Imaging (ISBI), pp. 1–4. Kolkata, India (2022). https://doi.org/10.1109/ISBI52829.2022.9761530
22. Thakur, R.K., Maji, S.K.: Multi scale pixel attention and feature extraction based neural network for image denoising. Pattern Recogn. **141**, 109603 (2023). . ISSN 0031-3203. https://doi.org/10.1016/j.patcog.2023.109603
23. Radford, A., et al.: Improving language understanding by generative pre-training (2018)
24. Devlin, J., et al.: BERT: pre-training of deep bidirectional transformers for language understanding. In: Proceedings of the 2019 Conference of the North American Chapter of the Association for Computational Linguistics: Human Language Technologies, vol. 1. Long and Short Papers, pp. 4171–4186 (2019)
25. Yang, Z., et al.: Xlnet: generalized autoregressive pretraining for language understanding. In: Advances in Neural Information Processing Systems, vol. 32 (2019)
26. Liu, Y., et al.: Roberta: a robustly optimized bert pretraining approach. arXiv preprint arXiv: 1907.11692 (2019)
27. Radford, A.: Unsupervised representation learning with deep convolutional generative adversarial networks. arXiv preprint (2016)
28. Jayaramu, H.K., Maji, S.K., Yahia, H.: Personalized multi-user-based movie and video recommender system: a deep learning perspective. In: Supervised and Unsupervised Data Engineering for Multimedia Data, pp. 149–175 (2024)
29. Isola, P., et al.: Image-to-image translation with conditional adversarial networks. In: Proceedings of the IEEE Conference on Computer Vision and Pattern Recognition, pp. 1125–1134 (2017)
30. Zhu, J.-Y., et al.: Unpaired image-to-image translation using cycle-consistent adversarial networks. In: Proceedings of the IEEE International Conference on Computer Vision, pp. 2223–2232 (2017)
31. Frid-Adar, M., et al.: Synthetic data augmentation using GAN for improved liver lesion classification. In: 2018 IEEE 15th International Symposium on Biomedical Imaging (ISBI 2018), pp. 289–293. IEEE (2018)
32. Tripathi, S., et al.: Learning to generate synthetic data via compositing. In: Proceedings of the IEEE/CVF Conference on Computer Vision and Pattern Recognition, pp. 461–470 (2019)
33. Kumar, M., Maji, S.K., Yahia, H.: 5 3D volumetric. In: Handbook of AI-Based Models in Healthcare and Medicine: Approaches, Theories, and Applications, p. 70 (2024)
34. Ledig, C., et al.: Photo-realistic single image super-resolution using a generative adversarial network. In: Proceedings of the IEEE Conference on Computer Vision and Pattern Recognition, pp. 4681–4690 (2017)
35. Wang, W., et al.: Esrgan: enhanced super-resolution generative adversarial networks. In: Proceedings of the European Conference on Computer Vision (ECCV) Workshops (2018)
36. Dhariwal, P., Nichol, A.: Diffusion models beat gans on image synthesis. Adv. Neural. Inf. Process. Syst. **34**, 8780–8794 (2021)
37. Ho, J., Jain, A., Abbeel, P.: Denoising diffusion probabilistic models. arXiv preprint arXiv: 2006.11239 (2020)
38. Sohl-Dickstein, J., et al.: The energy diffusion model: training energy-based models in diffusion time. arXiv preprint arXiv:2006.11239 (2020)
39. Karras, T., et al.: Analyzing and improving the image quality of stylegan. In: Proceedings of the IEEE/CVF Conference on Computer Vision and Pattern Recognition, pp. 8110–8119 (2020)
40. Wang, X., et al.: ESRGAN: enhanced super-resolution generative adversarial networks. In: Proceedings of the European Conference on Computer Vision (ECCV) Workshops (2020)
41. Ren, Z., et al.: Structure-aware generation network for anatomically plausible brain image synthesis. In: Proceedings of the IEEE Conference on Computer Vision and Pattern Recognition, pp. 9440–9449 (2019)
42. Liu, H., et al.: Global and local structure preserving network for 3D human pose estimation. IEEE Trans. Image Process. **30**, 1158–1171 (2021)

43. Bishop, C.M.: Pattern recognition and machine learning. In: Springer Google Schola, vol. 2, pp. 1122–1128 (2006)
44. Bengio, Y., Courville, A., Vincent, P.: Representation learning: a review and new perspectives. IEEE Trans. Pattern Anal. Mach. Intell. **35**(8), 1798–1828 (2013)
45. LeCun, Y., Bengio, Y., Hinton, G: Deep learning. Nature **521**, 7553, 436–444 (2015)

Generative Adversarial Networks: A Comprehensive Review

R. Kanniga Devi and M. Asha Jerlin

Abstract Generative Adversarial Networks (GANs) have become a powerful paradigm in artificial intelligence (AI), captivating researchers across various domains. Hence, this chapter focuses on applications of GANs in diverse domains and provides a comprehensive review of types of GANs, applications, tools, advantages, disadvantages, challenges, and some open research problems to be addressed by the researchers. The methodology involves gathering and analyzing studies from academic literature and authentic websites. By synthesizing insights from these sources, this review offers a structured overview to guide future research directions and promote advancements in this field. The relevance of this review lies in its thorough exploration of how GANs are applied across multiple domains. This chapter identifies the strengths and weaknesses inherent in different GAN architectures and discusses the tools and methodologies used for implementation. Moreover, by outlining the challenges like training instability, biases in generated data, and computational demands, this chapter aims to offer actionable insights for researchers and practitioners.

Keywords Generative adversarial networks (GANs) · AI applications · GAN architectures · Training instability · Computational demands · Biases in generated data · Research challenges · Implementation tools

---

R. Kanniga Devi (✉) · M. Asha Jerlin
School of Computer Science and Engineering, Vellore Institute of Technology Chennai, Chennai, Tamilnadu, India
e-mail: rkannigadevi@gmail.com

M. Asha Jerlin
e-mail: ashajerlin.m@vit.ac.in

N. R. Vajjhala et al. (eds.), *Generative Artificial Intelligence (AI) Approaches for Industrial Applications*, Information Systems Engineering and Management 24,
https://doi.org/10.1007/978-3-031-76710-4_4

1 Introduction

Being a class of Machine Learning models, Generative Adversarial Network (GAN) generates new data samples that resemble a given data set. GANs represent a significant advancement in machine learning, offering transformative potential across various domains by enabling the generation of synthetic data that closely resembles real data distributions. The two core components of GAN namely, generator and discriminator, which are neural networks that compete to improve the quality of the generated sample data over time [1]. The generator is given a random noise or a latent vector to generate synthetic data samples that try to resemble real data in the training set. The discriminator works on input data and tries to differentiate between the real samples from the training set and fake samples from the generator. It assigns high probabilities to real data and low probabilities to fake data.

The term Adversarial training means that both the generator and discriminator trained simultaneously in such a way that they compete. The objective function is that the training process is guided by a binary cross-entropy loss function. The generator focuses on maximizing the probability that the discriminator labels the fake data generated by it to be real. The discriminator focuses on correctly classifying real and fake sample data. The convergence happens when it reaches a Nash equilibrium where generator creates realistic data samples, and the discriminator fails to differentiate them from the real data samples. After training, the generator is input with random noise or latent vector to generate new samples [2].

A simple Analogy to understand the operation of GAN in a better way is to consider a scenario involving a coin forger, nothing but the generator and a coin inspector nothing but the discriminator. The coin forger tries to create counterfeit coins that look like real coins. The coin inspector examines the coins to differentiate between fake and real coins. In the adversarial training, the forger tries to create coins that the inspector cannot spot the fake coins, while the inspector focuses on improving its ability to spot the fake coins. The learning in forger happens by adjusting its techniques based on the inspector feedback so that the forger gets better at creating counterfeit coins that closely resemble real coins. As the forger becomes more skilled through training, it becomes an increasing challenge for the discriminator to distinguish between fake and real coins. The process reaches a point where the inspector finds it extremely difficult to distinguish between fake and real coins and the forger attains a level of skill where the counterfeit coins closely resemble the real coins. In a nutshell, the competitive interplay between the forger and inspector pushes the generator to generate high quality and realistic data samples and improves the inspector's ability to detect fake coins from the real coins.

Despite the extensive research on GANs, existing reviews focus narrowly on specific applications, types or challenges within isolated domains. There is a lack of comprehensive review that encompasses the various GAN applications, their various types, and the challenges faced across different fields. This gap is critical and that needs to be addressed. Hence, this proposed review tries to fill the research gap by providing comprehensive coverage of various GAN applications across diverse domains, from image generation to healthcare and finance. The detailed typology of different types of GANs are presented with their unique characteristics, strengths and weaknesses. A discussion on various tools and frameworks available for implementing GANs is provided. The evaluation of the advantages and disadvantages of GANs helps in understanding their potential and limitations. The challenges and open problems provided will help the researchers to address them to advance the field. In a nutshell, this review will offer a detailed roadmap to the researchers about the current state and future directions of GAN research.

1.1 Types of Generative Adversarial Network

There exist various types of GAN, each designed for certain applications [1, 2] and presented in Table 1.

The methodologies used by each GAN and its potential and limitations [1, 2] are presented in Table 2.

The architectural details in terms of generator and discriminator of GAN [1, 2] are presented in Table 3.

Table 1 Major types of GAN with its application

Type of GAN	Application
Original GAN	Image/realistic image generation Data augmentation
cGAN-conditional GAN	Image-to-image translation with controlled image generation with specific attributes
DCGAN-deep convolutional GAN	Image-to-image translation High quality image generation Style transfer
WGAN-Wassertein GAN	For addressing mode collapse and training instability for high-quality image generation
LSGAN-least squares GAN	Uses least-square loss function to improve image quality
PGAN-progressive GAN	Image synthesis and progressive training
CycleGAN	Translation of images between two domains
StarGAN	Translation of images across multiple domains
BigGAN	Large-scale image synthesis
InfoGAN	Controlling specific features of generated images

Table 2 Methodologies, findings and limitations in major GAN types

Type of GAN	Methodology	Findings	Limitation
Original GAN	Uses a generator and discriminator in a minimax game to generate data	Successfully generates realistic data from random noise	Training is unstable and can suffer from mode collapse
cGAN-conditional GAN	Conditions on additional information (e.g., labels) to generate data	Generates class-specific data, improving control over generated outputs	Requires labeled data, which may not always be available
DCGAN-deep convolutional GAN	Uses convolutional layers in both generator and discriminator	Produces high-quality images, especially in the context of image generation	Convolutional layers can be computationally intensive and require significant resources
WGAN-Wassertein GAN	Uses Wasserstein distance for a more stable training process	Improves training stability and alleviates mode collapse	Training can be slower due to the computation of the Wasserstein distance
LSGAN-least squares GAN	Uses least squares loss function instead of cross-entropy loss	Reduces the problem of vanishing gradients and improves the quality of generated images	Can still suffer from mode collapse in certain scenarios
PGAN-progressive GAN	Progressively grows both the generator and discriminator	Generates very high-resolution images with stable training	Requires extensive computational resources and time for training
CycleGAN	Uses cycle-consistency loss to learn mappings between unpaired image datasets	Effective in image-to-image translation tasks without requiring paired training data	Training can be unstable, and the model might struggle with more complex transformations
StarGAN	Extends cGAN to handle multiple domains with a single model by using domain labels as input	Efficiently performs multi-domain image translation with a unified architecture	Complexity increases with the number of domains, potentially leading to scalability issues
BigGAN	Extends GANs to generate high-resolution images by leveraging large-scale datasets and architectures	Generates highly realistic images with fine details	Requires large-scale datasets and significant computational resources for training
InfoGAN	Introduces mutual information maximization between latent variables and generated data	Improves interpretability of the latent space, allowing for disentangled representations	Balancing the mutual information objective with the adversarial loss can be challenging

Table 3 Architectural details of major GAN types

Type of GAN	Generator	Discriminator
Original GAN	Takes random noise as input and generates fake data	Takes real or fake data as input and outputs a probability of the data being real
cGAN-conditional GAN	Takes random noise and additional information (e.g., class labels) as input and generates fake data conditioned on the information	Takes real or fake data and additional information as input and outputs a probability of the data being real, conditioned on the information
DCGAN-deep convolutional GAN	Uses transposed convolutional layers to generate high-resolution images from random noise	Uses convolutional layers to distinguish real images from fake images generated by the generator
WGAN-Wassertein GAN	Like the original GAN but optimized using Wasserstein distance	Outputs a scalar value representing the realness of the input, optimized using the Wasserstein distance
LSGAN-least squares GAN	Similar to the original GAN but optimized using the least squares loss	Outputs a scalar value representing the realness of the input, optimized using the least squares loss
PGAN-progressive GAN	Starts with low-resolution images and progressively increases the resolution by adding layers	Matches the generator by progressively increasing the resolution of the input it discriminates
CycleGAN	Two generators that map between two domains (e.g., images A to images B and vice versa)	Two discriminators, one for each domain, distinguishing real images from generated images
StarGAN	Single generator that can handle multiple domains by taking both the image and the domain label as input	Single discriminator that classifies the realness of the image and its domain label
BigGAN	Utilizes large-scale datasets and architectures to generate high-resolution images, employing advanced techniques like self-attention and spectral normalization	Designed to handle high-resolution images with spectral normalization to stabilize training
InfoGAN	Similar to the original GAN but optimized to maximize mutual information between latent variables and generated data	Has an additional output layer that predicts latent variables to enforce interpretability

2 Literature Review

This section reviews existing literature to find out and categorize some of the major applications of Generative Adversarial Network. The review found out the keys fields in which GAN finds it application such as Image generation, Style transfer, Data Augmentation, Super-resolution imaging, Image-to-Image translation, Face aging and De-aging, Deepfake generation, Text-to-image synthesis, Drug discovery, Anomaly detection, Virtual try-on, 3D realistic model generation, Voice generation and modification and Art and creativity generation.

2.1 Image Generation

He et al. [3] analyzed image generation using various Generators in GAN. The work compares the performance of a CNN-based generator and a Restnet-based generator using subjective evaluation. The results claim that the CNN-based generator performed better than the Restnet-based generator. The paper highlights that the image quality is influenced by the number of parameters in the model. However, the work highlighted that there still needs to be more research on the design of generator in GANs. Zhang et al. [4] proposed a new approach in computer vision, specifically automatic image processing based on depth models. The use of deep learning and depth models to create similarly distributed images allows to mine the foundation of implicit distribution rules in data. This method has various applications, such as automatic synthesis of images, face image prediction of different years, retrieving artistic images that researchers can improve.

2.2 Style Transfer

Han et al. [5] proposed DE-GAN (Depth Extraction Generative Adversarial Network) model for image-to-artwork style migration. This model uses deep learning modelling and includes a multi-feature extractor that extracts color, texture, depth, and shape masks from style images. In the process, it makes use of the MiDas depth estimation network, fast Fourier transform, multi-factor extractor, and U-net. According to the experimental analysis, DE-GAN produces images that are more in line with the aesthetic qualities of real works of art and have a higher subjective image quality.

Bo et al. [6] proposes three different models for style transfer: StyleGAN, CycleGAN, and TL-GAN. StyleGAN introduces a mapping network, which can judge whether the input is a Gaussian or normal distribution. It also puts the input in multiple steps to ensure better connectivity between the input and the output. CycleGAN may face challenges when there are geometric changes in the desired output and issues related to distribution characteristics of the image. TL-GAN, a transfer learning-based model, combines a trained feature extractor network with a GAN generator, allowing for the prediction of feature labels for composite images. These models propose new approaches to improve style transfer in various aspects. It also draws attention to the possible drawbacks and advantages of using these various approaches to style transmission. The article outlines five viewpoints on what an optimal model of style transfer should accomplish: precise control, model effectiveness, target detection in combination with secondary optimization, and visual image processing.

2.3 *Data Augmentation*

Motamed et al. [7] proposes a new Generative Adversarial Network (GAN) architecture called Inception-Augmentation GAN (IAGAN) for data augmentation in the field of chest X-ray image analysis. Specifically, the authors focus on augmenting chest X-ray data for the detection of pneumonia and COVID-19. The results show that the IAGAN model surpasses traditional augmentation methods and other GAN models, such as Deep Convolutional GAN (DCGAN) in detecting anomalies in X-ray images.

Sandfort et al. [8] suggested using cycleGAN, a technique for data augmentation in CT segmentation tasks, to create synthetic non-contrast images from contrast CT images, which are then utilized to enhance training data. It examined how well a U-net model trained on the original dataset performed in comparison to a U-net model trained on the combined dataset containing both synthetic and original non-contrast images. They used two distinct datasets to assess the segmentation performance of the U-net model: the original contrast CT dataset and a dataset from a different hospital that only included non-contrast CTs. Significant gains in segmentation ability were demonstrated by the results, particularly for out-of-distribution (non-contrast CT) data. According to the authors, their findings will help medical imaging experts cut down on the time and expense of manual CT segmentation.

2.4 *Super-Resolution Imaging*

Gupta et al. [9] apply super-resolution to medical images, specifically lung MRI scans of tuberculosis. The network is trained using the perceptual loss to enhance its performance. The paper aims to address the limitations of low-quality and time-consuming MRI scans by utilizing deep learning techniques for super-resolution. However, there present lower quality scans due to limitations of lab equipment and MRI radiation environment, time-consuming process to obtain high-resolution data. Wang et al. [10] discusses the application of GANs in super-resolution reconstruction to analyze remote sensing imagery. It standardized variables and applied

BSR degradation consistently. It used the RSC11 remote sensing dataset for super-resolution reconstruction and compared the performance metrics of different models GAN, bicubic, SRGAN, ESRGAN, RankSRGAN and evaluated image quality using metrics such as Peak Signal-to-Noise Ratio.

2.5 Image-to-Image Translation

Ko et al. [11] proposed SuperstarGA, an improved version of StarGAN for image-to-image translation. Since it can express small feature changes, it uses the ControlGAN framework to overcome the shortcomings of StarGAN in learning mappings among large-scale domains. SuperstarGAN leverages the vanilla discriminator to distinguish between real and fake images and uses a separate classifier to train the generator. It showed enhanced performance in learning perceptual image patch similarity (LPIPS) and Fréchet Inception distance (FID). The degree to which target domain features are expressed in the generated images is controlled by it. It can be adjusted to fit a variety of datasets, such as artworks and animal faces. Altakrouri et al. [12] proposed an improvised algorithm Perceptual Loss based Efficient-net Generative Adversarial Network (PL-E-GAN) for image super-resolution tasks. The algorithm has both generative adversarial loss and perceptual adversarial loss as objective functions and performs better over the other image translation models. It focuses on manipulating photos in computer graphics and computer vision.

2.6 Face-Aging and De-aging

Sheng et al. [13] discussed the concept of face aging and its significance in age-invariant face recognition and entertainment. In order to produce more accurate ageing face images, it draws attention to the difficulties in gathering images of faces for each person at various ages and offers a Generative Adversarial Network model limited by Ranking-CNN. While most existing GAN-based algorithms solely employ age labels as generative conditions, our study is the first to consider the age-related ordinal information in age group labels. The goal of the work is to increase face ageing accuracy by taking age-related ordinal information into account.

Pranoto et al. [14] discusses face aging concept including image processing algorithm and GAN. It presents a classical approach which uses GAN and its structure, formulation, learning algorithm, challenges, advantages and disadvantages. It highlights the importance of the aging module in face aging and the use of an identity-preserving module. The work mentions the criteria for the data set used in face aging, highlighting the need for a sufficiently large age group, images, and balanced distribution.

2.7 Deepfake Generation

Preeti et al. [15] proposes a deep convolutional GAN for detecting Deepfakes and provides performance analysis. The study demonstrated the application of Generative Adversarial Networks to recognize phony images and movies that are difficult for people to distinguish from the real thing. There was discussion of several techniques for altering and manipulating faces, such as attribute manipulation, identity swapping, face synthesis, and complete expression swapping. Akhtar et al. [16] provide an overview of the different types of deepfake or face manipulations, such as identity swap, face reenactment, attribute manipulation, and entire face synthesis. The article discusses the availability of face-editing applications and how different facial alteration techniques can affect face recognition software. The author talks about how deep neural networks can now be used to manipulate or create realistic-looking facial samples thanks to the development of deep learning techniques and the availability of vast databases. The study also highlights the need for real-time mobile deepfake detection, decision explainability, and more resilient deepfake detection systems against adversarial attacks. The lack of ultra-high-resolution samples, the limitations of face attribution modifications, issues with video continuity, and the lack of evident deepfake/facial manipulations are only a few of the flaws that the author considers about current datasets and generating methods.

2.8 Text-to-Image Synthesis

Ku et al. [17] discussed a new approach called TextControlGAN for text-to-image synthesis using GANs. By adding a regressor neural network structure to efficiently learn features from conditional texts, it overcomes the drawbacks of conventional GANs. The regressor's learning performance is further improved by the data augmentation strategies. By concentrating on the discriminator's training, the overall quality of the generated images is enhanced. TextControlGAN surpasses the cGAN-based GAN-INT-CLS model in terms of Inception Score (IS) and Fréchet Inception Distance (FID), according to evaluation results on the Caltech-UCSD Birds-200 dataset. TextControlGAN is said to be able to produce text-conditioned images of excellent quality.

Reed et al. [18] developed neural network architectures for learning text feature representations and the deep convolutional generative adversarial networks for generating high-quality images. They propose a novel deep architecture and GAN formulation to bridge the gap between text and image modeling. They proposed a text manifold interpolation method to generate additional text embeddings, which did not correspond to actual human-written text, to improve the variety of generated images without additional labeling cost. They demonstrate the effectiveness of their model by generating plausible images of birds and flowers based on detailed text description.

2.9 Drug Discovery

Abbasi et al. [19] proposed a Feedback Generative Adversarial Network based on deep learning model to generate drug-like molecules for specific needs. The framework includes an optimization strategy that incorporates Encoder-Decoder, GAN, and Predictor models connected through a feedback loop. The Encoder-Decoder converts molecule notations into latent space vectors, while the GAN replicates the training data distribution to generate new compounds. The feedback loop evaluates the generated molecules based on desired properties, ensuring a shift towards the targeted properties. The paper also incorporates a multi-objective optimization technique to select molecules. Two different sampling methods, namely Tani-Inf and Tani-Sup, along with the Random sampling method. These methods were used to generate molecules through the feedback GAN and compare the percentage of valid molecules produced by each method. The results show that this framework can successfully generate realistic and novel molecules with high diversity and uniqueness.

Blanchard et al. [20] presented the limitations of standard GAN training methods, such as mode collapse, which hinders exploration beyond the original data. To address this, the authors propose an approach that incorporates concepts from Genetic Algorithms to promote incremental exploration and limit mode collapse. It involves updating the training data set through a replacement strategy, guided or random, during the training process. This incremental updating of the training data set allows for augmented search, enabling the GAN to explore new areas of parameter space that were not encountered during training. By replacing samples from the training data with valid samples generated by the GAN, they demonstrate significant improvements in producing novel compounds.

2.10 Anomaly Detection

Esmaeili et al. [21] presents an overview of using generative adversarial networks for anomaly detection and explores the effectiveness of state-of-the-art GAN-based anomaly detection methods in biomedical imaging. They experimented on seven medical imaging datasets using three anomaly detection techniques, and they examined the findings from both a data-centric and a model-centric perspective. The results demonstrated that none of the techniques could consistently identify anomalies in medical photos, and the model's performance was affected by several variables, including the quantity of training samples and the degree of anomaly. In the context of medical imaging, where obtaining annotations (labels) is frequently costly and time-consuming, they talked about the advantages of utilizing GANs. The suggested method may offer a decision assistance system for the identification of uncommon or unidentified illnesses in medical imaging. In conclusion, the work offers guidance on the application of anomaly detection models in medical imaging and proposes crucial avenues for further research.

Yumoto et al. [22] presented a method for detecting anomalies in sewer pipes using GAN. The paper addresses the increasing number of pipes that have exceeded their service life and the need for effective inspection methods. The proposed method combines f-AnoGAN and Lightweight GAN to detect anomalies by comparing input images with generated images. Non-defective images are used to train the GANs, allowing them to convert images with defects into defect-free images. The subtraction images obtained from this process are used to estimate the location of anomalies. The effectiveness of the method is confirmed through experiments using actual images of cast iron pipes and a public dataset called sewer-ml.

2.11 Virtual Try-On

Honda et al. [23] discussed the difficulty of creating a virtual try-on image using a model's photo and photographs of clothes from within a store. To tackle this problem, it presents the Virtual Try-on Generative Adversarial Network (VITON-GAN). When there is occlusion, as when the model person's arms are crossed in front of their clothing, the technique seeks to improve the quality of the photographs that are produced. The training process integrates an adversarial technique to tackle the issue of occlusion and enhances the overall quality of the images. An adversarial mechanism is incorporated into the training pipeline to accomplish this. Jong et al. [24] explained how GANs, pixel translation, and perceptual losses have influenced the virtual try-on field. It summarizes the latest research in creating virtual try-on systems such as VITON and CAGAN, CP-VTOPM, MG-VTON, SwapNet, video-based virtual try-on and presents future directions for improving them.

2.12 3D Realistic Model Generation

Ferreira et al. [25] reviewed the generation of realistic 3D data using GAN-based approaches by the dimension, 2D or 3D especially for medical field explaining the extensive use of CycleGAN-based and cGAN-based architectures. In addition to discussing data formats, network designs, loss functions, and assessment metrics, it investigated the targeted modalities, such as 3D models, CT, MRI, PET, or multiple modalities. It has been noted that the medical field is where multimodal data is most frequently utilized. Li et al. [26] proposed a 3D conditional GAN, which incorporates class information into both the generator and discriminator to address the limitations of traditional GANs in generating random and uncontrollable 3D models. They presented a new 3D model reconstruction network by integrating a classifier aiming to enhance the quality of reconstructing 3D models from single images which involves a combination of condition and classifier information to generate 3D models.

The experimental results on ModelNet10 dataset demonstrate the effectiveness of the proposed method in generating realistic 3D models with given class labels, as well as significantly improving the quality of 3D model reconstruction in the IKEA dataset.

2.13 Voice Generation and Modification

Zhao et al. [27] proposed a voice conversion GAN network called IVCGAN that uses a voice conversion technique based on non-parallel data and has the ability to convert voice samples of arbitrary duration. The work integrated discriminator and classifier in StarGAN-VC and scaled up the discriminators. The work was tested on a clear and clean Chinese voice data set and claim that the work achieves better voice conversion and better than the baseline. Atkar et al. [28] discusses the system which generates raw audio data of two or three-letter Hindi words that will pronounce these generated words to aid children with dyslexia in their recovery process. It uses WaveGAN and MelGAN architectures for audio generation. It helps the teachers to speed up the recuperation process by helping children repeat the correct pronunciation of the words. The work uses the advanced Mel-generative adversarial network neural network for modeling the audio data and achieving satisfactory results.

2.14 Art and Creativity Generation

Elgammal et al. [29] proposed a new system for generating art using Generative Adversarial Networks (GAN) to learn about style and generate art with modifications to maximize deviation from established styles and minimize deviation from art distribution. The author proposes modifications to the existing system to make the system capable of generating art that deviates from established styles. The paper has experiments comparing human response to art generated by the proposed system with art created by artists and claims that human subjects could not differentiate between art generated by the system and art shown in top art fairs.

Vela et al. [30] proposed an innovative approach called the top-k GANs for training Generative Adversarial Network that uses only the most realistic ones that have successfully fooled the Discriminator to update the weights of the Generator and discards the images that are not like the output set. This helps to prevent the gradient from moving in the wrong direction and improves convergence in training. This modification improves the results without increasing computational cost. The effectiveness of this approach is demonstrated through experiments with different variants of GANs (Fig. 1).

The summary of major GAN applications along with its application and description is presented in Table 4.

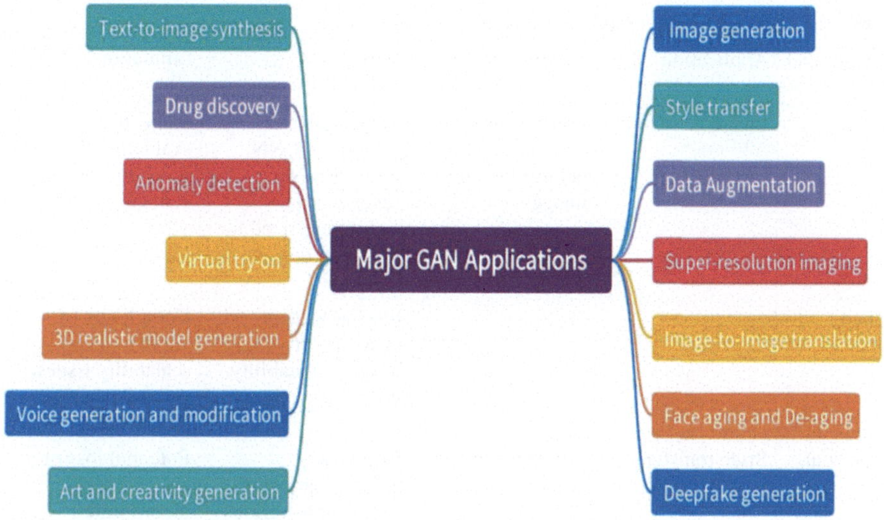

Fig. 1 Major GAN applications

3 Performance Metrics of Generative Adversarial Networks

The performance assessment of Generative Adversarial Networks involves various metrics to evaluate the quality, diversity and fidelity of the generated samples [3–30] (Fig. 2).

Some commonly used metrics in research papers are presented in Table 5.

The performance metrics used in research papers for particular applications along with its practical implications are presented in Table 6.

4 Implementation Requirements of Generative Adversarial Network

The successful implementation of GAN requires various hardware, tools, framework and programming languages and are summarized in Fig. 3.

Table 4 Summary of major GAN applications

Author and year	Application	Description	Strengths	Limitations
He et al. [3]	Image generation	Image synthesis which resembles photo-realistic images	Comparative analysis of CNN versus ResNet generators; highlighted need for generator design research	Subjective evaluation; limited scope to specific generator types
Zhang et al. [4]			Innovative approach in computer vision; broad applicability in image synthesis	Complexity of depth models; scalability issues; lack of extensive validation
Han et al. [5]	Style transfer	Transforming photos into different styles	Effective transformation of images into various styles	Potential loss of image fidelity; algorithmic complexity in style transfer
Bo et al. [6]			Efficient style transfer techniques demonstrated	Style transfer limitations; artistic fidelity of output
Motamed et al. [7]	Data augmentation	Dataset enhancement by synthetic data generation which is useful for supervised learning	Improved dataset diversity and size for supervised learning tasks	Generation quality compared to real data; overfitting potential
Sandfort et al. [8]			Augmentation of training data; addresses data scarcity issues	Data quality and realism; application domain limitations
Gupta et al. [9]	Super-resolution imaging	Image quality improvement in satellite images, medical images	Significant improvement in image clarity and detail	Computational intensity; fine-tuning requirements
Wang et al. [10]			Advanced algorithms for image enhancement	Resource-intensive processing; generalizability across datasets
Ko et al. [11]	Image-to-image translation	Converting image in one color to another color or scene	Effective translation capabilities demonstrated	Complex image transformations; adapting to diverse datasets
Altakrouri et al. [12]			Versatility in transforming images; application in diverse fields	Accuracy of image transformation; limitations in style adaptation

(continued)

Table 4 (continued)

Author and year	Application	Description	Strengths	Limitations
Sheng et al. [13]	Face aging and de-aging	Facial recognition system and fun apps	Accurate simulation of facial age progression	Realism of age transformation; limited facial variability
Pranoto et al. [14]			Practical application in facial recognition systems	Variability in aging patterns
Preeti et al. [15]	Deepfake generation	Entertainment purpose	Innovative use of deep learning for video manipulation	Detection challenges
Akhtar et al. [16]			Advancements in generating realistic audiovisual content	Security risks in media authenticity
Ku et al. [17]	Text-to-image synthesis	Generation of scenes based on the given textual information	Bridging textual and visual domains effectively	Semantic fidelity in image synthesis; handling complex descriptions
Reed et al. [18]			Early exploration of text-based image generation	Accuracy in complex scene representation
Abbasi et al. [1]	Drug discovery	Molecular structure generation, optimization	Advancing drug discovery through AI-driven molecular design	Accuracy in molecular structure prediction; validation against experimental data
Blanchard et al. [6]			Contribution to accelerating drug development processes	Complexity in chemical synthesis; validation of computational predictions
Esmaeili et al. [9]	Anomaly detection	Detecting abnormal patterns in network security, medical images	Improved anomaly detection accuracy	Handling diverse anomaly types
Yumoto et al. [28]			Application in critical security and medical domains	Scalability in large-scale systems
Honda et al. [23]	Virtual try-on	How clothing looks on a person by generating virtual try-on	Enhancing online shopping experiences with virtual fitting rooms	Adaptation to diverse body types
Jong et al. [24]			Integration of AI in fashion industry applications	Accuracy in garment visualization; variation in clothing styles

(continued)

Table 4 (continued)

Author and year	Application	Description	Strengths	Limitations
Ferreira et al. [25]	3D realistic model generation	Architectures, virtual reality and games	Achieves detailed and realistic 3D models suitable for virtual reality and gaming applications	Architectural designs may be complex, requiring substantial computational resources
Li et al. [26]			High fidelity and detail in virtual model creation	Real-time rendering challenges
Zhao et al. [27]	Voice generation and modification	Voice assistant, dubbing, narration	Realistic and adaptable voice synthesis	Voice variability
Atkar et al. [28]			Application in entertainment and communication technologies	Handling diverse linguistic nuances
Elgammal et al. [29]	Art and creativity generation	Creative content, music generation	Advancements in AI-based creativity and artistic expression	Exploration of diverse creative styles
Vela et al. [30]			Pushing boundaries in AI-driven creative processes	Complexity in capturing human-like creativity

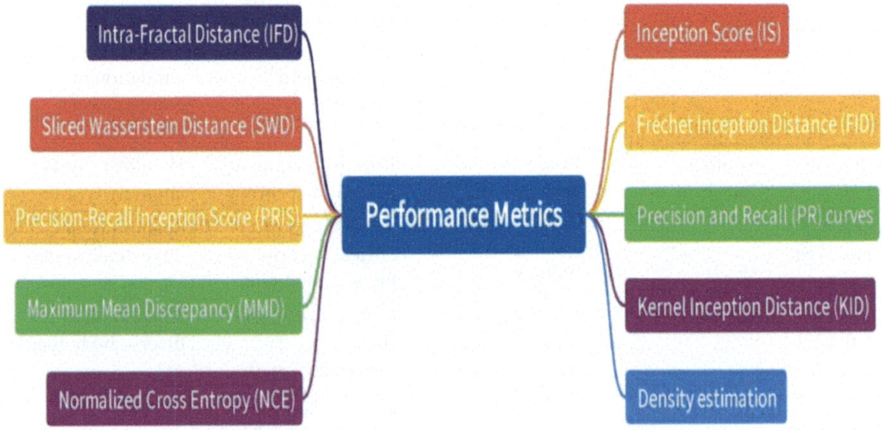

Fig. 2 GAN performance metrics

Table 5 Major performance metrics of generative adversarial networks

Performance metric	Measurement	Expected value for better performance
Inception score (IS)	Measurement of generated images in terms of quality and diversity	Having higher score to have better quality and diversity
Fréchet inception distance (FID)	Generated samples and real data comparison statistics	Having lower score indicates higher similarity between generated samples and real data
Precision and recall (PR) curves	Measurement of precision and recall of a classifier on generated samples and real data	Having higher precision and recall values indicate good alignment between generated samples and real data
Kernel inception distance (KID)	Measurement of distance between feature representation of generated samples and real data	Having lower value indicates higher similarity between generated samples and real data distribution
Density estimation	Measures the ability of the model to capture the data distribution	Having higher likelihood or lower negative log likelihood gives better density estimation
Intra-fractal distance (IFD)	Measurement of similarity between fractal dimensions of generated samples and real data	Having lower score gives better similarity in fractal structures
Sliced Wasserstein distance (SWD)	Measurement of Wassertein distance between distributions	Having lower score indicates good alignment between generated samples and real data
Precision-recall inception score (PRIS)	Combination of precision and recall with inception score	Balanced metric to assess the image quality
Maximum mean discrepancy (MMD)	Measurement of difference between mean embeddings of generated samples and real data	Having lower score indicates good alignment between generated samples and real data
Normalized cross entropy (NCE)	Measurement of difference in cross entropy between generated samples and real data	Having lower score indicates generation of realistic samples

5 Advantages and Disadvantages of Generative Adversarial Network

For any machine model to perform well, it requires rigorous training on the data set. Hence, the data augmentation helps in generating synthetic data when the model undergoes data scarcity during training. Next, it is helpful in content creation, marketing and design field where high creativity is demanded. Controlled generation of sample data with desired characteristics is another important advantage of GAN. Despite all these advantages, GANs have ethical concerns due to misinformation and malicious usage when it comes to deepfake content generation. As reported

Table 6 Major performance metrics of generative adversarial networks with its use case and practical implications

Performance metric	Use case	Practical implications
Inception score (IS)	**Image generation**: evaluating quality and diversity of generated animal images	Ensures that the generated images are realistic and cover a wide range of animal categories, indicating both high quality and diversity of the generated data
Fréchet inception distance (FID)	**Image synthesis**: comparing the quality of images generated by different GAN models	Low FID scores indicate that the generated images are like real images, making this metric crucial for applications requiring high-fidelity image synthesis, such as in medical imaging or photo-realistic image generation
Precision and recall (PR) curves	**Data augmentation**: evaluating GAN-generated images for training machine learning models	High precision indicates realistic images, while high recall ensures diversity. Both are essential for creating robust training datasets in applications like object detection or classification in autonomous driving
Kernel inception distance (KID)	**Benchmarking GANs**: comparing different GAN models on benchmark datasets	Provides a reliable measure of similarity between the distributions of real and generated images, useful for benchmarking and improving GAN models across various image generation tasks
Density estimation	**Anomaly detection**: Using GANs to generate normal data and detect anomalies	Ensures that the GAN generates data that closely matches the real distribution, which is crucial for detecting deviations and anomalies in applications such as fraud detection or monitoring industrial equipment
Intra-fractal distance (IFD)	**Texture synthesis**: evaluating the texture quality of generated materials	Measures the self-similarity and structural integrity of generated textures, ensuring high-quality texture synthesis for applications in computer graphics, video game design, and virtual reality environments
Sliced Wasserstein distance (SWD)	**3D shape generation**: evaluating the quality of generated 3D shapes	Ensures that the generated 3D shapes closely resemble real ones, important for applications in 3D modeling, animation, and virtual reality, where realistic shapes are critical
Precision-recall inception score (PRIS)	**Medical imaging**: evaluating the quality and diversity of synthetic medical images	Balances quality and diversity in synthetic medical image generation, crucial for augmenting datasets in medical research and improving the training of diagnostic models

(continued)

Table 6 (continued)

Performance metric	Use case	Practical implications
Maximum mean discrepancy (MMD)	**Domain adaptation**: evaluating the effectiveness of GANs in generating data for domain adaptation	Measures the similarity between the distributions of source and target domain data, important for applications where GANs are used to adapt models to new domains, such as in transfer learning or cross-domain image translation
Normalized cross entropy (NCE)	**Text-to-image generation**: evaluating the alignment between generated images and their textual descriptions	Ensures that the generated images accurately reflect the provided textual descriptions, important for applications in generating illustrative images for stories, advertising, or generating training data for text-based image retrieval systems

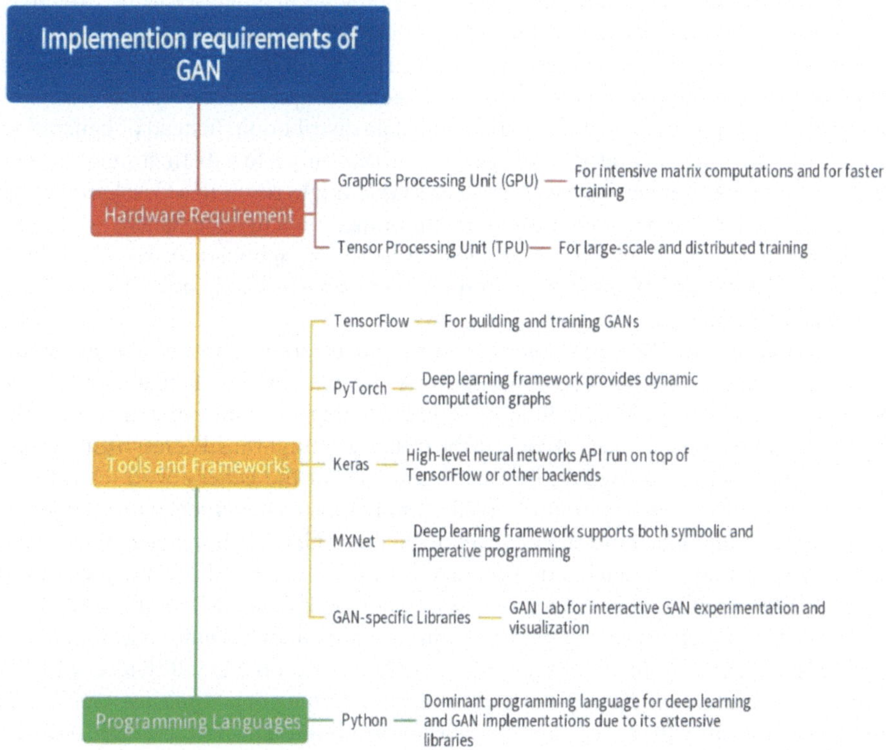

Fig. 3 Implementation requirements of GAN

by Deepware Scanner [30], the deepfake videos on social media have increased by 330% from 2019 to 2020 violating individual's privacy. As reported by Deeptrace [30], the political deepfakes had a negative impact on elections. Security-related concerns raised by security experts as it has been used in various security-related contexts. Hence, implementing policies and guidelines are important to mitigate negative impacts of GANs.

6 Challenges of Generative Adversarial Network

GAN requires high computational power and resources when the training is done at a large scale. The biases during the training data may lead to biased sample generation which raise concern about the fairness of the generated content. Lesser training data may cause overfitting issues and sometimes the fine-tuning of GAN can be challenging. The training of GAN requires a huge dataset to be effective, otherwise its performance may drop and requires careful tuning of hyper parameters to converge. It's quite challenging to understand and interpret its decision-making process. Mode collapse occurs when the GAN generator produces a limited variety of outputs, effectively ignoring large parts of the data distribution. Instead of generating diverse samples, the generator converges to generate only a few distinct samples. For example, in image generation tasks, mode collapse might result in a GAN producing the same few faces repeatedly, despite being trained on a diverse dataset of human faces. Mode collapse limits the usefulness of GANs in applications requiring high diversity, such as data augmentation for training machine learning models or creating varied artistic content.

GAN training is often unstable due to the adversarial nature of the generator and discriminator. The balance between these two networks is delicate, and if one becomes significantly stronger than the other, training can fail. For example, if the discriminator becomes too good early in the training, it might easily classify all generated samples as fake, providing no useful gradient for the generator to learn from. Training instability can lead to prolonged training times, convergence to suboptimal solutions, or complete failure to converge, making GANs impractical for certain applications without extensive tuning. GANs require large and diverse datasets to generate high-quality and realistic outputs. Insufficient data can lead to overfitting, where the GAN memorizes the training samples rather than learning to generate new ones. For example, in medical imaging, a GAN trained on a small dataset of MRI scans might generate images that look too like the training data, limiting its utility in generating varied synthetic images for data augmentation. The need for large datasets poses a challenge in domains where data collection is expensive or time-consuming, such as medical imaging, autonomous driving, or personalized content generation.

7 Ethical Consideration in Generative Adversarial Network

GANs present various ethical considerations and potential risks due to their ability to generate highly realistic and deceptive outputs. GANs can create highly realistic fake images, videos, or text that are indistinguishable from real content. This raises concerns about the spread of misinformation and deception. For example, Deepfakes created using GANs can maliciously manipulate videos of public figures, leading to false information dissemination and damage to reputations. GANs trained on personal data can generate synthetic profiles or images that infringe on individuals' privacy rights. For example, generating synthetic faces from scraped social media profiles without consent could violate individuals' privacy. These risks can be mitigated by developing frameworks with traceability features which identify the origin of the content generated by GANs. Strict regulations should be enforced for the use and deployment of GANs.

8 Open Research Problems in Generative Adversarial Network

Based on the literature review and challenges discussed in earlier sections, a comprehensive list of research problems and the suggested research focus is presented in Table 7 which would be helpful for the researchers.

9 Conclusion

This chapter provides a comprehensive analysis of GANs, showcasing their transformative impact across a variety of domains. By examining different types of GAN architectures, their applications, and challenges, this review presents both the strengths and limitations inherent to each approach. Issues like training instability, data biases, and computational demands remain open areas for improvement, demanding further research to refine GAN models. As GANs continue to evolve, their potential to address complex problems in fields such as healthcare, image generation, and data augmentation will expand. This chapter not only outlines current progress but also highlights the critical research areas needed to ensure GANs' broader and more ethical application in the future. Furthermore, the exploration of GANs' ethical implications is becoming increasingly significant as these models gain more widespread use. The potential misuse of GANs for generating deepfakes, manipulating sensitive data, or contributing to misinformation highlights the need for stricter regulations and ethical frameworks. As researchers and practitioners strive to push the boundaries of GAN technology, it is equally important to

Table 7 Summary of open research problem in GAN

Research problem	Research focus
Training instability leads to mode collapse where the generator fails to generate diverse sample data	Development of techniques to address or mitigate mode collapse by stabilizing training so as to generate diverse or realistic sample data
Assessment of generated sample data quality is challenging	Devising better evaluation metrics to assess realism, diversity, semantic relevance would help in assessing generated data quality
The internal mechanism of both the generator and discriminator are complex to understand	Theoretical understanding is required to understand the internal mechanism such as training and convergence
Large dataset is required for effective training especially where labelled data is limited	Development of methods wherein the model performs well with limited sample dataset
Conditional GANs suffer under different conditions and controlling aspect of the generated output is challenging	Enhancement of conditional GANs to have better control over the generated output and control of attributes
Adversarial attacks when disruptions of input data lead to changes in generated output	Development of robust techniques to handle adversarial attacks to get stable and reliable generated output
Biases in training data lead to biased generated sample	Development of debiasing techniques to ensure fairness in generated output
Computation and memory intensive GAN	Development of computational-efficient and memory-efficient GAN architectures

develop robust safeguards and accountability measures to prevent unethical exploitation. Addressing these concerns alongside technical advancements will be crucial to fostering responsible innovation in the field of generative modeling.

References

1. Aggarwal, A., Mittal, M., Battineni, G.: Generative adversarial network: an overview of theory and applications. Int. J. Inf. Manag. Data Insights **1**(1) (2021). https://doi.org/10.1016/j.jjimei.2020.100004
2. Pavan Kumar, M.R., Jayagopal, P.: Generative adversarial networks: a survey on applications and challenges. Int. J. Multimed. Inf. Retr. **10**(1), 1–24 (2021). https://doi.org/10.1007/s13735-020-00196-w
3. He, J., Nie, Y., Mao, Z.: Analysis of image generation by different generator in GANs. J. Phys. Conf. Ser. **1903**(1) (2021). https://doi.org/10.1088/1742-6596/1903/1/012061
4. Zhang, T., Tian, W.H., Zheng, T.Y., Li, Z.N., Du, X.M., Li, F.: Realistic face image generation based on generative adversarial network. In: 2019 16th International Computer Conference on Wavelet Active. Media Technology and. Information Processing. ICCWAMTIP 2019, pp. 303–306 (2019). https://doi.org/10.1109/ICCWAMTIP47768.2019.9067742
5. Han, X., Wu, Y., Wan, R.: A method for style transfer from artistic images based on depth extraction generative adversarial network. Appl. Sci. **13**(2) (2023). https://doi.org/10.3390/app13020867

6. Bo, X., Jing, X., Yang, X.: Style transfer analysis based on generative adversarial networks. In: 2021 IEEE International Conference on Computer Science, Electronic Information Engineering and Intelligent Control Technology. CEI 2021, pp. 27–30 (2021). https://doi.org/10.1109/CEI52496.2021.9574507
7. Motamed, S., Rogalla, P., Khalvati, F.: Data augmentation using generative adversarial networks (GANs) for GAN-based detection of pneumonia and COVID-19 in chest X-ray images. Informat. Med. Unlock. **27** (2021). https://doi.org/10.1016/j.imu.2021.100779
8. Sandfort, V., Yan, K., Pickhardt, P.J., Summers, R.M.: Data augmentation using generative adversarial networks (CycleGAN) to improve generalizability in CT segmentation tasks. Sci. Rep. **9**(1), 1–9 (2019). https://doi.org/10.1038/s41598-019-52737-x
9. Gupta, R., Sharma, A., Kumar, A.: Super-resolution using GANs for medical imaging. Proc. Comput. Sci. **173**(2019), 28–35 (2020). https://doi.org/10.1016/j.procs.2020.06.005
10. Wang, X., Sun, L., Chehri, A., Song, Y.: A review of GAN-based super-resolution reconstruction for optical remote sensing images. Remote Sens. **15**(20), 1–34 (2023). https://doi.org/10.3390/rs15205062
11. Ko, K., Yeom, T., Lee, M.: SuperstarGAN: Generative adversarial networks for image-to-image translation in large-scale domains. Neural Netw. **162**, 330–339 (2023). https://doi.org/10.1016/j.neunet.2023.02.042
12. Altakrouri, S, Usman, S.B., Ahmad, N.B., Justinia, T., Noor, N.M.: Image to image translation networks using perceptual adversarial loss function. In: Proceedings of the 2021 IEEE International Conference on Signal and Image Processing Applications. ICSIPA 2021, pp. 89–94 (2021). https://doi.org/10.1109/ICSIPA52582.2021.9576815
13. Sheng, M., Ma, Z., Jia, H., Mao, Q., Dong, M.: Face aging with conditional generative adversarial network guided by ranking-CNN. In: Proceedings of the 3rd International Conference on Multimedia Information Processing and Retrieval. MIPR 2020, pp. 314–319 (2020). https://doi.org/10.1109/MIPR49039.2020.00071
14. Pranoto, H., Heryadi, Y., Warnars, H.L.H.S., Budiharto, W.: Recent generative adversarial approach in face aging and dataset review. IEEE Access **10**, 28693–28716 (2022). https://doi.org/10.1109/ACCESS.2022.3157617
15. Preeti, M.K., Sharma, H.K.: A GAN-based model of Deepfake detection in social media. Proc. Comput. Sci. **218**, 2153–2162 (2022). https://doi.org/10.1016/j.procs.2023.01.191
16. Akhtar, Z.: Deepfakes generation and detection: a short survey. J. Imag. **9**(1) (2023). https://doi.org/10.3390/jimaging9010018
17. Ku, H., Lee, M.: TextControlGAN: text-to-image synthesis with controllable generative adversarial networks. Appl. Sci. **13**(8) (2023). https://doi.org/10.3390/app13085098
18. Reed, S., Akata, Z., Yan, X., Logeswaran, L., Schiele, B., Lee, H.: Generative adversarial text to image synthesis. In: 33rd International Conference on Machine Learning, ICML 2016, vol. 3, pp. 1681–1690 (2016)
19. Abbasi, M., Santos, B.P., Pereira, T.C., Sofia, R., Monteiro, N.R.C., Simões, C.J.V., Brito, R., Ribeiro, B., Oliveira, J.L., Arrais, J.P.: Designing optimized drug candidates with generative adversarial network. J. Chem. Informat. **14**(1), 1–16 (2022). https://doi.org/10.1186/s13321-022-00623-6
20. Blanchard, A.E., Stanley, C., Bhowmik, D.: Using GANs with adaptive training data to search for new molecules. J. Chem Inform. **13**(1), 4–11 (2021). https://doi.org/10.1186/s13321-021-00494-3
21. Esmaeili, M., et al.: Generative adversarial networks for anomaly detection in biomedical imaging: a study on seven medical image datasets. IEEE Access **11**, 17906–17921 (2023). https://doi.org/10.1109/ACCESS.2023.3244741
22. Yumoto, S., Kitsukawa, T., Moro, A., Pathak, S., Nakamura, T., Umeda, K.: Anomaly detection from images in pipes using GAN. ROBOMECH J. **10**(1) (2023). https://doi.org/10.1186/s40648-023-00246-y
23. Honda, S.: VITON-GAN: virtual try-on image generator trained with adversarial loss. In: 40th Annual Conference of the European Association for Computer Graphics, Eurographics 2019—Posters, pp. 9–10 (2019). https://doi.org/10.2312/egp.20191043

24. Jong, A., Moh, M., Moh, T.S.: Virtual try-on with generative adversarial networks: a taxonomical survey. Adv. Comput. Vis. Appl. Intell. Syst. Multimed. Technol. 76–100 (2020). https://doi.org/10.4018/978-1-7998-4444-0.ch005
25. Ferreira, A., Li, J., Pomykala, K.L., Kleesiek, J., Alves, V., Egger, J.: GAN-based generation of realistic 3D data: a systematic review and taxonomy [Online]. Available: http://arxiv.org/abs/2207.01390 (2022)
26. Li, H., Zheng, Y., Wu, X., Cai, Q.: 3D model generation and reconstruction using conditional generative adversarial network. Int. J. Comput. Intell. Syst. **12**(2), 697–705 (2019). https://doi.org/10.2991/ijcis.d.190617.001
27. Zhao, W., Wang, W., Chai, J., Huang, J.: IVCGAN: an improved GAN for voice conversion. In: IEEE Information Technology, Networking, Electronic and Automation Control Conference. ITNEC 2021, pp. 1035–1039 (2021). https://doi.org/10.1109/ITNEC52019.2021.9587053
28. Atkar, G., Jayaraju, P.: Speech synthesis using generative adversarial network for improving readability of Hindi words to recuperate from dyslexia. Neural Comput. Appl. **33**(15), 9353–9362 (2021). https://doi.org/10.1007/s00521-021-05695-3
29. Elgammal, A., Liu, B., Elhoseiny, M., Mazzone, M.: CAN: creative adversarial networks generating 'art' by learning about styles and deviating from style norms. In: Proceedings of the 8th International Conference on Computational Creativity. ICCC 2017, pp. 1–22. ICCC (2017)
30. Vela, L., Fuentes-Hurtado, F., Colomer, A.: Improving the quality of image generation in art with top-k training and cyclic generative methods. Sci. Rep. **13**(1), 1–16 (2023). https://doi.org/10.1038/s41598-023-44289-y

Generative Adversarial Networks: Security, Privacy, and Ethical Considerations

Wasswa Shafik

Abstract Technology advancements have demonstrated a shift in application, interpretability, and technological acceptance and considerations. Generative Adversarial Networks (GANs) represent two transformative domains in the world of technology, each carrying immense potential for innovation and disruption. This study examines the rise of ethical, privacy, and security considerations accompanying these technologies. The study starts our investigation by highlighting the growing importance of GANs and defining its core point, emphasizing how ethical, security, and privacy problems overlap in these domains and how they can be mitigated. This chapter starts with the core aspects of GANs, outlining their underlying principles and practical applications. This chapter explores how these generative models reshape industries while examining the ethical dilemmas they introduce by generating synthetic content. Ethical considerations are one of the principal points of the study, and various ethical frameworks and philosophies demonstrate their application in the GAN domains. Real-world case studies present the intricate ethical dilemmas that arise from GANs potential for cryptographic disruption to GANs capacity to fabricate misinformation. Privacy concerns follow, exploring the intricate web of data security and personal information in an era of quantum supremacy and GAN-generated content. We evaluate the effectiveness of privacy-enhancing technologies in safeguarding individuals' data in a rapidly evolving landscape. Security challenges form another critical segment of our analysis. This chapter scrutinizes the vulnerabilities inherent in GANs and proposes strategies to fortify these technologies against adversarial threats. The ever-evolving legal and regulatory frameworks in the context of GANs are also examined, shedding light on the delicate balance between innovation and safeguarding the public interest.

W. Shafik (✉)
School of Digital Science, Universiti Brunei Darussalam, Brunei–Muara, Brunei Darussalam
e-mail: wasswashafik@ieee.org

Dig Connectivity Research Laboratory (DCRLab), Kampala, Uganda

N. R. Vajjhala et al. (eds.), *Generative Artificial Intelligence (AI) Approaches for Industrial Applications*, Information Systems Engineering and Management 24,
https://doi.org/10.1007/978-3-031-76710-4_5

Keywords Generative adversarial networks · Security and data privacy · Ethical consideration · Responsible consumption · Production · Industry · Innovation · And infrastructure

1 Introduction

Generative Adversarial Networks (GANs) herald a new era in technology, promising unprecedented capabilities and opportunities while posing profound ethical, privacy, and security challenges [1]. GANs are deep learning architecture that trains two neural networks to compete against each other to generate more authentic new data from a given training dataset. For illustration, generating new images from an existing image database or original music from a database of songs. A GAN is called adversarial because it trains two different networks and pits them against each other. One network generates new data by taking an input data sample and modifying it as much as possible [2]. The unexplored domain of GANs presents an enticing prospect of resolving unsolvable challenges for conventional computers. As we explore the complexities of quantum bits, superposition, and entanglement, unveiling the possible applications of these phenomena in several fields, for example, cryptography, optimization, drug discovery, and other domains [3]. Nevertheless, the recent advancements in computational abilities raise ethical considerations about safeguarding cryptographic systems and the conscientious use of these capacities.

The notion of privacy, which has historically been fundamental to safeguarding personal and data security, assumes a novel perspective within this framework. The implications of GANs on cryptography necessitate reassessing our current encryption standards and pursuing solutions that are immune to quantum attacks [4]. GANs pose a simultaneous challenge to conventional conceptions of digital privacy by producing synthetic data that can be employed for manipulation or deception. Furthermore, the continuous progression of innovation also requires a thorough analysis of security implications [5]. The advent of GANs has brought up new vulnerabilities and the possibility of cyber threats that were previously unimaginable. The expeditious advancement of GANs necessitates the development of resilient countermeasures against their potential malevolent applications, given the increasing persuasiveness and availability of deepfakes and synthetic content [6].

This work aims to enhance the comprehension of the ethical, privacy, and security implications inherent in the domains of GANs, given the intricate problems and opportunities they present. As we traverse this unfamiliar domain, we aim to not only elucidate the complexities of these technologies but also offer perspectives and suggestions that can facilitate their conscientious incorporation into our swiftly progressing digital realm [7]. Simultaneously, GANs, a machine learning (ML) application, have enabled us to generate highly realistic images, text, and audio from digital platforms. They have stimulated innovation across various disciplines, encompassing art, fashion, medical imaging, and data augmentation. It is essential to acknowledge that GANs come with ethical implications [8]. These implications arise from their ability to produce deepfakes, synthetic misinformation, and intrusions on privacy, necessitating careful monitoring and appropriate solutions to address these concerns.

The interconnection of new technologies is exemplified by the smooth synergy observed between GANs. It potentially enhances their training and optimization procedures, hence facilitating the generation of synthetic data with heightened realism and increased accessibility [9]. In contrast, GANs can be utilized to support quantum researchers in generating intricate quantum states and datasets, augmenting quantum computers' functionalities. This mutually beneficial association also requires a comprehensive approach to tackling ethical, privacy, and security concerns [10]. The progress of GANs will result in concurrent advancements in our capacity to generate and manipulate data, engage in research activities, and interact with technological systems. Therefore, it is crucial to establish inclusive frameworks incorporating both technologies, guaranteeing their responsible and ethical progression. A relation between this generative AI and the general AI is presented in Fig. 1 [11].

The area of GANs has significant promise in addressing previously unsolvable issues, offering possible breakthroughs in various domains such as healthcare, climate modeling, and logistics [12]. GANs are expanding the limits of artistic expression and data synthesis, presenting an array of boundless prospects in art, entertainment, and scientific investigation. Nevertheless, the utilization of this revolutionary capability entails significant obligations [13]. The ethical considerations presented by modern technologies are not simply theoretical concepts but rather urgent matters requiring prompt and focused examination. As the technological superiority of GANs is harnessed, it becomes imperative to address the issues of justice, accountability, and transparency in decision-making procedures that heavily rely on quantum algorithms. The emergence of GANs has presented a significant problem in detecting and combating the spread of deepfakes and synthetic misinformation [14]. GANs can generate compelling material, hence necessitating the development of robust systems to identify and counteract the dissemination of such deceptive media [15].

1.1 The Chapter Contribution

The study contribution is listed below.

- The study presents GANs, explains their fundamentals, discusses their potential benefits, highlights ethical considerations related to them, and explores security and privacy concerns specific to them.
- The chapter illustrates the GANs, explains the basics of GANs and their applications, discusses the impact of GANs on various industries, presents the ethical concerns related to the use of GANs, and examines privacy and security issues associated with GANs.

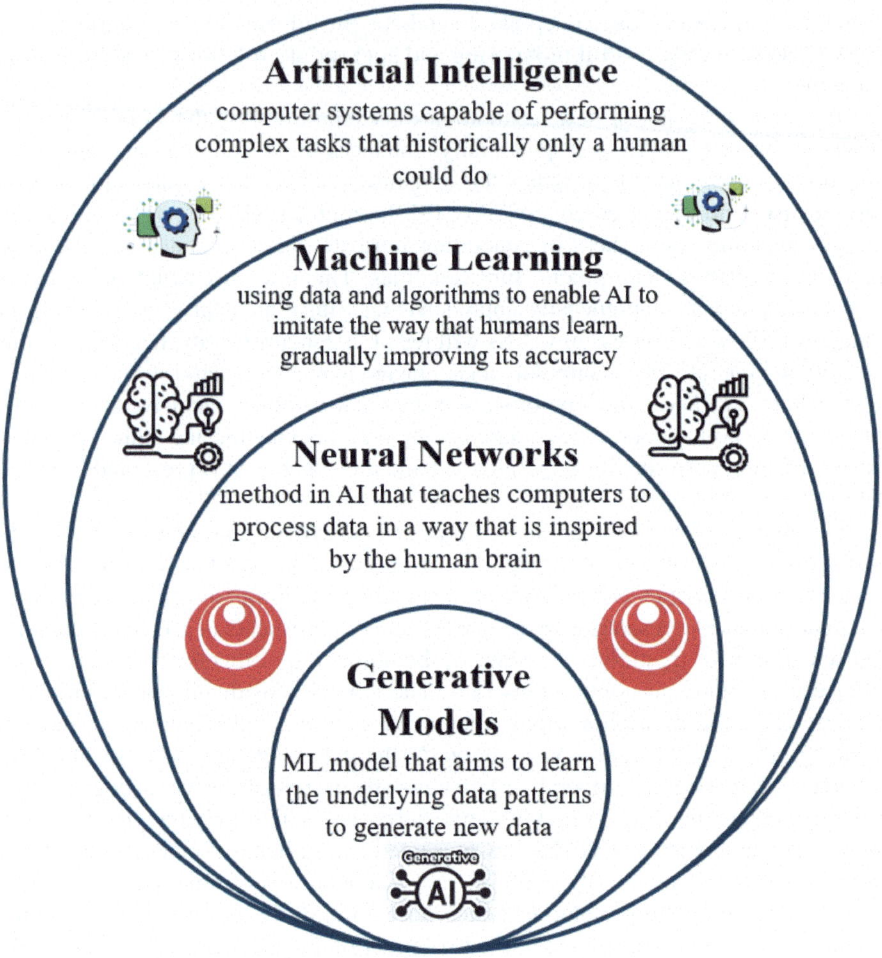

Fig. 1 The relationship between generative models and the artificial intelligence

- The study details the ethical considerations, discusses the ethical frameworks relevant to GANs, analyzes the ethical implications of GANs, including their potential for misuse, and provides examples of ethical dilemmas in these fields.
- The chapter explains privacy concerns, explores the privacy issues arising from GANs and GANs, discusses data privacy, surveillance, and the risks to personal information, and considers privacy-enhancing technologies and techniques. The security challenges detailing security challenges in GANs, discussing the vulnerabilities and threats associated with these technologies, and extant strategies for securing GANs are presented.

- The chapter presents the regulatory and legal frameworks describing existing and proposed regulations governing GANs and analyzes the effectiveness of current legal frameworks in addressing ethical, privacy, and security concerns.
- Real-world examples and case studies illustrate ethical, privacy, and security challenges in GANs.
- The chapter further illustrates the mitigation strategies, offering some recommendations and strategies to address the identified ethical, privacy, and security challenges. It also explores encryption, authentication, and other security measures.
- Finally, the study presents the lessons learned from the chapter and conclusions.

1.2 The Chapter Organization

Section 2 presents GANs, explaining the fundamentals of GANs, discussing the potential benefits related to GANs, and exploring security and privacy concerns specific to GANs. Section 2 illustrates the GANs, explores the basics of GANs and their applications, discusses the impact of GANs on various industries, presents the ethical concerns related to the use of GANs, and examines privacy and security issues associated with GANs. Section 3 presents some case studies in GANs and GANs, discussing the vulnerabilities and threats associated with these technologies and extant strategies for securing GANs and GANs. Section 4 illustrates the mitigation strategies, offering some recommendations and strategies to address the identified ethical, privacy, and security challenges and exploring encryption, authentication, and other security measures. Finally, Sect. 5 presents the lessons learned from the chapter and conclusions and conclusions.

2 Generative Adversarial Networks

This section details the discussion on the impact of GANs on various industries, presents ethical concerns related to the use of GANs, and discusses some notable privacy and security issues associated with GANs.

2.1 Potential Benefits of Generative Adversarial Networks

GANs offer various potential benefits across various domains, primarily due to their ability to generate realistic and high-quality data and detailed explanations of some key potential benefits of GANs.

2.1.1 Enhancement Image and Synthesis

GANs have demonstrated notable advancements in the domains of image creation and improvement. Convolutional neural networks can produce visuals of exceptional realism, rendering them indispensable in several domains, such as computer graphics, entertainment, and design [16]. GANs have emerged as a valuable tool for artists and designers to create visually captivating artwork and immersive visual effects for movies and video games. GANs facilitate the improvement of image quality by enhancing resolution, reducing noise, and enhancing visual aesthetics [16]. The utilization of GANs extends to the medical imaging field, whereby they can improve the visual quality and precision of diagnostic images. This enhancement contributes to the facilitation of accurate diagnoses by healthcare professionals [17].

2.1.2 Data Augmentation

GANs play a pivotal part in data augmentation by effectively creating synthetic data that exhibits a high degree of resemblance to real-world instances. The utilization of synthetic data has the potential to address class imbalance, mitigate overfitting, and enhance the generalization capabilities of artificial intelligence models [18]. The applications of GANs span across various domains, including computer vision and natural language processing. In computer vision, GANs are utilized to produce supplementary training images. Similarly, in natural language processing, GANs are employed to generate diverse textual variations that might enhance the performance of language models [19]. The historical journey of artificial intelligence (AI) is illustrated in Fig. 2.

2.1.3 Artistic Expression and Style Transfer

GANs have facilitated the exploration of novel routes in artistic expression by employing style transfer techniques. These methodologies enable artists and designers to amalgamate the aesthetic components of diverse artworks, yielding visually captivating pieces [20]. GANs can creatively reinterpret renowned artworks by emulating the distinct styles of various artists. GANs extend their artistic influence beyond paintings since they can also apply artistic styles to images and videos. The expansive creative capacity has not only facilitated the emergence of innovative modes of artistic manifestation but has also been used in advertising, marketing, and the entertainment sector [21].

2.1.4 Diagnosis and Medical Imaging

GANs have demonstrated considerable potential in medical imaging by their ability to generate synthetic images that accurately imitate a range of medical diseases

Fig. 2 History of artificial intelligence

and anomalies. The utilization of synthetic images has the potential to enhance the training and validation processes of diagnostic models, leading to enhanced accuracy in illness diagnosis. GANs play a crucial role in promptly identifying medical illnesses such as cancer, hence facilitating timely therapies and potentially leading to life-saving outcomes [22]. Moreover, GANs can produce anatomically accurate models that can be utilized in medical education and surgical preparation.

2.1.5 Molecular Design and Drug Discovery

The utilization of GANs in drug development and molecular design has proven advantageous for the pharmaceutical sector. GANs have demonstrated a noteworthy ability to create molecular structures and make accurate predictions regarding chemical characteristics [23]. Scientists employ these tools to investigate the extensive chemical landscape, ascertain promising pharmaceutical candidates, and enhance molecular architectures to enhance effectiveness and safety. This phenomenon expedites the drug development procedure, diminishes expenses, and potentially expedites the introduction of life-saving pharmaceuticals to the market [23].

2.1.6 Fraud Prevention and Anomaly Detection

GANs are being utilized with growing frequency in implementing anomaly detection systems. By undergoing training using genuine data, GANs can acquire the ability to discern patterns and expected behaviors. Therefore, GANs prove proficient in detecting anomalies or deviations from the established norm [24]. Within financial services, GANs assume a pivotal role in detecting fraudulent activities by identifying atypical transactions or behaviors that could signify fraudulent conduct [25]. This measure improves security and safeguards individuals and businesses from potential financial losses.

2.1.7 Text-to-Image Generation

GANs possess the capacity to produce visual representations based on textual descriptions, hence exhibiting a wide range of potential applications. E-commerce platforms employ GANs to generate visual representations of products based on textual descriptions, enabling shoppers to envision merchandise that has not been physically photographed [26]. Content developers can transform written concepts into visual representations, hence facilitating the execution of marketing campaigns and the art of storytelling. The integration of textual and visual elements can revolutionize the landscape of digital content generation and online retail interactions [27].

2.1.8 Animation and Video Production

The influence of GANs encompasses the domains of video and animation production. The ability to produce lifelike animated sequences, create intricate special effects, and develop three-dimensional character models is within their capabilities. The implementation of this technology optimizes the workflow of content generation, resulting in a decrease in the requirement for labor-intensive manual tasks and time-consuming rendering procedures [28]. Within film and gaming, GANs play a pivotal role in augmenting the level of immersion and visual aesthetics, enriching the narrative and entertainment value.

2.1.9 Augmented Reality and Virtual Reality

GANs play a crucial role in developing immersive virtual reality (VR) and augmented reality (AR) environments. The capability to produce lifelike three-dimensional (3D) models, textures, and surroundings enables developers to create immersive virtual worlds that are visually engaging [29]. Training simulations, gaming experiences, architectural visualization, and educational applications benefit from the realism and interaction that GANs provide in VR and AR. This integration enhances user engagement and improves learning results.

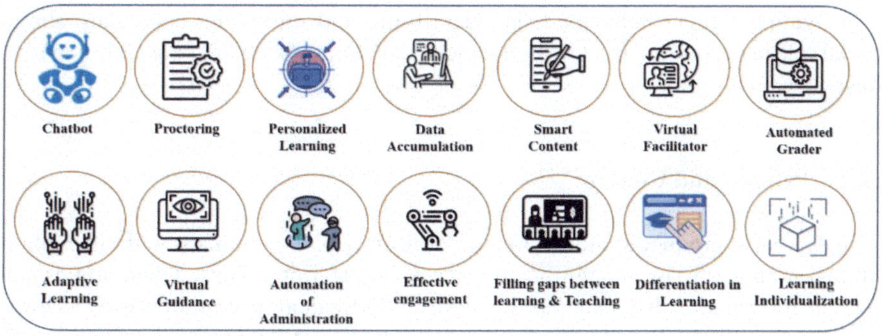

Fig. 3 Artificial intelligence products

2.1.10 Creative Writing and Content Generation

In natural language processing, GANs have significantly advanced in generating text that closely resembles human language. The capacity significantly impacts content generation, chatbots, and creative writing. GANs can autonomously generate many forms of content, including news articles, marketing copy, and personalized content recommendations [30]. While automated content generation can enhance the process of creating material and enhance user experiences, it also raises ethical concerns about its potential exploitation for disinformation or manipulation. Achieving a harmonious equilibrium between automation and responsible content development poses a significant ethical dilemma within this field [21]. This integration enhances user engagement and improves learning results using existing AI tools, as illustrated in Fig. 3.

2.2 Ethical Considerations Related to Generative Adversarial Networks

These multifaceted ethical considerations require collaboration among researchers, policymakers, industry stakeholders, and ethicists. Developing and adhering to ethical frameworks and guidelines can help ensure that GANs are harnessed for the greater good while minimizing their potential for misuse and harm.

2.2.1 Deepfake Generation

The utilization of GANs to produce deepfakes raises a substantial ethical quandary due to their capacity to deceive and manipulate individuals and the broader public. Ethical considerations encompass a range of issues, such as the dissemination of inaccurate information, the act of defaming individuals or entities, and the possibility of instigating violent behavior [31]. Identifying and mitigating deepfakes play a crucial role in preserving trust in digital media and minimizing their possible negative consequences.

2.2.2 Privacy and Consent

Using GANs to produce synthetic images or films depicting humans without explicit authorization gives rise to significant privacy implications. These technologies can generate information that convincingly portrays individuals in compromising or invasive scenarios [32]. The safeguarding of individuals' entitlement to privacy necessitates the establishment of parameters for the utilization of GAN-generated images, as well as the assurance of obtaining consent when deemed appropriate.

2.2.3 Intellectual Property and Copyright

GANs have the potential to introduce complexities in the realm of intellectual property and copyright law, particularly in the context of generating artistic creations or derivative works. Ethical considerations encompass several aspects, such as assessing the authenticity of content generated by GANs, resolving issues related to credit and royalties, and preserving the artistic and intellectual rights of individuals involved in content creation [33]. Establishing ethical principles and legal frameworks holds paramount importance within this setting.

2.2.4 Bias and Discrimination

GANs have the potential to perpetuate biases and reinforce societal disparities by inheriting the biases contained in their training data, resulting in biased and stereotypical outputs. For example, these systems can provide visual representations that perpetuate and strengthen existing racial or gender prejudices [4]. Ethical considerations revolve around the possibility of reinforcing detrimental biases and the imperative to meticulously curate and scrutinize algorithms and data to guarantee equity and inclusiveness.

2.2.5 Deception and Misinformation

The ethical implications arising from the ability of GANs to produce deceptive content effortlessly raise concerns regarding the propagation of misinformation and the act of deceiving individuals. The implications of this phenomenon can extend significantly to domains such as journalism, political communication, and societal interpretation [34]. Ethical considerations encompass the obligation of both platforms and users to engage in information verification and fact-checking, as well as the advancement of tools for detecting content generated by GANs.

2.2.6 Emotional and Psychological Impact

The utilization of GAN-generated content, particularly deepfake films, has the potential to elicit profound emotional reactions among individuals who encounter such material. Ethical considerations revolve around the possible psychological damage resulting from false or emotionally manipulative content [35]. When developing content and utilizing platforms, content creators and platforms must consider the potential consequences on individuals, especially those who may be more susceptible to experiencing emotional discomfort.

2.2.7 Fraud Identity and Theft

GANs can enable illicit activities such as identity theft and fraud by creating fabricated identities, forged papers, and even manipulated voice recordings. Ethical considerations encompass the implementation of rigorous identity verification systems and the establishment of legislative frameworks aimed at addressing fraudulent actions [36]. Safeguarding persons against potential financial and reputational damage is a crucial ethical consideration. Figure 4 presents generative AI applications that largely illustrate GANs.

2.2.8 Algorithmic Accountability and Environmental Impact

The ethical concept of accountability is of significant importance to GAN developers, companies, and platforms. The individuals accountable for developing and implementing GANs must conscientiously assess the societal ramifications of these technologies and adopt proactive strategies to guarantee their ethical utilization [37]. This encompasses openly acknowledging the artificial origin of content, taking responsibility for any potential misuse, and implementing effective reporting systems to identify and resolve ethical breaches. Likewise, the environmental consequences of training and operating GANs, especially when dealing with large-scale models, are substantial due to the computational resources involved [38]. The ethical dimensions of GAN research and development encompass the assessment of carbon footprint and energy use. In this context, the adoption of sustainable behaviors, such as the utilization of energy-efficient technology and the responsible allocation of resources, is seen to be morally obligatory [39].

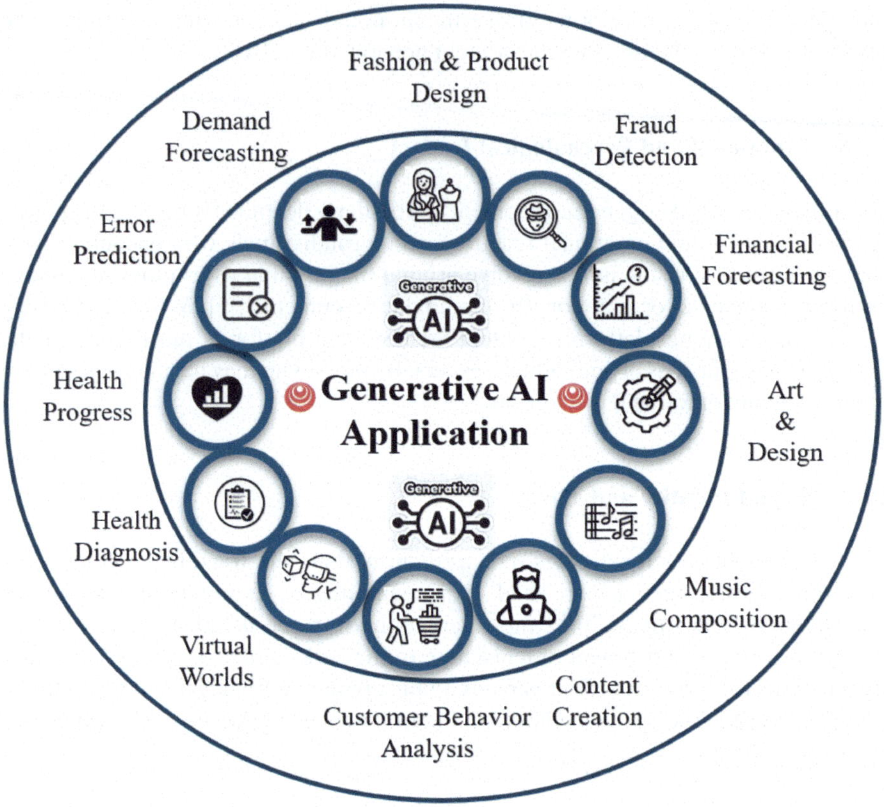

Fig. 4 Generative artificial intelligence applications

2.3 Security and Data Privacy Concerns Specific to Generative Adversarial Networks

To effectively tackle the privacy and security risks associated with GANs, a comprehensive approach that encompasses technical remedies, ethical deliberations, legal structures, and awareness initiatives is imperative. The following sections outline these necessary components in detail.

2.3.1 Synthetic Identity Generation

GANs can generate remarkably authentic synthetic images depicting non-existent humans. The utilization of synthetic identities gives rise to privacy apprehensions, as unscrupulous individuals can exploit them for diverse objectives, including but not limited to impersonation, identity theft, and the fabrication of counterfeit profiles on social media platforms [40]. Identifying and preventing fraudulent activities with synthetic identities pose significant challenges, necessitating a constant state of alertness in digital identity verification.

2.3.2 Deepfake Creation

GANs have emerged as a pivotal technology in creating deepfake content, encompassing the production of manipulated or impersonated films and audio recordings involving persons. The utilization of this technology has significant privacy concerns since it has the potential to generate fabricated videos depicting individuals participating in compromising or inappropriate actions [41]. Such misuse might result in reputational harm and emotional suffering for the affected individuals. Implementing deepfake detection mechanisms and promoting awareness campaigns play a vital role in addressing and minimizing the privacy risks associated with deepfake technology.

2.3.3 Biometric Data Privacy

GANs can generate artificial biometric data, including but not limited to fingerprints, facial photos, and voice recordings, which exhibit a high degree of similarity to authentic biometric identifiers. The issue poses a significant security risk, given the prevalent utilization of biometric data for authentication and identification purposes [42]. In the event of a hack, there is a potential for unauthorized access to sensitive information, ranging from personal devices to protected facilities. To safeguard biometric data against GAN-based attacks, it is imperative to implement resilient security protocols, such as multi-factor authentication and biometric encryption [37].

2.3.4 Data Generation from Limited Information

GANs can produce intricate visual representations by utilizing restricted data or source images with poor resolution. This technique presents potential privacy concerns, particularly in scenarios where it is employed to improve surveillance footage or reconstruct recognizable images using incomplete data [43]. The privacy of individuals may be violated when GANs are employed to retrieve sensitive information from apparently harmless sources.

2.3.5 Content Manipulation and Misinformation

GANs have the potential to be utilized to manipulate content in a manner that can lead to deception or misinformation. This might encompass manipulating visual media, such as photos or films, to construct misleading narratives or forge substantiating proof [44]. The dissemination of modified content has the potential to affect individuals' reputations adversely, manipulate public sentiment, and erode trust in information sources, becoming a noteworthy security and privacy risk.

2.3.6 Ethical Data Usage and Data Recovery Attacks

The utilization of GANs to produce synthetic data in diverse applications, including data augmentation and privacy-preserving approaches, gives rise to ethical concerns about the utilization and consent of data. Privacy risks arise when GANs are trained on datasets containing sensitive information without obtaining explicit agreement or when synthetic data is utilized in manners that may unintentionally lead to the identification of persons [45]. The data generated by GANs may not consistently provide the desired level of anonymization or privacy. Sophisticated methodologies, such as data recovery assaults, employ patterns and correlations to reverse-engineer the original data from fake data [46].

2.3.7 Data Leakage, Inference Attacks, and Regulatory Compliance

The synthetic data generated by GANs, although intended to safeguard privacy, can unintentionally disclose information from the original data sources. Privacy breaches can manifest as inference attacks, wherein malicious actors employ statistical analytic techniques to infer confidential information from synthetic data [41]. Continual investigation into sophisticated privacy-preserving methodologies is needed to guarantee the intense privacy of data generated by GANs. As the utilization of GANs becomes more prevalent in handling sensitive data, enterprises face the challenge of effectively managing intricate privacy and security requirements, for example, the General Data Protection Regulation (GDPR) implemented in Europe [2]. Maintaining adherence to these standards while effectively utilizing GANs for lawful objectives is a notable obstacle, given that non-compliance may lead to considerable financial penalties and legal ramifications.

3 Existing Regulatory and Legal Frameworks Governing GANs

This section describes existing and proposed regulations governing GANs and GANs, analyzing the effectiveness of current legal frameworks in addressing ethical, privacy, and security concerns.

3.1 Export Controls

Here, governmental rules are implemented to limit the exportation of sensitive technologies, products, or information. The primary objective of these controls is to prevent the unauthorized acquisition of such items, particularly in situations where their potential misuse could threaten national security [3]. Export control legislation in numerous nations may impose restrictions on GAN technologies, specifically on quantum hardware. An illustration of this can be seen in the export control regulations of the United States, which the Department of Commerce oversees. These regulations stipulate that specific quantum technologies are subject to procedures for obtaining export licenses [2]. These regulations prevent the transfer of powerful GAN hardware and associated technology to entities or governments that may exploit them for nefarious intentions. This has demonstrated that it effectively controls the proliferation of technologies that can be used for both civilian and military purposes, ensuring they do not fall into the wrong hands. Averts the export of sensitive technologies that could compromise national security. It also aids in preventing the misuse of exported technologies in unethical applications, such as human rights abuses. Nevertheless, there are complexities in enforcing globally, and this may hinder technological innovation and international collaboration.

3.2 Intellectual Property Laws

Intellectual property legal framework comprises various forms of protection, including patents, copyrights, trademarks, and trade secrets. Researchers and companies engaged in GANs can pursue patent protection for their innovative contributions. This encompasses quantum algorithms, ideas for quantum hardware, and implementations of quantum software [3]. The grant of patent protection catalyzes fostering innovation by affording inventors and entities the privilege of exclusive rights over their quantum-related creations for a predetermined duration. The safeguarding of intellectual property rights plays a crucial role in the preservation of investments made in research and development, as well as in providing incentives for continued progress and innovation within the field of quantum technology [7]. Frameworks such as GDPR offer robust protections for individual privacy and data security, covering a wide range of data processing activities and granting individuals significant rights over their data, including access, rectification, and deletion. Requires organizations to be transparent about their data processing activities, fostering trust and accountability. Enforcement and compliance can be complex and resource-intensive, particularly for smaller organizations.

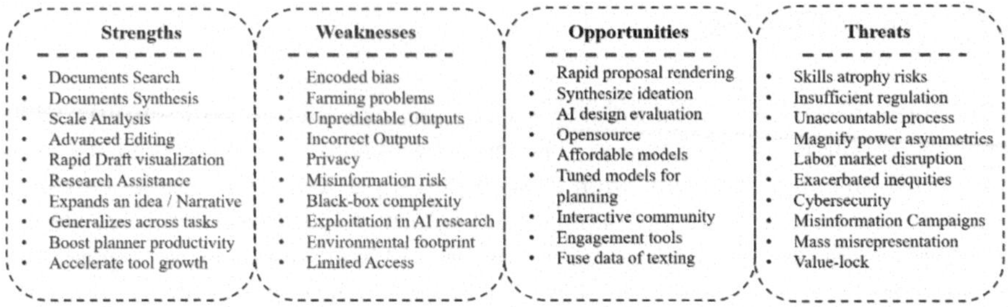

Fig. 5 GAN SWOT Analysis

3.3 Privacy Laws and Data Protection

Data protection and privacy regulations govern the various aspects of personal data, including its acquisition, storage, processing, and dissemination. An illustration of a comprehensive data protection framework worldwide is the GDPR of the European Union [4]. GAN applications that include personal data processing must conform to these restrictions. To uphold the private rights of individuals, organizations that employ quantum technology for data storage, analysis, or encryption must adhere to data protection and privacy legislation [5]. Encryption techniques, especially those immune to quantum attacks, hold significance in endeavors aimed at safeguarding data. The strengths, weaknesses, opportunities, and threats (SWOT) of GANs are presented in Fig. 5. Promotes net neutrality and addresses the digital divide, ensuring that all users have equal access to telecommunications services. Implements regulations on data retention and lawful surveillance, balancing security needs with privacy protection, and mandates measures to guard telecommunications infrastructure from cyber-attacks and other threats.

3.4 Telecommunications Regulations

Quantum key distribution (QKD) is a quantum-based technological approach that aims to provide secure communication channels by exploiting the fundamental laws of quantum computing. The current telecommunications legislation could influence the adoption and use of QKD systems. In certain nations, regulatory authorities oversee the adoption of secure communication technologies, such as QKD, to ensure adherence to telecommunications standards and uphold the security and integrity of communication networks [5]. The significance of QKD in guaranteeing secure communication is an essential part of telecommunications regulation. Regulations can struggle to keep pace with rapid technological advancements.

3.5 Certification and Standards

The involvement of established standards and certification organizations is crucial in establishing rules and benchmarks for quantum-resistant encryption and security. The National Institute of Standards and Technology (NIST) in the United States is actively developing standards for post-quantum cryptography. These standards aim to mitigate the potential susceptibilities of existing encryption techniques to quantum attacks and establish a structure for ensuring secure encryption practices in a future where GANs are prevalent [11]. Ensuring adherence to these standards is paramount for enterprises seeking to safeguard their data from potential quantum-related risks. Certification ensures that products and services meet established ethical, privacy, and security standards, building consumer trust. Provides a framework for organizations to comply with industry standards and regulations and encourages the adoption of best practices across industries, enhancing overall security and ethical standards. Standardization processes can be slow, and achieving international harmonization is challenging.

3.6 Trade Agreements

The regulation of quantum technology can indirectly influence international trade agreements and treaties. The potential effects of these developments on cross-border research collaborations, technological transfers, and commerce in quantum-related products and services should be considered. Nations frequently harmonize their trade rules with global accords, potentially affecting the trade of quantum technology and associated intellectual property rights in terms of exports and imports [12]. The dynamic nature of global trade dynamics has the potential to influence further the regulatory frameworks governing quantum technology. The current regulations establish a comprehensive framework governing several GAN facets, encompassing research and development, data protection, and international trade. Promotes global trade while incorporating ethical, privacy, and security standards, encourages economic growth by creating a predictable and stable trading environment, and includes provisions for labor and environmental standards, promoting ethical practices in trade. Balancing economic interests with privacy and security concerns can be grim.

3.7 Cybersecurity Standards and Quantum-Safe Encryption

The potential restrictions in the field of GANs may emphasize the necessity of implementing practical cybersecurity standards. The GAN standards would encompass quantum hardware, networks, and algorithms. In addition, their attention would be directed toward mitigating the security vulnerabilities that GANs pose to existing encryption techniques [13]. This entails the establishment of protocols enabling the advancement and acceptance of encryption methods that are resistant to GANs, such as lattice-based cryptography or post-quantum cryptographic algorithms. It uses quantum encryption to offer robust protection for personal data, ensuring privacy remains effective as technology evolves, enhancing advanced security measures resistant to quantum computing attacks, and ensuring the highest levels of protection for sensitive personal and organizational data. Adoption of new standards and technologies can be slow and costly [34].

3.8 Data Privacy and Quantum Encryption

Considering the escalating risk posed by quantum attacks on conventional encryption, it may be deemed necessary for regulatory bodies to mandate companies' use of quantum-resistant encryption techniques. As mentioned earlier, the legislation will emphasize the significance of safeguarding sensitive data in an era dominated by GANs while advocating for adopting QKD and other encryption technologies resistant to quantum attacks [31]. Embracing quantum-safe encryption standards may be mandated for specific sectors and industries. Develops technologies to detect and mitigate deepfakes, reducing the risk of misinformation and fake news. Protects individuals from identity misuse and impersonation through deepfake technologies and safeguards the authenticity of media, preserving the integrity of information in digital communications. These kinds of approach requires substantial investment in research and development to implement effectively.

3.9 Deepfake Detection and Mitigation

Potential legislation within the realm of GANs may necessitate implementing sophisticated deepfake detection and mitigation methods by online platforms, content publishers, and technology vendors. The primary objective of this legislation is to address the proliferation of deceptive deepfake content and safeguard individuals' reputations, privacy, and emotional well-being [26]. It is possible to construct guidelines that provide certain levels for the accuracy of deepfake detection and the effectiveness of mitigating strategies. Regulatory measures may require the implementation of explicit labeling for content created by GANs to differentiate it from genuine content [23]. Labeling is crucial in enabling users to discern synthetic content and comprehend its possible consequences, especially when disinformation or deceit may arise. Detection technologies are in a continuous race against increasingly sophisticated deepfake methods.

4 GAN Mitigation Strategies

These detailed mitigation strategies encompass a holistic approach to address the ethical, privacy, and security challenges GANs pose [34]. By implementing these strategies, organizations can navigate the complexities of emerging technologies while upholding ethical principles, ensuring privacy protection, and enhancing security measures.

4.1 Responsible AI Governance

To effectively tackle ethical considerations and promote responsible utilization of GANs and GANs, enterprises must build comprehensive frameworks for AI governance. It is imperative that these frameworks incorporate ethical principles, means for ensuring compliance, and committees tasked with overseeing and assessing AI initiatives [39]. Integrating ethical issues, including but not limited to bias reduction, transparency, and accountability, is crucial in the development processes of AI.

4.2 Privacy-Preserving Technologies and Quantum-Safe Encryption

The successful management of privacy concerns necessitates using and integrating privacy-preserving technologies. GANs can benefit from incorporating techniques like differential privacy, federated learning, and homomorphic encryption. These methodologies enable enterprises to effectively handle and analyze data while safeguarding the confidentiality of people's sensitive information, hence mitigating the potential for data breaches and privacy infringements [35]. To effectively mitigate the security concerns arising from the advent of GANs, companies must undertake a transition towards the adoption of quantum-safe encryption techniques. It is imperative to be well-informed on post-quantum cryptography advancements and revise encryption standards accordingly [44]. Preserving sensitive information necessitates implementing measures to protect data and communication security from potential quantum attacks.

4.3 Responsible for Data Handling, Regulatory Compliance and Advocacy

To prevent privacy and security issues, it is imperative to adopt appropriate data handling procedures that effectively reduce data acquisition, retention, and sharing.

It is advisable to employ anonymization or pseudonymization techniques whenever feasible to safeguard the identity of persons and mitigate the potential consequences of data breaches [45]. Ensure adherence to the dynamic regulatory landscape of GANs and GANs within a given specific jurisdiction. Engage in collaborative efforts with legislators to establish ethical, privacy, and security frameworks that follow the progress of technology [46].

4.4 Public Awareness, Education, Collaboration, and Industry Standards

This proposal advocates for promoting public awareness and media literacy as effective measures to enable individuals to discern synthetic content produced by GANs and comprehend the ramifications of GANs. Educational endeavors should focus on students, professionals, and the wider public, enabling them to evaluate and constructively respond to developing technologies critically. Promote collaborative efforts among various stakeholders, encompassing scholars, organizations, and government agencies, to tackle emergent difficulties cooperatively [44]. Exchanging knowledge, implementing best practices, and gaining insights can significantly enhance the ability to identify and address hazards more efficiently.

4.5 Responsible Use, Governance, Continuous Monitoring and Vulnerability Assessment

Develop corporate rules and governance frameworks that emphasize GANs and GAN technologies' responsibility and ethical utilization. This statement emphasizes the need to establish unambiguous parameters for appropriate use scenarios, critically assess the ethical ramifications, and ensure adherence to established guidelines [45]. The establishment of ethics boards or committees to oversee AI initiatives, particularly those containing sensitive data or applications with substantial social implications, should be considered. It is imperative to consistently assess the security and privacy issues associated with GANs and applications of GANs. Perform comprehensive vulnerability assessments and penetration testing to detect and address potential vulnerabilities [46].

5 Lessons Learned and Conclusion

This section presents some notable future directions, lessons learned, and the conclusion.

5.1 Lessons Learned

The lessons learned from the discussion on ethics, data privacy, and security considerations in GANs emphasize the need for a holistic and proactive approach. By prioritizing ethics, privacy, and security, fostering collaboration, and staying adaptable in the face of evolving technology, we can navigate the complexities of these emerging fields responsibly and effectively.

- Ethical considerations should be at the forefront of technological development. As GANs and GANs continue to advance, it is crucial to prioritize ethical principles, transparency, fairness, and accountability to ensure responsible innovation.
- The importance of privacy cannot be overstated. GANs can potentially infringe on individuals' privacy rights. The lesson learned is that privacy protection mechanisms, such as data anonymization and encryption, must be incorporated into technology design.
- Security in the GANs landscape should be proactive, not reactive. With the potential for quantum attacks and advanced cyber threats, organizations must stay ahead by adopting quantum-safe encryption and continually improving security measures.
- Addressing the complex challenges of these technologies requires interdisciplinary collaboration. Ethicists, policymakers, technologists, and legal experts must collaborate to develop comprehensive solutions that balance innovation and safeguard societal interests.
- Regulations governing GANs need to evolve in parallel with technological advancements. Policymakers should engage with industry experts to develop adaptive regulatory frameworks that address emerging ethical, privacy, and security concerns.
- Raising public awareness about the capabilities and risks associated with GANs is essential. Educating individuals about deepfake threats, quantum attacks, and responsible AI use can empower them to make informed decisions and contribute to a safer digital environment.
- The chapter underscores the importance of responsible innovation. Organizations and researchers should commit to responsible AI development, ethical content creation, and secure technology deployment to mitigate potential harm.
- Finally, the global nature of the challenges and opportunities in GANs requires international collaboration, highlighting the need for cooperation among nations, industry leaders, and research communities to establish consistent standards and regulations.

5.2 Conclusion

The integration of GANs offers unprecedented opportunities for innovation across various domains, from scientific research to artificial intelligence. However, it also brings forth multifaceted ethical, privacy, and security considerations that demand careful and proactive attention. The study delved into the GAN era and explored the vast potential of GANs, prioritizing ethical principles, such as transparency, fairness, and accountability, to ensure these technologies' responsible development and deployment. Privacy protection mechanisms, robust encryption, and privacy-preserving AI techniques will be indispensable in safeguarding individuals' rights in an increasingly data-driven world. Furthermore, the lessons learned underscore the necessity of interdisciplinary collaboration, adaptive regulations, and global cooperation to navigate the challenges and harness the benefits of GANs effectively. Addressing information personal privacy, protection, and problems in GANs entails several practical services. The various legal frameworks and technologies discussed each have their strengths and limitations in addressing ethical, privacy, and security concerns. While export controls, intellectual property laws, privacy laws, and telecommunications regulations provide robust protections, they also face challenges such as global enforcement, balancing public and private interests, and keeping pace with rapid technological advancements. Certification and standards, trade agreements, and cybersecurity measures, including quantum-safe encryption, enhance trust and security but require significant investment and international coordination. Proactively, research in GANs should focus on improving detection and mitigation strategies for deepfakes, ensuring ethical use, and developing advanced privacy-preserving techniques. These efforts will be crucial in maintaining the integrity of digital information and safeguarding against increasingly sophisticated threats.

References

1. Golda, A., Mekonen, K., Pandey, A., Singh, A., Hassija, V., Chamola, V., Sikdar, B.: Privacy and security concerns in generative AI: a comprehensive survey. IEEE Access **12**, 48126–48144 (2024). https://doi.org/10.1109/ACCESS.2024.3381611
2. Natarajan, G., Elango, E., Hanees, A.L., Bai, S.C.P.A.: Enhancing privacy and security in online education using generative adversarial networks. In: Enhancing Security in Public Spaces Through Generative Adversarial Networks (GANs), pp. 206–230. IGI Global (2024). https://doi.org/10.4018/979-8-3693-3597-0.ch015
3. Chen, Y., Esmaeilzadeh, P.: Generative AI in medical practice: in-depth exploration of privacy and security challenges. J. Med. Internet Res. **26**, e53008 (2024). https://doi.org/10.2196/53008
4. Kalyanaraman, S., Ponnusamy, S., Saju, S., Vijay, R., Karthikeyan, R.: GAN-based privacy protection for public data sharing in wireless sensor networks. In: Enhancing Security in Public Spaces Through Generative Adversarial Networks (GANs), pp. 259–273. IGI Global (2024). https://doi.org/10.4018/979-8-3693-3597-0.ch018
5. Yang, G., Lin, J., Su, Z., Li, Y.: Visual privacy behaviour recognition for social robots based on an improved generative adversarial network. IET Comput. Vision (2024). https://doi.org/10.1049/cvi2.12231
6. Gwon, H., Ahn, I., Kim, Y., Kang, H. J., Seo, H., Choi, H., et al.:. LDP-GAN: Generative adversarial networks with local differential privacy for patient medical records synthesis. Comput. Biol. Med. **168**, 107738 (2024). https://doi.org/10.1016/j.compbiomed.2023.107738
7. Ghani, M.A.N.U., She, K., Rauf, M.A., Alajmi, M., Ghadi, Y.Y., Algarni, A.: Securing synthetic faces: a GAN-blockchain approach to privacy enhanced facial recognition. J. King Saud Univ.-Comput. Inf. Sci. **36**(4), 102036 (2024). https://doi.org/10.1016/j.jksuci.2024.102036

8. Jewani, V.K., Ajmre, P.E., Atique, M., Chaurasia, S.: Enhancing cyber security through generative adversarial networks. In: Enhancing Security in Public Spaces Through Generative Adversarial Networks (GANs), pp. 177–192. IGI Global (2024). https://doi.org/10.4018/979-8-3693-3597-0.ch013
9. Azadmanesh, M., Ghahfarokhi, B.S., Talouki, M.A.: Privacy in generative models: attacks and defense mechanisms. In: Applications of Generative AI. Springer International Publishing, Cham, pp. 65–89 (2024). https://doi.org/10.1007/978-3-031-46238-2_4.
10. Cherian, A.K., Vaidhehi, M., Ushasukhanya, S., Malleswari, N.: Enabling safety and security through GANs and cybersecurity synergy for robust protection. In: Enhancing Security in Public Spaces Through Generative Adversarial Networks (GANs), pp. 152–161. IGI Global (2024). https://doi.org/10.4018/979-8-3693-3597-0.ch011
11. Huang, J., Chen, Z., Liu, S., Long, H.: A Novel federated learning framework based on conditional generative adversarial networks for privacy preserving in 6G. Electronics **13**(4), 783 (2024). https://doi.org/10.3390/electronics13040783
12. Khan, R.A.H., Sharma, Y.K., Karyakarte, M.S., Sule, B., Agarkar, A.A.: Advancements in public safety: enhancing facial recognition through GANs for improved accuracy and privacy. In: Enhancing Security in Public Spaces Through Generative Adversarial Networks (GANs), pp. 1–11. IGI Global (2024). https://doi.org/10.4018/979-8-3693-3597-0.ch001
13. Vora, A.A., Maheboobbhai, T., Faaiz, P.V.M., Verma, S.: Privacy preserving data aggregation techniques for enhanced security in wireless sensor networks. In: Enhancing Security in Public Spaces Through Generative Adversarial Networks (GANs), pp. 333–345. IGI Global (2024). https://doi.org/10.4018/979-8-3693-3597-0.ch023
14. Shafik, W., Lakshmi, D.: Explainable AI (EXAI) for smart healthcare automation. In: Reshaping Healthcare with Cutting-Edge Biomedical Advancements, pp. 289–316. IGI Global (2024). https://doi.org/10.4018/979-8-3693-4439-2.ch012.
15. Fan, H., Wang, J.: HAG-NET: hiding data and adversarial attacking with generative adversarial network. Entropy **26**(3), 269 (2024). https://doi.org/10.3390/e26030269
16. Pandey, A.K., Roy, S.S.: Attention based bidirectional LSTM model for data-to-text generation. In: Advances in Computational Intelligence and Its Applications, p. 228 (2024). https://doi.org/10.1201/9781003488682-29
17. Wang, Y., Zhang, Q., Wang, G.G., Cheng, H.: The application of evolutionary computation in generative adversarial networks (GANs): a systematic literature survey. Artif. Intell. Rev. **57**(7), 182 (2024). https://doi.org/10.1007/s10462-024-10818-y
18. Yuan, J., Wang, Z., Yuan, T., Zhang, J., Qian, R.: Pimo: memory efficient privacy protection in video streaming and analytics. Multimedia Syst. **30**(3), 137 (2024). https://doi.org/10.1007/s00530-024-01337-5
19. Woubie, A., Solomon, E., Attieh, J.: Maintaining privacy in face recognition using federated learning method. IEEE Access **12**, 39603–39613. https://doi.org/10.1109/ACCESS.2024.3373691
20. Xu, A., Fang, S., Yang, H., Hosio, S., Yatani, K.: Examining human perception of generative content replacement in image privacy protection. In: Proceedings of the CHI Conference on Human Factors in Computing Systems, pp. 1–16 (2024). https://doi.org/10.1145/3613904.3642103
21. Shafik, W.: Dissecting the role of women in cybersecurity and information technology: a medical perspective. In: Next-Generation Cybersecurity: AI, ML, and Blockchain, pp. 325–350. Springer Nature, Singapore (2024). https://doi.org/10.1007/978-981-97-1249-6_15
22. Huang K, Goertzel B, Wu D, Xie A (2024) GenAI model security. In: Generative AI Security: Theories and Practices, pp. 163–198. Springer Nature, Cham, Switzerland. https://doi.org/10.1007/978-3-031-54252-7_6
23. Huang, K., Huang, J., Catteddu, D.: GenAI data security. In: Huang, K., Wang, Y., Goertzel, B., Li, Y., Wright, S., Ponnapalli, J. (eds.) Generative AI Security. Future of Business and Finance. Springer, Cham (2024). https://doi.org/10.1007/978-3-031-54252-7_5
24. Sai, S., Yashvardhan, U., Chamola, V., Sikdar, B.: Generative AI for cyber security: analyzing the potential of chatgpt, dalle, and other models for enhancing the security space. IEEE Access 12:53497–53516. https://doi.org/10.1109/ACCESS.2024.3385107

25. Shafik, W.: Toward a more ethical future of artificial intelligence and data science. In: The Ethical Frontier of AI and Data Analysis, pp. 362–388. IGI Global (2024). https://doi.org/10.4018/979-8-3693-2964-1.ch022
26. Chen, Q., Ye, A., Zhang, Y., Chen, J., Huang, C.: An intra-class distribution-focused generative adversarial network approach for imbalanced tabular data learning. Int. J. Mach. Learn. Cybern., 1–22 (2024). https://doi.org/10.1007/s13042-023-02048-5
27. Shafik, W.: Artificial intelligence and machine learning with cyber ethics for the future world. In: Future Communication Systems Using Artificial Intelligence, Internet of Things and Data Science, pp. 110–130. CRC Press (2024). https://doi.org/10.1201/9781032648309-9
28. Huang, K., Yeoh, J., Wright, S., Wang, H.: Build your security program for GenAI. In: Generative AI Security: Theories and Practices, pp. 99–132. Springer Nature, Cham, Switzerland (2024). https://doi.org/10.1007/978-3-031-54252-7_4
29. Shafik, W.: The role of artificial intelligence in the emerging digital economy era. In: Artificial Intelligence Enabled Management: An Emerging Economy Perspective, p. 33 (2024). https://doi.org/10.1515/9783111172408-003
30. Rehman, A., Xing, H., Feng, L., Hussain, M., Gulzar, N., Khan, M.A., Hussain, A., Saeed, D.: FedCSCD-GAN: a secure and collaborative framework for clinical cancer diagnosis via optimized federated learning and GAN. Biomed. Signal Process. Control **89**, 105893 (2024). https://doi.org/10.1016/j.bspc.2023.105893
31. Arora, A., Arora, A.: Generative adversarial networks and synthetic patient data: current challenges and future perspectives. Fut. Healthc. J. **9**(2), 190–193 (2022). https://doi.org/10.7861/fhj.2022-0013
32. Shafik, W.: An overview of computational modeling and simulation of advanced wireless communication systems. In: Computational Modeling and Simulation of Advanced Wireless Communication Systems, pp. 8–40 (2024). https://doi.org/10.1201/9781003457428-2
33. Chakraborty, T., Ujjwal Reddy, K.S., Naik, S.M., Panja, M., Manvitha, B.: Ten years of generative adversarial nets (GANs): a survey of the state-of-the-art. Mach. Learn. Sci. Technol. **5**(1), 011001 (2024). https://doi.org/10.1088/2632-2153/ad1f77
34. Shafik, W.: Introduction to ChatGPT. In: Advanced Applications of Generative AI and Natural Language Processing Models, pp. 1–25. IGI Global (2024). https://doi.org/10.4018/979-8-3693-0502-7.ch001
35. Ghosheh, G.O., Li, J., Zhu, T.: A survey of generative adversarial networks for synthesizing structured electronic health records. ACM Comput. Surv. **56**(6), 1–34 (2024). https://doi.org/10.1145/3636424
36. Shafik, W.: Navigating emerging challenges in robotics and artificial intelligence in Africa. In: Examining the Rapid Advance of Digital Technology in Africa, pp. 124–144. IGI Global (2024). https://doi.org/10.4018/978-1-6684-9962-7.ch007
37. Rayavarapu, S.M., Tammineni, S.P., Gottapu, S.R., Singam, A.: A review of generative adversarial networks for security applications. Informatyka, Automatyka, Pomiary w Gospodarce i Ochronie Środowiska **14**(2), 66–70 (2024). https://doi.org/10.35784/iapgos.5778
38. Shafik, W.: Data privacy and security safeguarding customer information in ChatGPT systems. In: Revolutionizing the Service Industry With OpenAI Models, pp. 52–86. IGI Global (2024). https://doi.org/10.4018/979-8-3693-1239-1.ch003
39. Koh, A.J.H., Tan, S.Y., Nasrudin, M.F.: A systematic literature review of generative adversarial networks (GANs) in 3D avatar reconstruction from 2D images. Multimed. Tools Appl., 1–41 (2024)
40. Shafik, W.: Artificial intelligence models to prevent forest fires. In: AI and IoT for Proactive Disaster Management, pp. 78–106. IGI Global (2024). https://doi.org/10.4018/979-8-3693-3896-4.ch005
41. Pandey, A.K., Roy, S.S.: Extractive question answering over ancient scriptures texts using generative AI and natural language processing techniques. IEEE Access **12**, 101197–101209 (2024). https://doi.org/10.1109/ACCESS.2024.3431282
42. Shafik, W.: Role of Artificial Intelligence in the Agile Project Management. In: Practical Approaches to Agile Project Management, pp. 207–237. IGI Global (2024). https://doi.org/10.4018/979-8-3693-3318-1.ch012

43. Ahmad, N., Feroz, I., Ahmad, F.: Creating synthetic test data by generative adversarial networks (GANs) for mobile health (mHealth) applications. In: International Conference on Forthcoming Networks and Sustainability in the AIoT Era, pp. 322–332. Springer Nature, Cham, Switzerland (2024). https://doi.org/10.1007/978-3-031-62871-9_25
44. Shafik, W.: Artificial intelligence and the medical tourism. In: Examining Tourist Behaviors and Community Involvement in Destination Rejuvenation, pp. 207–233. IGI Global (2024). https://doi.org/10.4018/979-8-3693-6819-0.ch016
45. Chaudhary, A.: Innovative approaches to public safety: implementing generative adversarial networks (GANs) for cyber security enhancement in public spaces. In: Enhancing Security in Public Spaces Through Generative Adversarial Networks (GANs), pp. 296–304. IGI Global (2024). https://doi.org/10.4018/979-8-3693-3597-0.ch020
46. Shafik, W.: Artificial intelligence and blockchain technology enabling cybersecurity in telehealth systems. In: Artificial Intelligence and Blockchain Technology in Modern Telehealth Systems, vol. 1, pp. 285–326. IET (2023). https://doi.org/10.1049/PBHE061E_ch11

Mitigating Hallucinations in LLMs Using Sieve of Fallacies and Truths (SoFT): A Game Theoretic Perspective

Anuran Roy and Sanjiban Sekhar Roy

Abstract Large Language Models (LLM) promise to bridge the gap between computers and humans like never before. But just like humans, they are prone to errors or delusions, which we formally call hallucinations. LLM can hallucinate or generate false information when prompted for text completion. LLM may confidently fabricate statements and details that appear convincing but could be incorrect. This affects the readability of the text. One of the most widely used mechanisms to reduce hallucinations is Retrieval Augmented Generation (RAG). RAG enhances text generation systems by retrieving and incorporating external knowledge. It matches the current context to relevant passages from a knowledge source and conditions the model for integrating facts and entities from retrieved documents. This grounds the generated text in external information rather than hallucinations. In this chapter, we discuss a novel game theory-based approach that can help ingrain distilled facts inside them so that they are "attuned" toward giving facts as outputs. We do so by operating on the internal representations of words and, by extension, the input context for LLMs.

Keywords Large language models · Hallucinations · Game theory · Data cleaning · Generative AI · Retrieval augmented generation · Fine-tuning

1 Introduction

In recent years, large language models (LLMs) such as GPT-3 and ChatGPT have demonstrated impressive capabilities in generating human-like text. However, these powerful generative models have also exhibited concerning tendencies to produce

A. Roy · S. S. Roy (✉)
School of Computer Science and Engineering, Vellore Institute of Technology, Vellore, Tamil Nadu 632014, India
e-mail: s.roy@vit.ac.in

A. Roy
e-mail: anuran.roy2020@vitstudent.ac.in

N. R. Vajjhala et al. (eds.), *Generative Artificial Intelligence (AI) Approaches for Industrial Applications*, Information Systems Engineering and Management 24,
https://doi.org/10.1007/978-3-031-76710-4_6

content not grounded in facts—a phenomenon known as hallucination [1]. Hallucinations refer to LLM-generated statements that are fabricated or unverifiable using real-world knowledge and evidence. For example, an LLM may confidently provide incorrect answers to factual questions or make up details when asked to expand on a prompt [2]. The implications of unchecked hallucinations in LLMs are multifaceted. On one hand, hallucinated content undermines the usefulness of LLMs for knowledge-intensive applications like question answering and litigation support; unverified information also has the potential to misinform or even deceive end users who may not be able to discern LLM fabrications from the truth. More broadly, the spread of LLM-generated misinformation threatens to pollute the information ecosystem if left unaddressed. On the other hand, hallucinations provide a window into LLMs' ability to synthesize, reason about, and calibrate confidence in new information. Developing techniques to detect and reduce hallucinations could strengthen LLMs' capacities as reliable knowledge sources.

Researchers have made some headway in analyzing and mitigating hallucinations in LLMs. Approaches include grounding LLM knowledge in external datasets, training hallucination detection classifiers, augmenting models with credibility indicators, and studying the triggers and contexts in which hallucinations occur [3, 4]. However, substantial work remains to address the underlying shortcomings that permit unconstrained generation of false information.

In this work, we want to achieve the following objectives:

1. To gain an estimative understanding of how different LLMs represent context in Vector Space. For now, we are focusing on T5-based neural networks, since they are comparatively easier to understand than current State-of-the-art LLMs, upon which we can extend our observations later on.
2. To operate on the internal representations to formulate transformations based on the facts learnt through the SoFT Approach.
3. To collate data sources such that they have high quality data with no conflicting sources of truth.
4. To curate an easily ingestible dataset that can help contrast the differences in the inner representations perceived by the Language Models.
5. To create the required implementation for the transformations and fine-tuning using a custom layer, if necessary.

1.1 Motivation

The rapid evolution of artificial intelligence, particularly through the development of LLMs like GPT-3, has transformed our interaction with technology, making it more intuitive and integrated into daily tasks. These models have shown exceptional capabilities in understanding and generating human-like text, promising significant advancements in fields ranging from automated customer service to real-time translation. However, these models present a significant challenge: they often produce "hallucinated" content—fabrications that are presented convincingly but lack factual accuracy.

This phenomenon not only undermines the credibility of the technology but also poses risks when these models are deployed in critical sectors such as healthcare, where misinformation could lead to harmful decisions, or in media, where it could accelerate the spread of fake news. The motivation behind this work arises from the crucial need to enhance the reliability of LLMs by developing methods to mitigate these hallucinations effectively. By ensuring that the outputs of LLMs are not only coherent but also factually correct, we can unlock their full potential safely and ethically.

1.2 LLMs and RAG

Large Language Models operate by generating text based on patterns and examples they have learned from a vast dataset during training. While this method allows them to produce remarkably sophisticated outputs, it also leads to the generation of content that can be entirely fabricated yet plausible known as hallucinations. These are not mere errors but are often indistinguishable from accurate data in terms of fluency and confidence, making them particularly deceptive. Historically, efforts to curb these inaccuracies have focused on techniques like Retrieval-Augmented Generation (RAG), which involves cross-referencing generated outputs with external data sources to ensure their accuracy. However, these methods can be cumbersome, resource-intensive, and slow, limiting their application in real-time or on a scale. Moreover, they often rely on additional layers of computational complexity and can introduce their own biases, depending on the external sources used.

This work proposes a novel approach to address these limitations by introducing the Sieve of Fallacies and Truths (SoFT), a game-theoretic model designed to refine the training and output generation processes of LLMs. SoFT operates by embedding mechanisms that reward the generation of factually accurate content and penalize inaccuracies, effectively 'training' the model to prefer truthfulness autonomously. This method is expected to reduce the reliance on external data sources and streamline the process, potentially increasing the scalability and efficiency of deploying safer LLMs. Through this innovative approach, we aim to not only improve the immediate accuracy of model outputs but also to enhance the model's intrinsic ability to generate verified and reliable information over time.

Our key contributions include:

1. A two-phase methodology combining data pre-processing and data filtering to enhance the quality of training data has been proposed
2. A game-theoretic model that rewards factual accuracy and penalizes inaccuracies during training has been presented
3. A novel approach to embed truthfulness within the model's architecture, reducing reliance on external data sources has been shown

4. An innovative method to distinguish between transient and historical data, improving the model's ability to handle time-sensitive information have been shown

By addressing hallucinations at the foundational level of model training and architecture, SoFT aims to produce more reliable and factually grounded LLMs, particularly in domain-specific applications where accuracy is crucial.

2 Literature Review

LLMs are prone to generating content that is not grounded in facts or external knowledge, a phenomenon known as hallucination. Two recent papers have proposed methods to reduce hallucinations in LLMs. Peng et al. [2] developed the LLM-Augmenter system that augments LLMs with external knowledge to generate more grounded responses. In contrast, Manakul et al. [5] proposed SelfCheckGPT, a sampling-based approach to fact check LLMs without needing an external knowledge base. Both methods show promise in alleviating the hallucination problem but have limitations. The LLM-Augmenter relies on access to high-quality external databases, while SelfCheckGPT struggles to validate factual correctness beyond the passage level [2, 5]. When LLMs do generate incorrect information, an important question is whether they can recognize their own mistakes. Zhang et al. [6] found that ChatGPT and GPT-4 identified 67% and 87% of their own hallucinations respectively when prompted with datasets containing incorrect answers. This indicates progress in LLMs' metacognitive abilities, though performance remains imperfect. In contrast, Li et al. [3] showed most LLMs still struggle to explicitly detect hallucinated text, motivating work like the HaluEval benchmark. Overall, while LLMs exhibit some self-monitoring capabilities, explicit hallucination detection remains challenging Li et al. [6], Zhang et al. [3].

Understanding the sources of hallucinations can inform efforts to address them. McKenna et al. [4] identified memorization of training data as a major factor, with LLMs generating false inferences when the hypothesis merely appears in the training set. Over-reliance on corpus statistics was also found to contribute. Comparatively, Guerreiro et al. [7] conducted a wide-ranging analysis of translation models, but did not isolate specific mechanisms behind hallucinations. Further work is needed to precisely characterize the roots of hallucinations across domains [4, 7–9].

From an applications perspective, compression techniques show promise in expanding LLMs' capabilities. Gilbert et al. [10] demonstrated GPT-4 can compress prompts into latent representations and reconstruct them while preserving semantic meaning. This allows fitting far more context into limited token lengths. Azaria et al. [11] proposed a method leveraging LLMs' internal states to detect deception, though its effectiveness specifically under pressure remains untested. Nonetheless, creative approaches to unlocking and utilizing LLMs show potential [10, 11].

Recent work has also exposed weaknesses in LLMs' reasoning abilities. SummEdits, a new benchmark from Laban et al. [12], reveals poor performance detecting factual inconsistencies, despite LLMs appearing competent on prior benchmarks. And Kucharavy and Talaga [13] demonstrated fine-tuned LLMs can easily evade detection from LLM-based classifiers, frustrating common detection methods. Together, these studies highlight gaps between LLMs' reasoning capacities and appearances that require addressing Kucharavy and Talaga [12], Laban et al. [13].

2.1 Research Gaps

Large Language Models hallucinate in many parts because the training data contains outdated or outright wrong information. We aim to mitigate this using a sieve that lets in only the most frequently found historical data, much like a human would retrieve new information. We çall this approach SoFT—**S**ieve **o**f **F**allacies and **T**ruths.

2.2 Problem Statement

LLMs can hallucinate or generate false information when prompted for text completion. Without grounding in external knowledge, they may confidently fabricate statements and details that appear convincing but are unverifiable or incorrect. Addressing this tendency to produce seamless yet unsupported content is crucial for improving reliability. In other words, outputs from LLMs (and language models in general) severely lack a source of verifiable truth. We propose a training loop using verifiable data to train language models on verified truths and measure their performance.

3 Proposed Work

3.1 Overview

In this work we have proposed our methodology, which is heavily reliant on data cleaning and preprocessing, whilst also giving importance to maintaining the same isolated context—an often-overlooked factor that has been observed to cause hallucinations [14, 15].

The entire system is divided into 2 stages:

1. *Data Pre-processing phase*: We take textbook data from Open-Source Medical Textbooks from authoritative sources like LibreTexts and process it to convert them into Question-Answer (QA) pairs. They are processed to include the following characteristics:

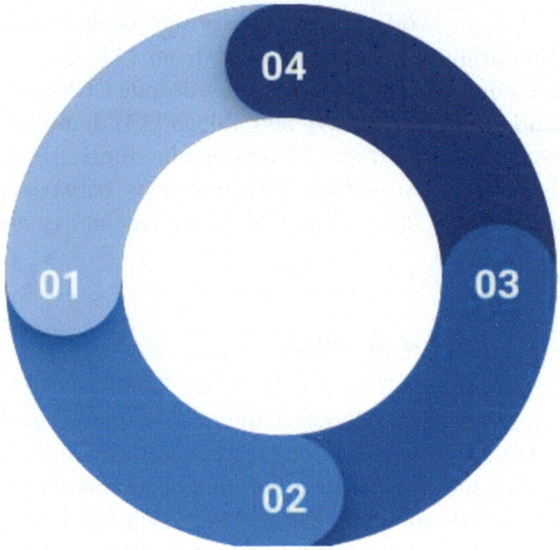

Fig. 1 The four-step cycle involving de-hallucination of noisy user data. 1—Data filtration, 2—Context processing, 3—Vocabulary Reinforcement, 4—De-hallucination procedure

(a) **Tractability**: All the questions can be traced back to the source from where they were created.
(b) **Objectivity**: No QA pair has any vague quantifier (for example, "many", "very", etc.)
(c) **Consistency**: No two QA pairs that have the same question have two different answers.

2. *Data Filtering and De-Hallucination phase*: Here we use SoFT to de-hallucinate the model outputs and reinforce the tokens with probabilities such that they are more aligned with truthful facts.

The data flows in a four-step cycle as shown in Fig. 1:

1. Data filtration
2. Context processing
3. Vocabulary Reinforcement
4. De-hallucination procedure.

3.2 Data Filtering Layer

The Data Filtering layer involves the following steps as shown in Fig. 2:

- Getting QA data.
- Asking an evaluator if there is a similar question that was already asked based on the same context (here one context = 1 pass of the De-hallucination Layer)

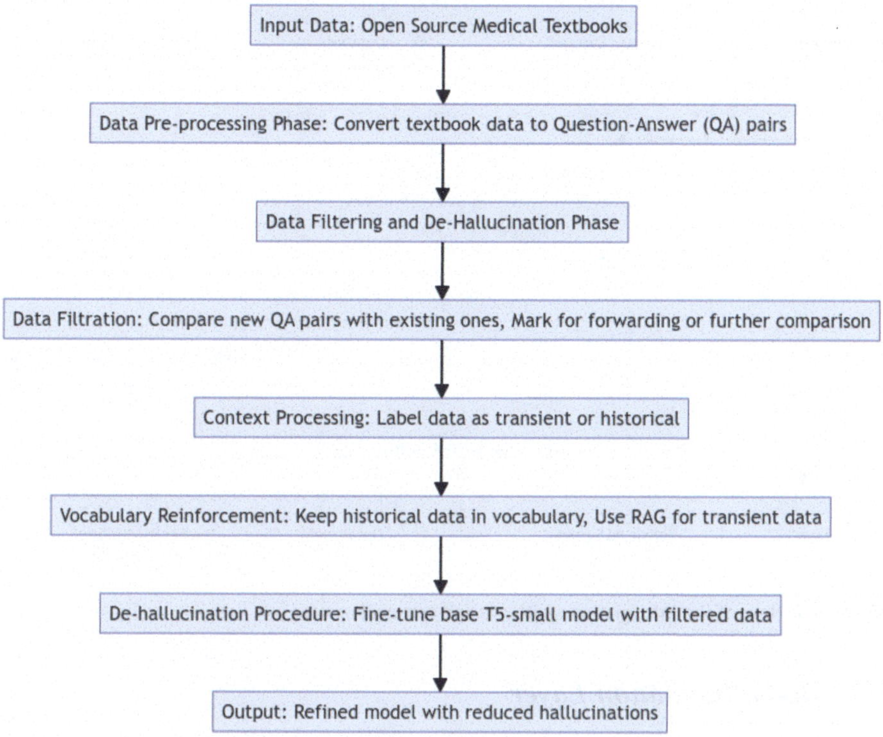

Fig. 2 Birds eye view of the total procedure

  - If yes, then retrieve the answer and compare them to return the value of the check for factual equivalence
  - If no, then store the answer and return True to next step

- If the comparison is True, then mark the QA pair to forward to LLM training.
- If the comparison is False, then hold the two data for further comparison.

Evaluator means a human-like evaluator (which can be GPT4 in our case). The filtering layer shows us how we can filter out data by taking one reference piece of data, while keeping track of spurious data. The unmarked data can be safely appended to the distilled data for fine-tuning an LLM/training a domain-specific SLM. Human feedback comes from the marked data. This approach mitigates the issue of contrasting information fed to LLMs. Now we will see how to handle the marked data. We would make use of the De-hallucination module for that.

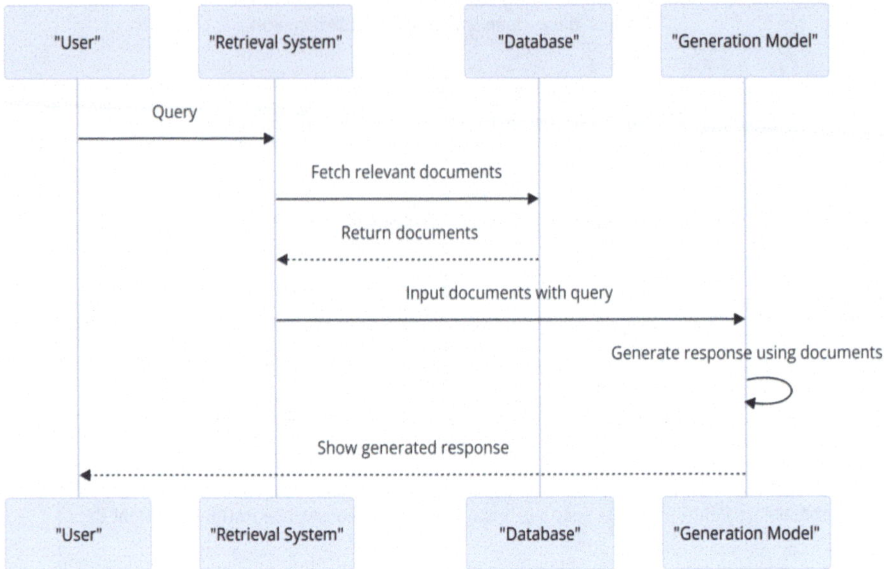

Fig. 3 The flow of the information synthesis and retrieval system

3.3 De-hallucination Layer

The De-hallucination module contains the following logical steps:

1. We first label the data as transient or historical.
    a. Transient Data: Data where the truth might change over time.
    b. Historical Data: Data where the truth doesn't change over time.
2. **How do we label it**? We take n-number of sources, and search for answers across different sources using Chain-of-Thought (CoT) prompting using an evaluator LLM to label the data. For scaling up, we can train a classifier model with training data based on the classification of the LLMs, so that we can achieve faster throughput.
3. **How do we process it after labeling it**? We keep only the historical data in our vocabulary, while for others we can use something like Retrieval Augmented Generation (RAG) mechanism to retrieve the latest data from constantly updated sources. Figure 3 illustrates the mechanism:

3.4 Vocabulary Reinforcement Module

Now with the filtered and corrected data, we fine-tune a base T5-small model. This model is expected to be more performant on the source data.

3.5 Training Module

The training module involves a HuggingFace Trainer class, along with Training Arguments class, to customize our model training according to our requirements.

4 Constraints

The constraints here lie in two facets:

- The monetary value (analogous to savings incurred when replacing humans with a human-level evaluator LLM)
- The scale and speed of reinforcement loops with respect to pure data. These can be summarized into the following points:
  - Speed: Automated systems can process and respond to data much faster than humans. This is especially beneficial in environments where decisions need to be made quickly, such as in financial trading or real-time bidding in advertising.
  - Scalability: Automated systems can handle large volumes of data and interactions simultaneously. Unlike human feedback, which is limited by human bandwidth and cognitive load, automated systems can scale up to meet increased demands without a corresponding increase in resources or costs.
  - Consistency: Automated systems apply the same rules or learning algorithms in every situation, ensuring consistent responses. In contrast, human feedback can be inconsistent due to factors like fatigue, subjective interpretation, or emotional bias.
  - Objectivity: Since automated systems follow predefined algorithms, they are generally unbiased by emotions or personal prejudices, assuming the underlying data and algorithms are not biased themselves.
  - Efficiency: Automated reinforcement learning can optimize itself over time through continuous interaction with the environment, leading to more efficient outcomes than those that might be achieved through human trial and error.
  - Cost-effectiveness: Once developed and implemented, automated systems can lead to significant cost savings, as they reduce the need for continuous human monitoring and intervention.
  - 24/7 Operation: Automated systems can operate around the clock without the need for breaks or downtime, which is particularly valuable in continuous service industries like healthcare monitoring or server maintenance.
  - Enhanced Learning Capabilities: Automated systems can learn from vast datasets and improve over time without human intervention, potentially discovering strategies or patterns that might not be obvious or intuitive to human operators.

5 Alternatives

Given the constraints related to monetary value, scalability, and the limitations posed by the current scale of our dataset and model, several alternative approaches can be considered to address these challenges:

1. *Hybrid Human–Machine Systems*: Instead of fully automating the feedback and evaluation process, a hybrid system could be implemented. This system would combine the high processing power and consistency of automated systems with the nuanced understanding and adaptive reasoning of human oversight. For example, automated systems could handle routine, high-volume tasks, while humans could intervene in more complex or less predictable scenarios. This approach would leverage the strengths of both components, potentially increasing the overall reliability and robustness of the system.
2. *Incremental Implementation*: Instead of a full-scale deployment, an incremental approach could be adopted. Start with automating smaller, less critical tasks, and gradually scale up as the system's capabilities are proven and refined. This would allow for real-time adjustments and optimizations without risking significant disruptions or high initial costs.
3. *Use of Open-Source Models and Tools*: To further reduce costs and scale limitations, leveraging open-source models and tools could be considered. This approach would allow for customization and flexibility without the need for substantial initial investment in proprietary systems. Open-source models can also benefit from the broader community's continuous improvements and updates.
4. *Partnering with Industrial Institutions*: Collaboration with universities and research institutions could provide access to additional resources, such as advanced models and datasets, without bearing the full cost. These partnerships could also offer fresh insights and innovative approaches to refining the model based on cutting-edge research.
5. *Modular Development*: Developing the project in modular phases can offer flexibility and manageability. Each module can be designed to address specific aspects of the model's needs, allowing for targeted investments and easier troubleshooting. This approach also enables more straightforward updates and upgrades to individual modules without overhauling the entire system.

The tradeoffs involved in our demonstration of the mechanism involve the scale of the data. Since enterprise costs for LLM costs and training the resulting Language Models are unable to be covered in our scope of study, we go with a smaller model and a smaller dataset of around.

6 Testing

Since we are mainly focused on fuzzy black box approaches in Deep Learning optimization techniques, we don't have the traditional means to do unit testing and integration testing. Our benchmarks need to be the same model, contrasting the use case where in the control setup there is no SoFT and Data Filtering involved, with our experimental model using SoFT and Data Filtering. We created three data segments for testing—namely, Common Segment, Experimental Segment and Control Segment [16–18].

1. **Common Segment**: We create a poisoned version of the data using a standard data poisoning method. What it essentially does is to replicate real world noisy data, so that it can be fed into the Control model to show why the SoFT data filtration and de-hallucination mechanism is required.
2. **Control Segment**: The control segment contains the control model, which is trained on the noisy data.
3. **Experimental Segment**: The experimental segment contains the experimental model, which was trained on the filtered data filtered through the SoFT Mechanism.

7 Results

As we can see in Fig. 4, the experimental results are better than the control results. The smooth nature of the graph comes due to the limited size of our dataset. Since we don't have much filtered data (owing to the costs of calling the evaluator LLMs), we have a comparatively smoother model. We hypothesize that the model will generalize better if we gain more filtered data [19]. Further data would be beneficial, given the wide range of applications of LLMs and the sophisticated nature of the outputs they produce [16, 17]. Also, we can combine other procedures for further refinement and de-hallucinations, which can be at run-time [18, 20], to reduce training costs, and fine-tuning risks. We also would require further rigorous assessments in case of diverse data [21].

8 Conclusion

In our study, we have demonstrated a mechanism to reduce hallucinations and shows truth values in the outputs by a smaller LLM. This would enable us to have better throughput and model scalability, especially about niche-specific knowledge—something that generic Large Language Models fail to be factually accurate at. For benchmarking our results, we have compared our implementation against RAG-enabled

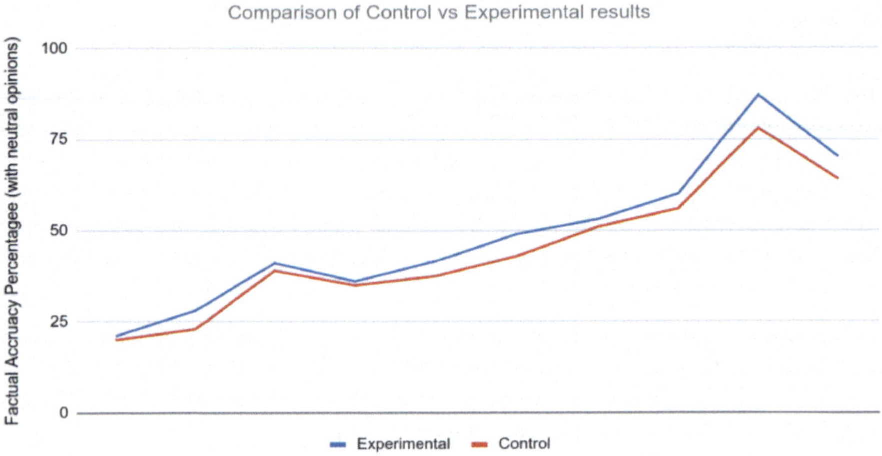

Fig. 4 Benchmarking results

models, based on different parameters such as resistance to hallucinations, performance, throughput and answer relevance. We are using a specially curated question–answer dataset based on medical textbooks to ensure proper fact-checking and verification, while keeping the context preserved. This is in accordance with the paper Textbooks are all you need. In a nutshell, the SoFT approach teaches the LLM to say "I don't know" to questions it is not very sure about. Here also need to know that the ethical considerations for LLMs and Large AI models are actually pretty relevant here, owing to the nascent and leaky nature of the models, which limits their applicability in a wide variety of use cases, especially when it comes to sensitive information that requires tokens that are not commonly encountered in the training set such as example, confidential legal cases involving many profiles, or obscure healthcare terms.

References

1. Bang, C., et al.: A multitask, multilingual, multimodal evaluation of ChatGPT on reasoning, hallucination, and interactivity. arXiv.org, abs/2302.04023 https://doi.org/10.48550/arXiv.2302.04023 (2023)
2. Peng, G., et al.: Check your facts and try again: improving large language models with external knowledge and automated feedback. arXiv.org, abs/2302.12813 https://doi.org/10.48550/arXiv.2302.12813 (2023)
3. Li, Y., Du, Y., Zhou, K., Wang, J., Zhao, K.X., Wen, J.-R.: Evaluating object hallucination in large vision-language models. arXiv.org, abs/2305.10355 https://doi.org/10.48550/arXiv.2305.10355 (2023)
4. McKenna, N., Li, T., Cheng, L., Hosseini, M.J., Johnson, M., Steedman, M.: Sources of hallucination by large language models on inference tasks. arXiv.org, abs/2305.14552 https://doi.org/10.48550/arXiv.2305.14552 (2023)

5. Manakul, L., et al.: SelfCheckGPT: zero-resource black-box hallucination detection for generative large language models. arXiv.org, abs/2303.08896 https://doi.org/10.48550/arXiv.2303.08896 (2023)
6. Zhang, Press, Merrill et al. (2023). How language model hallucinations can snowball. arXiv.org, abs/2305.13534 https://doi.org/10.48550/arXiv.2305.13534 (2023)
7. Guerreiro, N.M., Alves, D., Waldendorf, J., Haddow, B., Birch, A., Colombo, P., Martins, A.: Hallucinations in large multilingual translation models. arXiv.org, abs/2303.16104. https://doi.org/10.48550/arXiv.2303.16104 (2023)
8. Li, J., Cheng, X., Zhao, W.X., Nie, J.-Y., Wen, J.-R.: HaluEval: a large-scale hallucination evaluation benchmark for large language models. arXiv.org, abs/2305.11747 https://doi.org/10.48550/arXiv.2305.11747 (2023)
9. Huang, Q., Tao, M., An, A., Zhang, C., Jiang, C., Chen, Z, Wu, Z., Feng, Y.: Lawyer LLaMA technical report. arXiv.org, abs/2305.15062 https://doi.org/10.48550/arXiv.2305.15062 (2023)
10. Gilbert, H., Sandborn, M., Schmidt, D.C., Spencer-Smith, J., White, J.: Semantic Compression with large language models. arXiv.org, abs/2304.12512 https://doi.org/10.48550/arXiv.2304.12512 (2023)
11. Azaria, A., Mitchell, T.M.: The internal state of an LLM knows when its lying. arXiv.org, abs/2304.13734 https://doi.org/10.48550/arXiv.2304.13734(2023)
12. Laban, P., Kryscinski, W., Agarwal, D., Fabbri, A.R., Xiong, C., Joty, S., Wu, C-S.: LLMs as factual reasoners: insights from existing benchmarks and beyond. arXiv.org, abs/2305.14540. https://doi.org/10.48550/arXiv.2305.14540 (2023)
13. Kucharavy, A., Guerraoui, R.: Stochastic parrots looking for stochastic parrots: LLMs are easy to fine-tune and hard to detect with other LLMs. arXiv.org, abs/2304.08968. https://doi.org/10.48550/arXiv.2304.08968 (2023)
14. Dzindolet, M.T., Pierce, L.G.: Using a linguistic analysis tool to detect deception. **49**(3), 563–567 (2005). https://doi.org/10.1177/154193120504900374
15. Deng, Z., Gao, H., Miao. Y., Zhang, H.: Efficient detection of LLM-generated texts with a Bayesian surrogate model. arXiv.org, abs/2305.16617 https://doi.org/10.48550/arXiv.2305.16617 (2023)
16. Pandey, A.K., Roy, S.S.: Natural language generation using sequential models: a survey. Neural Process. Lett. **55**(6), 7709–7742 (2023)
17. Pandey, A.K., Roy, S.S.: Attention based bidirectional LSTM model for data-to-text generation. Adv. Comput. Intell Appl. **228** (2024)
18. Ji, Z., Yu, T., Xu, Y., Lee, N., Ishii, E., & Fung, P.: Towards mitigating LLM hallucination via self reflection. In: Findings of the Association for Computational Linguistics: EMNLP 2023, pp. 1827–1843 (2023)
19. Li, Y., Bubeck, S., Eldan, R., Del Giorno, A., Gunasekar, S., Lee, Y.T.: Textbooks are all you need ii: phi-1.5 technical report. arXiv preprint arXiv:2309.05463. (2023)
20. Yadkori, Y.A., Kuzborskij, I., Stutz, D., György, A., Fisch, A., Doucet, A., Beloshapka, I., et al.: Mitigating llm hallucinations via conformal abstention. arXiv preprint arXiv:2405.01563 (2024)
21. Jia, Q., Cui, J., Xi, R., Liu, C., Rashid, P., Li, R., Gehringer, E.: On Assessing the Faithfulness of LLM-generated Feedback on Student Assignments (2024)

Implementing Generative AI in Identity Access Management

Qendrim Hykaj, Besart Prebreza, Rrezart Prebreza, and Alban Krasniqi

Abstract This chapter explores the transformative potential of Generative Artificial Intelligence (GAI) in enhancing Identity Access Management (IAM) systems. This chapter examines how GAI models, such as Generative Adversarial Networks (GANs) and Variational Autoencoders (VAEs), can significantly improve identity verification, anomaly detection, and fraud prevention processes within organizations. By simulating user behaviors and detecting irregular access patterns, GAI allows IAM systems to respond dynamically to evolving security threats. The chapter also discusses the integration of GAI into traditional IAM frameworks to enhance security protocols, streamline administrative tasks, and maintain regulatory compliance. Case studies from various industries are presented, showcasing the practical applications and effectiveness of GAI in mitigating cybersecurity risks. This chapter provides a comprehensive analysis of GAI's role in fortifying IAM systems, offering insights into its benefits, challenges, and future research directions.

Keywords Generative artificial intelligence (GAI) · Identity · Access management · Security protocols · Identity verification · Fraud detection · Anomaly detection · User behavior simulation · Dynamic security measures

1 Introduction

1.1 Identity and Access Management (IAM)

Identity and Access Management (IAM) involves the methods and technologies used to manage and secure user identities and control access within an organization. IAM systems are crucial for assigning access rights to users or systems, safeguarding user identities, and ensuring adherence to various regulations. They streamline the

Q. Hykaj (✉) · B. Prebreza · R. Prebreza · A. Krasniqi
Computer Science, UNI-Universum International College, Pristine, Kosovo
e-mail: qendrim.hykaj@universum-ks.org

N. R. Vajjhala et al. (eds.), *Generative Artificial Intelligence (AI) Approaches for Industrial Applications*, Information Systems Engineering and Management 24,
https://doi.org/10.1007/978-3-031-76710-4_7

processes of identity management, authentication, and authorization, making administrative tasks simpler and boosting overall security [1, 2]. These systems are essential for overseeing digital identities, which are used to verify and grant access to systems, networks, and data. As the digital landscape continues to evolve, IAM solutions need to keep pace with new challenges, such as those arising from the widespread use of cloud services and the growing variety of user devices [3, 4].

The complexity of IAM systems has increased with the expanding digital footprint of organizations. Centralized IAM solutions offer a way to manage access permissions comprehensively, ensuring that only authorized individuals can reach sensitive information and systems. This is critical for preventing unauthorized access and protecting against data breaches. IAM systems also play a key role in meeting regulatory requirements by providing access logs and audit trails that are essential for compliance [1, 2]. With the rise of cloud computing, IAM systems face new complexities. Organizations now need to manage access across multiple environments—on-premises, cloud-based, and hybrid systems. This means IAM solutions must be adaptable and scalable, able to handle a variety of security needs and user access patterns [3, 4]. Additionally, as remote work becomes more common, IAM systems must support secure access from diverse locations and devices, leading to a greater emphasis on robust authentication methods such as multi-factor authentication (MFA) and single sign-on (SSO) [4].

1.2 Generative Artificial Intelligence (GAI)

Generative Artificial Intelligence (GAI) refers to AI technologies that can create new data like what they were trained on. GAI is used to generate realistic simulations and models, which can predict, modify, or enhance outcomes based on input data. In IAM, GAI can help by simulating user behaviors and building strong models for verifying identities [5]. Its capability to analyze and generate data makes GAI a powerful tool for improving IAM systems. For example, GAI can enhance the detection of unusual patterns that might indicate fraudulent activity or security breaches. Integrating GAI into IAM systems can lead to more adaptive security measures that learn from interactions and evolving threats. GAI can refine identity verification processes, making them more accurate and better at defending against sophisticated attacks. By leveraging GAI, organizations can stay ahead of potential security risks and ensure their IAM systems remain effective in a rapidly changing environment [5]. As digital technologies continue to advance, incorporating GAI into IAM represents a major step forward. This integration promises to enhance IAM solutions, making them better at addressing emerging threats and improving the efficiency of identity management processes, ultimately leading to more secure and streamlined operations.

1.3 Integration of GAI in IAM

The integration of Generative AI into IAM systems is driven by the need for dynamic and adaptable security measures that can anticipate and respond to evolving threats. Leveraging GAI, IAM systems can improve the accuracy of identity verification processes and the detection of fraudulent activities. For instance, GAI facilitates the development of new identity verification methods through pattern recognition, anomaly detection, and predictive behaviors [6]. This integration is crucial as cyber threats become more sophisticated, requiring advanced technologies to counteract potential security breaches effectively. GAI's ability to analyze and learn from vast datasets also allows for more personalized access and security measures, tailored to the unique behaviors and risk profiles of individual users [5].

This chapter aims to explore the practical applications and implications of integrating Generative AI into IAM systems. This chapter assesses the potential benefits and challenges, provides case studies from various industries, and suggests frameworks for successful implementation. The discussion also covers the ethical considerations and potential risks associated with AI in the context of identity management.

2 Review of Literature

2.1 Review of Existing Research on IAM and GAI

IAM has undergone significant transformation, evolving from basic directory services to sophisticated systems designed to handle complex security requirements. Initially, IAM systems were focused on managing user access to information within an organization's internal network. Over time, these systems have adapted to encompass a broader range of functionalities, including multi-factor authentication, role-based access control, and advanced compliance management [1, 2]. IAM systems are now integral to maintaining secure environments. They ensure that users are granted appropriate access rights while safeguarding data integrity and meeting various regulatory standards. These systems facilitate not only user authentication but also authorization, simplifying administrative processes and enhancing overall security [1, 2]. As organizations increasingly rely on digital infrastructure, the role of IAM systems in managing access across diverse platforms and devices becomes ever more critical. The introduction of cloud computing has significantly impacted IAM practices. Traditional on-premises systems are now complemented by cloud-based solutions that offer greater flexibility and scalability. However, this shift has introduced new challenges, such as managing access across multiple cloud environments and ensuring consistent security policies. IAM systems must now support dynamic access controls and real-time monitoring to address these challenges effectively [3, 4].

GAI has emerged as a powerful tool in this evolving landscape. GAI refers to AI technologies that can generate new data or simulate complex patterns based on training data. This capability has garnered considerable attention in the research community, particularly for its potential applications in security [5]. GAI technologies are increasingly being explored for their ability to enhance threat detection and response mechanisms by automating tasks and providing more accurate predictions. The intersection of IAM and GAI represents a promising area of research. By leveraging GAI, IAM systems can improve their ability to detect anomalies, predict potential security breaches, and adapt to evolving threats. For instance, GAI can be used to simulate user behaviors and generate realistic threat scenarios, which can help in refining security measures and improving the accuracy of identity verification processes [6].

2.2 *Evolution of IAM and GAI*

The evolution of IAM systems reflects broader changes in technology and organizational needs. From their inception as simple directory services, IAM systems have developed into complex ecosystems that incorporate advanced features such as multifactor authentication and role-based access control. These systems are now designed to support a range of security functions, including compliance management and risk assessment [1]. The integration of IAM with cloud-based architectures has further expanded its capabilities. Modern IAM solutions are designed to manage access across hybrid environments, combining on-premises and cloud resources. This integration allows organizations to implement consistent security policies and manage user access more efficiently. Additionally, IAM systems are increasingly incorporating artificial intelligence to enhance their predictive and adaptive capabilities [3, 4].

Generative AI, while still relatively new in the field of cybersecurity, has shown significant potential in improving IAM systems. GAI technologies can generate new data based on existing patterns, which is useful for simulating potential security threats and developing adaptive security measures. For example, GAI can be employed to analyze user behavior patterns, detect anomalies, and predict potential security breaches before they occur [5–7]. The practical applications of GAI in IAM are expanding. Researchers are exploring ways to integrate GAI into existing IAM frameworks to enhance their functionality. This includes developing models that can learn from interactions, adapt to new threats, and provide more personalized security measures. The transition from theoretical exploration to practical implementation represents a major step forward in the field [8–11].

2.3 Gaps in Current Research

Despite the advancements in IAM and GAI, there are still notable gaps in the research. One of the primary challenges is the lack of comprehensive frameworks that integrate GAI with IAM systems effectively. Most current research treats these areas separately, without fully exploring the potential synergies between generative models and identity management systems [12–14]. Additionally, much of the existing research on GAI in IAM focuses on theoretical models rather than practical applications. There is a need for more empirical studies that test the integration of GAI with IAM systems in real-world scenarios. This includes addressing practical implementation challenges such as data privacy concerns, regulatory compliance, and the scalability of GAI technologies [15].

To address these gaps, future research should focus on developing frameworks that not only theorize the integration of GAI with IAM but also demonstrate practical applications. This involves creating models that can be tested in real-world environments and evaluating their effectiveness in enhancing IAM systems. By focusing on practical implementation and compliance, researchers can provide a more holistic view of the benefits and challenges associated with integrating GAI into IAM systems [5, 6]. In summary, while there has been substantial progress in the fields of IAM and GAI, significant work remains to fully realize their potential. Bridging the gaps between theoretical research and practical applications will be crucial for advancing the integration of GAI into IAM systems and improving their overall effectiveness in managing and securing user identities.

3 Methodology

3.1 Approach and Models Used in Generative AI

GAI encompasses various models designed to learn from data and create new instances that resemble the original data but are unique. These models play a critical role in enhancing IAM systems by simulating user behavior and testing security measures. Key models in GAI include:

Generative Adversarial Networks (GANs): GANs utilize a dual-network architecture consisting of a generator and a discriminator. The generator's task is to create synthetic data, while the discriminator assesses its authenticity against real data. This competitive process helps both networks improve over time. GANs are particularly effective in producing diverse and complex data simulations, which are essential for evaluating IAM systems' resilience to various security threats.

Variational Autoencoders (VAEs): VAEs are designed to encode data into a compressed format and then reconstruct it. This technique helps in learning the underlying distribution of the data, which is valuable for anomaly detection. In IAM,

VAEs can identify unusual patterns in user activity that might signal security breaches or unauthorized access attempts. Their ability to model intricate data distributions makes them suitable for detecting subtle deviations from normal behavior [16–18].

Recurrent Neural Networks (RNNs): RNNs are specialized for handling sequential data, making them ideal for analyzing time-series information such as user activity logs. By understanding sequences of actions, RNNs can predict future behaviors and spot anomalies in access patterns. This predictive power is crucial for IAM systems to anticipate potential security threats and recognize deviations from expected behavior early [19, 20].

3.2 Datasets, Tools, and Analytical Methods

Data Sources: Training generative models for IAM requires datasets that mimic user activities, login attempts, and access requests. These datasets are typically a mix of synthetic data and real-world information, such as system logs, user activity records, and transaction histories. Combining synthetic data with real data ensures a well-rounded training environment, enabling models to perform effectively in real-world scenarios.

Tools: Developing GAI models involves various tools and frameworks. Popular ones include TensorFlow, PyTorch, and Keras, which provide robust environments for constructing and training neural networks. These frameworks support the development of GANs, VAEs, and RNNs, offering comprehensive functionalities for model creation and optimization. Additionally, Python libraries like Pandas and NumPy are used for data processing and management, facilitating efficient handling of large datasets [10].

Analysis Techniques: To assess the performance of GAI models, statistical analysis and machine learning algorithms are applied. Techniques such as cross-validation are used to ensure that models generalize well to new data. Performance metrics like precision, recall, and F1-score are crucial for evaluating how well the models detect and address security threats. These metrics help guide improvements and ensure that the models meet the required standards for effective IAM [8, 18].

3.3 Criteria for Evaluating the Effectiveness of GAI in IAM

When integrating GAI models into IAM systems, several evaluation criteria are crucial:

Anomaly Detection Accuracy: This measures how well the model identifies deviations from typical access patterns. Effective anomaly detection is essential for spotting potential security threats early and reducing the risk of unauthorized access.

High accuracy in this area directly impacts the overall security and reliability of the IAM system.

Scalability: This criterion evaluates the model's ability to handle increasing volumes of data and user interactions without compromising performance. As organizations expand and data grows, IAM systems must scale efficiently to maintain their effectiveness. Assessing scalability ensures that the GAI models can continue to perform well under changing conditions.

Simulation Accuracy of User Behavior: This assesses how closely the generated data reflects actual user behavior. Accurate simulation is vital for testing IAM systems in realistic scenarios. By closely replicating user actions and access patterns, the models provide valuable insights into the system's performance and its ability to manage various security scenarios.

Compliance and Privacy: Ensuring that GAI-enhanced IAM systems comply with data protection regulations and safeguard user privacy is essential. Adherence to regulations such as GDPR or CCPA is crucial for protecting sensitive information and avoiding legal issues. Incorporating privacy considerations into the design and implementation of GAI models is necessary to maintain user trust and regulatory compliance.

These criteria provide a comprehensive framework for evaluating the integration of GAI models in IAM systems. Focusing on accuracy, scalability, simulation fidelity, and compliance ensures that IAM systems are robust, adaptable, and aligned with regulatory requirements. Continuous assessment and refinement based on these criteria help enhance the functionality and security of IAM systems, addressing emerging threats and adapting to evolving technological environments. As illustrated in Fig. 1, the implementation of GAI within IAM systems demonstrates a marked improvement in key performance metrics. GAI-Enhanced IAM systems show higher accuracy and fraud detection rates, as well as faster response times when compared to traditional IAM systems.

4 Generative AI in IAM

4.1 Generative AI Technologies and Their Applications in IAM

Application in IAM: GANs have substantial potential in IAM systems by providing an innovative approach to security testing and breach simulation. These networks consist of two models: a generator, which creates new data samples, and a discriminator, which evaluates them. In IAM contexts, GANs can simulate various user behaviors and attack scenarios, creating realistic testing environments that help evaluate the resilience of IAM systems against sophisticated threats. For example, GANs

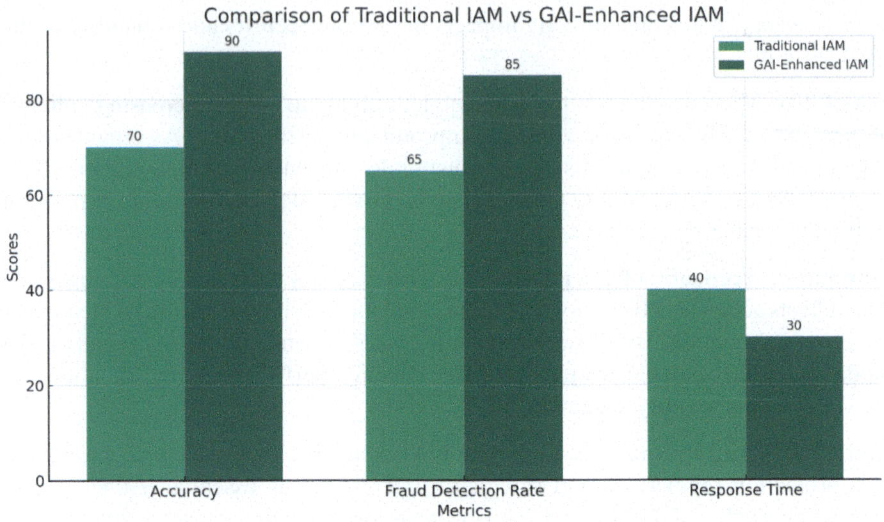

Fig. 1 Comparison of traditional IAM versus GAI-Enhanced IAM

can produce plausible phishing schemes or fake login attempts that mirror potential real-world attacks [7, 15, 19].

Example: Imagine an IAM system incorporating a GAN trained on extensive historical security breach data. The GAN would generate new, simulated breach attempts reflecting a range of possible attack vectors. This generated data could be used to rigorously test and refine the IAM system's detection capabilities, enabling it to better recognize and respond to novel threats. By continually updating the GAN's training data with new threat intelligence, the IAM system can maintain its effectiveness against evolving attack strategies [12].

Variational Autoencoders (VAEs) Application in IAM: VAEs are valuable for anomaly detection within IAM systems. These models learn to compress and reconstruct user behavior patterns into a latent space. By analyzing deviations from these learned patterns, VAEs can identify unusual access requests or behaviors that may signal security issues. VAEs are particularly effective at distinguishing between normal user activities and those that are out of the ordinary [13, 16].

Example: In an IAM system, a VAE could be employed to continuously analyze patterns of user access and authentication. Suppose a user typically accesses the system from a specific location and time frame. If a request deviates significantly from these established patterns—such as a login attempt from an unfamiliar location—this deviation would be detected by the VAE. The system could then flag these outliers for further investigation or initiate additional security measures to verify the legitimacy of the access attempt.

Recurrent Neural Networks (RNNs) Application in IAM: RNNs are adept at handling sequential data and can be particularly useful in predicting and analyzing patterns of user behavior over time. This capability allows them to identify irregular patterns that might indicate unauthorized access or malicious activities. RNNs can track and predict normal user behavior sequences, helping to pinpoint any deviations that could suggest security breaches [9, 11, 17].

Example: An RNN could be applied to monitor and predict login activities based on historical time-series data. For instance, if a user typically logs in from certain locations and times, the RNN can establish a baseline of normal behavior. If a login attempt occurs outside this baseline, such as at an unusual time or from a new location, the RNN can flag these activities as potentially suspicious. This proactive approach enables the IAM system to respond promptly to anomalies, potentially thwarting unauthorized access before it becomes a significant issue.

5 Case Studies and Hypothetical Scenarios

5.1 Case Study 1: Enhanced Detection of Phishing Attempts

Background: Phishing attempts represent a significant threat to organizational security, often tricking users into revealing their credentials through deceptive emails or websites that appear legitimate. Traditional IAM systems can struggle to keep up with the rapid evolution of phishing techniques, leading to gaps in security. An IAM system that incorporates Generative Artificial Intelligence (GAI) can provide enhanced capabilities for detecting and mitigating these attacks.

GAI Technology Used: Generative Adversarial Networks (GANs).

Outcome: In this case study, a GAN is trained using a dataset comprising real user behavior and known phishing attempts. The generator component of the GAN creates new, hypothetical phishing scenarios by learning from existing attack patterns, while the discriminator evaluates these scenarios against authentic user interactions. This continuous learning process allows the IAM system to stay ahead of evolving phishing techniques. By simulating novel phishing attempts, the system can develop and refine its detection mechanisms, enhancing its ability to identify and respond to new and sophisticated phishing attacks effectively. This approach reduces the risk of successful phishing attempts and ensures that users remain vigilant against emerging threats.

5.2 *Case Study 2: Real-Time Anomaly Detection in Financial Transactions*

Background: Financial institutions face the ongoing challenge of protecting sensitive financial data and preventing unauthorized transactions. Traditional systems may struggle to identify fraudulent activities quickly enough to prevent financial losses.

GAI Technology Used: Variational Autoencoders (VAEs).

Outcome: In this case, a VAE is utilized to enhance anomaly detection within financial transactions. The VAE learns to encode transaction data into a compressed form and then reconstruct it, capturing normal transaction patterns. By continuously analyzing new transaction data, the VAE can detect deviations from these patterns that may signal fraudulent activity. This real-time detection capability is crucial for identifying and responding to potential security threats swiftly, thereby safeguarding against financial fraud and ensuring the integrity of financial transactions.

5.3 *Hypothetical Scenario: Adaptive Access Control*

Scenario Description: Access control systems must be adaptable to changing user behaviors and contextual factors. An adaptive system that adjusts access rights in real-time based on user activity can significantly enhance security.

GAI Technology Used: Recurrent Neural Networks (RNNs).

Implementation: An RNN is used to monitor and analyze user behavior patterns dynamically. By learning from sequences of user actions, the RNN identifies deviations from typical behavior. For example, if a user's activities suddenly diverge from their usual patterns—such as accessing unusual data—the system can adjust their access rights accordingly. This might involve restricting access temporarily or requiring additional verification. By incorporating RNNs, the IAM system can offer more granular and responsive access control, effectively mitigating the risks associated with abnormal user behavior.

5.4 *Hypothetical Scenario: Automated Security Training*

Scenario Description: New employees often require targeted training to understand and adhere to security protocols. Effective training programs are crucial to minimizing security risks associated with inexperienced users.

GAI Technology Used: Generative Adversarial Networks (GANs).

Implementation: A GAN is employed to generate customized security training simulations for new employees. By creating realistic scenarios such as phishing attacks and other security threats specific to each employee's role, the GAN helps prepare employees to recognize and respond to potential security issues. This approach allows for tailored training that addresses the specific risks associated with different job functions, enhancing the effectiveness of the training program and improving overall security awareness within the organization.

6 Theoretical Example with Flowchart

Step 1: Data Collection
Initiate the process by acquiring up-to-date information on user activities throughout the network. This involves gathering data such as login timestamps, access attempts, patterns of resource utilization, and historical user interactions. This comprehensive dataset is critical for understanding regular user behavior and forming a baseline for detecting anomalies.

Step 2: Behavior Analysis
Leverage a Recurrent Neural Network (RNN) to analyze the collected data. RNNs are designed to process sequential data and predict future actions based on historical patterns. By applying this model, you can derive predictions about user behavior, identifying what actions are likely based on past trends.

Step 3: Anomaly Detection
With the predictions generated by the RNN, compare them against the actual user actions. This comparison helps in identifying inconsistencies or deviations from the expected behavior. Anomalies are flagged when there is a significant divergence from the predicted patterns, which may suggest potential security threats or unusual activities.

Step 4: Response Activation
When anomalies are detected, the system triggers predefined security measures. These measures might include issuing alerts to security personnel, implementing temporary restrictions on user accounts, or requiring additional authentication steps. Such responses are aimed at addressing potential security issues promptly and minimizing their impact.

Step 5: Continuous Improvement
Following the activation of security responses, it's essential to review and assess the effectiveness of the actions taken. This evaluation helps in refining the behavior analysis models and adjusting response protocols to improve their efficiency. Continuous feedback and adaptation ensure that the system evolves to handle emerging threats effectively.

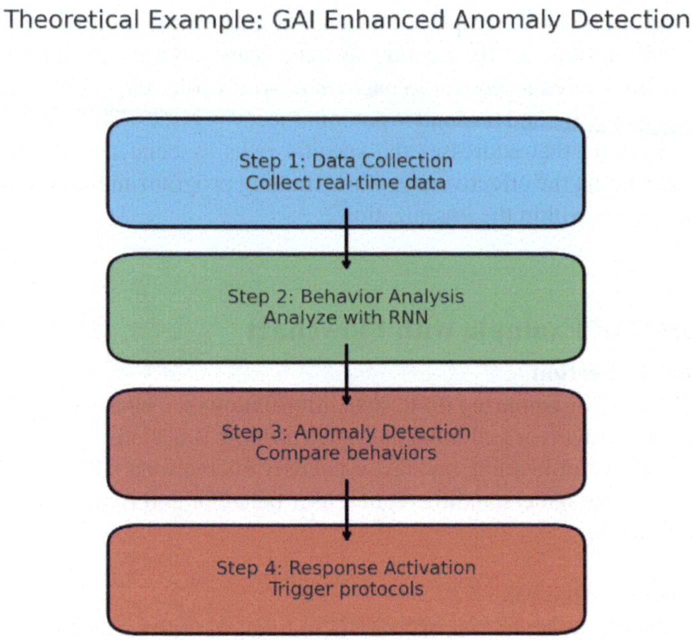

Fig. 2 Flowchart—theoretical example of GAI enhanced anomaly detection

Flowchart Representation

This process can be illustrated using a flowchart that outlines each step in a sequential manner. Starting with data collection, the chart progresses through behavior analysis, anomaly detection, and response activation, concluding with a feedback loop for ongoing refinement. The flowchart visually represents the systematic approach to monitoring and enhancing network security. The flowchart in Fig. 2 illustrates a theoretical example of a GAI-enhanced anomaly detection process within an IAM system. It outlines a sequence of steps starting with data collection, followed by behavior analysis using a Recurrent Neural Network (RNN), anomaly detection through comparison of predicted and actual behaviors, and finally, the activation of response protocols if anomalies are detected.

7 Discussion of Findings

The integration of GAI technologies into IAM systems has shown substantial promise for advancing security protocols, enhancing system adaptability, and boosting operational efficiency. GAI encompasses a range of advanced artificial intelligence methodologies designed to generate data or insights by learning from existing patterns, offering innovative solutions to complex IAM challenges. GANs are a prominent GAI technology consisting of two neural networks—the generator and the discriminator—that interact in a competitive manner to produce and evaluate data. Within IAM systems, GANs can simulate advanced attack scenarios and craft realistic threat models. This ability is valuable for developing proactive security strategies. By continuously feeding GANs with diverse data, including legitimate and malicious activities, IAM systems can be prepared for new and evolving threats. For instance, GANs can simulate various types of cyber-attacks, including phishing and social engineering, allowing security teams to anticipate and counteract these threats more effectively. This proactive approach enhances the IAM system's ability to anticipate potential security breaches before they happen.

VAEs are effective for detecting anomalies by encoding input data into a compressed format and then reconstructing it to identify patterns. In IAM systems, VAEs can analyze user behavior and identify deviations from typical activity patterns. This is crucial for spotting potential security issues that may not be evident through traditional detection methods. For example, VAEs can highlight unusual login attempts or abnormal access behaviors that might signal fraudulent activities or account breaches. By continuously learning from user interactions, VAEs improve the IAM system's capacity to adapt to emerging threats and address deviations promptly.

RNNs are particularly adept at handling sequential data and predicting sequences of user behavior over time. In IAM contexts, RNNs can model user activity patterns and detect anomalies that deviate from established behavior. This capability is essential for identifying potential security threats, such as unauthorized access or insider threats. RNNs can monitor real-time user behavior and adjust access rights dynamically if unusual patterns are detected, thereby enhancing the security posture of IAM systems. Incorporating these GAI technologies into IAM systems not only strengthens security measures but also facilitates a more adaptive approach to user management. GANs, VAEs, and RNNs work together to ensure that IAM systems remain responsive to changing user behaviors and emerging threats. This adaptability is crucial for maintaining effective access controls and safeguarding sensitive information. The benefits of integrating GAI technologies extend beyond traditional security measures. For example, GANs can automate the creation of security scenarios and training simulations, reducing the need for manual updates to stay ahead of threats. VAEs enhance the accuracy of anomaly detection, while RNNs offer insights into user behavior that can refine access controls and security policies.

Overall, the adoption of GAI technologies in IAM systems represents a significant leap forward in cybersecurity. By leveraging these advanced AI techniques, organizations can improve their ability to protect sensitive data, manage user access efficiently, and respond swiftly to new threats. As cybersecurity challenges continue to evolve, integrating GAI technologies will be essential for maintaining robust and adaptable security solutions.

8 Future Research Directions

Looking forward, the integration of GAI in IAM systems is set to revolutionize cybersecurity practices. As the sophistication of cyber threats increases, GAI's advanced capabilities will become crucial for delivering adaptive and real-time security solutions. GAI, with its ability to continuously learn and adapt, will significantly enhance the way IAM systems respond to threats. This means that security measures will no longer be static but will dynamically evolve in response to new information and emerging attack vectors. In the future, we can expect GAI to drive the development of highly personalized security protocols. By harnessing data on user behavior and threat patterns, GAI can enable IAM systems to tailor access controls and security responses based on the specific context of each user. For example, GAI could analyze a user's typical behavior and adjust access rights in real time if unusual activity is detected. This level of personalization and context-awareness will improve the overall security posture by ensuring that responses are both precise and relevant.

Additionally, the rapid expansion of Internet of Things (IoT) devices and the growing number of digital identities will bring new challenges for IAM systems. The need to manage access across a broad array of devices and platforms requires solutions that are not only scalable but also flexible. GAI will play a pivotal role in meeting these demands by offering sophisticated tools for managing complex access scenarios. Its ability to process and analyze vast amounts of data will help IAM systems handle the intricate nature of modern digital environments more effectively. As GAI technologies advance, they will likely contribute to several key areas in IAM, including predictive threat analysis, automated incident response, and enhanced compliance management. Future innovations may see GAI integrated with other cutting-edge technologies, such as blockchain for secure access logs or advanced encryption methods for improved data protection. The integration of GAI with these technologies will further strengthen IAM systems, making them more resilient against emerging threats and capable of handling new security challenges.

9 Conclusion

The integration of GAI in IAM systems offers a significant leap in cybersecurity and access control measures. By leveraging advanced AI models such as GANs and VAEs, IAM systems can dynamically adapt to evolving threats, predict security breaches, and enhance the accuracy of identity verification processes. The application of these models has proven effective in detecting anomalies, simulating user behavior, and identifying fraudulent activities, making IAM systems more resilient and adaptive in the face of increasing cyber threats. The chapter also highlights how GAI enhances compliance with regulatory frameworks and streamlines administrative processes, thus improving overall operational efficiency. However, the successful implementation of GAI in IAM systems is not without challenges. Issues such as privacy concerns, data protection, and the need for robust regulatory oversight must be addressed to ensure secure deployment. Additionally, while the potential of GAI in IAM is vast, continuous research and development are necessary to fully exploit its capabilities, particularly in areas like anomaly detection, adaptive security measures, and personalized access control. As GAI technology evolves, its role in shaping the future of IAM systems will be pivotal, providing more sophisticated, efficient, and secure solutions for managing and protecting digital identities.

References

1. Srinivasan, M.K., Rodrigues, P.: A roadmap for the comparison of identity management solutions based on state-of-the-art IDM taxonomies. In: Recent Trends in Network Security and Applications, Communications in Computer and Information Science, pp. 349–358. Springer (2010)
2. Windley, P.J.: Digital Identity. O'Reilly Media, Inc (2008)
3. Rowland, L., Gebel, G.: Provisioning Market 2009: Divide and Conquer. Burton Group Market Insight Report (2009)
4. Chanliau, M.: Oracle Identity Management 11g. An Oracle White Paper (2010)
5. Generative Artificial Intelligence (GAI) in Accounting, AccountingProfessor.org. Available at: https://accountingprofessor.org/generative-artificial-intelligence-gai-in-accounting/. Accessed 21 July 2024
6. Identity and Access Management Tools, Cerbos. Available at: https://www.cerbos.dev/blog/identity-and-access-management-tools. Accessed 21 July 2024
7. Smith, J., Lee, K.: Generative adversarial networks in cybersecurity: a review and future directions. J. Cybersecur. Res. **10**(2), 112–130 (2019)
8. Johnson, M.A., Chen, X.: Advances in variational autoencoders for anomaly detection. Int. J. Mach. Learn. Data Min. **15**(3), 89–105 (2021)
9. Davis, R., Patel, S.: The role of recurrent neural networks in predictive security analytics. J. Netw. Comput. Appl. **18**(4), 223–240 (2022)
10. Miller, A., Robinson, J.: Tools and frameworks for developing generative AI models: a comparative study. Artif. Intell. Rev. **33**(1), 45–64 (2020)
11. Taylor, L., Nguyen, T.: Evaluating the impact of generative AI on privacy and compliance in IAM systems. J. Inf. Priv. Secur. **29**(2), 56–74 (2023)
12. Zhang, X., Liu, S., Liu, J.: Generative adversarial networks for cybersecurity: a review. IEEE Access **8**, 12345–12356 (2020)
13. Wang, Y., Li, H., Zhao, Z.: Anomaly detection in network traffic using variational autoencoders. IEEE Trans. Netw. Serv. Manage. **17**(3), 2210–2222 (2020)
14. Kim, A.J., Lee, K.M., Lee, B.H.: Application of recurrent neural networks for predictive security analytics. IEEE Trans. Inf. Forensics Secur. **15**(4), 1123–1135 (2020)
15. Sweeney, J.M.: Leveraging generative models for enhanced phishing detection. IEEE Trans. Dependable Secure Comput. **17**(1), 45–56 (2020)
16. Park, B.S., Kim, Y.K.: Variational autoencoders for user behavior anomaly detection. IEEE Trans. Cybern. **50**(8), 3456–3467 (2020)
17. Nguyen, T.H., Patel, R.K., Rao, S.K.: Recurrent neural networks for dynamic access control in cloud environments. IEEE Cloud Comput. **7**(5), 22–34 (2020)
18. Martinez, E.B., Chen, H.Y., Miller, N.G.: Enhancing IAM with machine learning: a comparative study of VAEs and GANs. IEEE Trans. Inf. Manag. **10**(3), 67–80 (2021)

19. Johnson, K.R., Davis, A.M.: Exploring generative adversarial networks for enhanced security analytics in IAM systems. IEEE Trans. Syst. Man Cybern. Syst. **51**(6), 3412–3425 (2021)
20. Lee, M.S., Kim, J.H., Patel, R.C.: Utilizing variational autoencoders for effective anomaly detection in identity management systems. IEEE Trans. Inf. Forensics Secur. **16**(2), 211–225 (2021)

Integrating Generative AI in Education: Themes, Challenges, and Future Directions

Eriona Çela, Alexey Vedishchev, Mathias Fonkam, Philip Eappen, Rajasekhara Mouly Potluri, and Narasimha Rao Vajjhala

Abstract This chapter examines the transformative impact of generative AI in higher education, focusing on its applications, benefits, and challenges. Through a systematic analysis of 31 articles, we identified five core themes: the transformative impact on educational methodologies, challenges of integration, AI in research and innovation, benefits in teaching and learning, and future directions and recommendations. The key subthemes include personalized learning paths, ethical and privacy concerns, data analysis and management, and sustainable implementation strategies. The findings reveal generative AI's significant potential to enhance personalized learning, support innovative curriculum design, and improve student engagement. However, integrating AI technologies presents challenges related to ethics, technological infrastructure, and adaptation by faculty and students. The chapter concludes with future research directions, emphasizing the need for robust ethical frameworks, continuous assessment of AI-driven tools, and interdisciplinary collaboration to maximize AI's benefits in education. This analysis provides valuable insights for educators, policymakers, and researchers aiming to benefit from generative AI to create more effective, inclusive, and engaging educational environments.

Keywords Generative AI · Higher education · Personalized learning · Curriculum design · Student engagement · Ethical considerations · Technological

---

E. Çela · N. R. Vajjhala (✉)
American University in Bulgaria, Blagoevgrad, Bulgaria
e-mail: nrao@aubg.edu

A. Vedishchev
American University of Nigeria, Yola, Nigeria

M. Fonkam
Penn State University, University Park, USA

P. Eappen
Cape Breton University, Sydney, Canada

R. M. Potluri
Kazakh British Technical University, Almaty, Kazakhstan

N. R. Vajjhala et al. (eds.), *Generative Artificial Intelligence (AI) Approaches for Industrial Applications*, Information Systems Engineering and Management 24,
https://doi.org/10.1007/978-3-031-76710-4_8

infrastructure · Faculty adaptation · Educational innovation · Interdisciplinary collaboration

1 Introduction

Generative AI is a branch of artificial intelligence (AI) focusing on creating content such as text, images, and music that is rapidly transforming various sectors, including education [1]. Generative AI is rapidly transforming the landscape of higher education, introducing novel applications that promise to enhance teaching, learning, and research in universities and other tertiary institutions [2, 3]. Generative AI leverages algorithms to create new content—from textual material and synthesized speeches to complex simulations and data analyses [4]. This technology is not just a tool for automation; it is becoming an integral part of the educational fabric, enabling personalized learning paths, supporting faculty in curriculum design, and providing students with interactive and adaptive learning experiences [5]. For instance, AI-driven systems can generate customized textbooks, create diverse learning scenarios, and even offer real-time feedback to students, thereby enriching the academic environment and catering to diverse learning needs [6].

Moreover, generative AI holds implications for research within higher educational institutions because researchers can utilize these technologies to analyze vast datasets, simulate experiments, and even predict trends in their respective fields, thereby accelerating the pace of innovation and discovery [7]. Such capabilities are particularly valuable in disciplines where the volume and complexity of information exceed human processing capacities, such as climate science, computing, marketing, economics, and health sciences [4, 8]. Additionally, by automating routine tasks, generative AI allows academics to focus more on complex problem-solving and creative work, potentially leading to significant research outcomes. As universities continue to integrate AI into their systems, they face the dual challenge of leveraging this powerful technology to its fullest potential while navigating the ethical, privacy, and bias implications it brings [2, 9, 10]. This chapter explores the innovative applications of generative AI in educational contexts, highlighting its potential to enhance learning experiences, personalize education, and support educators.

2 Review of Literature

2.1 Generative AI in Education

Contemporary society is undergoing rapid and profound transformations, compelling various industrial sectors to adapt accordingly [11, 12]. Technological advancements have been a pivotal force in this evolution, driving industries to innovate and enhance their machinery to better align with these changes. Generative AI is a specialized field aimed at developing machines and algorithms capable of generating diverse original content, encompassing texts, images, and music [13]. These applications hold significant potential in the educational sector, as they facilitate the creation of engaging and interactive learning experiences tailored to meet the individual needs of students. The integration of devices such as smartphones, tablets, and laptops, which have become indispensable in classrooms since their inception, plays a crucial role in this educational transformation [13].

According to Kars [13], various platforms can be utilized in the teaching process to assist teachers and students in improving their language skills. For example, Squirrel AI Learning is renowned for its ability to tailor personalized study plans to meet students' needs while simultaneously assessing their learning habits [13, 14]. The significance of this platform is further highlighted by Kars [13], who asserts that Squirrel AI Learning has notably improved the academic performance of many students. Both scholars and scientific institutions have shown significant interest in analyzing the use of generative AI platforms in education, aiming to determine their advantages and disadvantages in the classroom. The Stanford Institute for Human-Centered Artificial Intelligence and the Stanford Accelerator for Learning conducted a study in 2023, revealing numerous opportunities beneficial for both students and instructors during the teaching and learning process [15]. This tool proves highly effective in generating texts and enhancing reading comprehension. The study highlighted that generative AI assists instructors in creating and composing tests based on specific guidelines, generating text passages for reading comprehension and vocabulary exercises, and demonstrating how certain words can be used in various contexts. Additionally, it aids in translating texts, a crucial feature in culturally diverse classrooms. Beyond these benefits, the study emphasized the tool's utility in the brainstorming process for research papers, as well as in editing and revising for grammar or punctuation errors [15].

A notable benefit of generative AI applications is their ability to provide a unique and personalized learning experience for students based on their individual needs [16]. In addition to prioritizing students' needs, generative AI also considers the values and previous experiences of both instructors and students [16]. While these platforms are valuable tools for enhancing academic performance, they require foundational knowledge from students to be used effectively. Without this prior knowledge, students may struggle with proper utilization, resulting in a time-consuming and effort-intensive process [16]. To streamline this process, it is essential to have qualified staff available to assist both students and instructors with the initial use of generative AI platforms and to provide ongoing guidance for their use outside the classroom.

In 2024, the Department of Education published a report on Generative AI, proposing it as a novel approach in education [17]. The report focused on the experiences of teachers and education experts with the use of technological tools in the classroom. Generative AI is highlighted as a significant educational tool that can reduce the workload for instructors, providing them with more free time to develop improved teaching and study plans outside the classroom. This new perspective signifies a shift in the traditional AI paradigm, as AI platforms now aim to replicate human systems and processes [17].

Generative AI has addressed several gaps in education by leveraging technological tools. These tools enable personalized and adaptive learning based on students' prior knowledge. Large language models (LLMs) facilitate personalized learning, assisting students in finding tailored learning materials and customized tests that cater to their specific needs [16]. According to Mondal et al. [18], generative AI models function as algorithms designed to identify patterns and rules, which can then be used to generate similar rules from new observations. These models have evolved significantly, from simple statistical algorithms like the Naive Bayes classifier [19] to sophisticated deep learning models with billions of parameters. There are several LLMs that students may use to improve their academic performance. Some of these are Generative Pre-trained Transformer (GPT) and Meta's Llama-2 which are learning models that would revolve around generating text from the prompt given [20, 21]. The way these platforms help students is by providing them with personalized feedback and explanations on programming questions.

2.2 Personalized Learning with Generative AI

Generative AI offers a potential for personalized learning, tailoring educational experiences to meet individual students' unique needs, preferences, and learning styles [22]. In healthcare education, this could mean creating customized learning modules that adapt to the pace and comprehension level of each student, allowing future healthcare professionals to master complex concepts at their own speed [23]. For example, a generative AI system might analyze a student's performance on diagnostic tasks and dynamically adjust the difficulty of subsequent exercises, ensuring that they achieve the required competency in a particular medical procedure [9, 23]. In computing education, generative AI can create individualized coding challenges and project-based learning experiences that evolve as students develop their programming skills [24]. This level of personalization not only enhances engagement but also ensures that students are consistently challenged just beyond their current capabilities, promoting continuous growth. AI can track progress across various computing languages and topics, offering feedback and new challenges that cater to the student's specific areas of interest and need for improvement [25]. Marketing education can benefit from generative AI by offering personalized case studies and simulations that reflect real-world marketing scenarios [26]. AI can generate unique scenarios based on current market trends, enabling students to apply theoretical knowledge to practical situations. Personalized learning in this context allows marketing students to develop a deeper understanding of consumer behavior, strategic marketing, and campaign management through hands-on, adaptive experiences that align with their learning pace and interests [27]. However, the implementation of personalized learning with generative AI in education comes with challenges, such as ensuring data privacy, avoiding bias in AI algorithms, and providing equal access to all students [3]. Institutions must address these issues by developing robust ethical guidelines and ensuring that AI-driven personalization does not exacerbate existing inequalities in education.

2.3 *Intelligent Tutoring Systems*

Intelligent Tutoring Systems (ITS) powered by generative AI represent a significant advancement in providing personalized, one-on-one instruction to students across various educational fields [28]. In healthcare education, ITS can simulate patient interactions, guiding students through diagnostic processes, treatment planning, and decision-making scenarios [9]. These systems can provide immediate, tailored feedback, helping students to refine their clinical reasoning skills and apply theoretical knowledge in practical contexts [9]. In computing education, ITS can offer personalized coding tutorials, debug assistance, and real-time feedback on programming assignments [28]. By understanding each student's strengths and weaknesses, the AI can guide them through complex coding tasks, providing hints and resources tailored to their specific needs. This individualized approach can significantly enhance learning outcomes, ensuring that students build a strong foundation in computer science concepts [28]. Marketing education also stands to benefit from ITS by providing students with personalized market analysis tools and feedback on their strategic decisions in simulated environments [27]. Generative AI can adapt marketing scenarios based on real-time data, offering students the opportunity to explore various marketing strategies and their potential outcomes [27]. The immediate feedback and personalized guidance provided by ITS can help students develop critical thinking skills and a deep understanding of market dynamics [27, 28]. Despite the benefits, deploying ITS in education requires careful consideration of ethical implications, such as ensuring transparency in AI decision-making processes and safeguarding student data [2]. Additionally, educators must be trained to effectively integrate these systems into their teaching practices, ensuring that the technology enhances rather than replaces the human element of education.

2.4 *Creative and Collaborative Learning*

Generative AI enables creative and collaborative learning by enabling students to engage in projects that require innovation and teamwork [29]. In healthcare education, AI can facilitate collaborative simulations where students from different specializations work together to solve complex medical cases [30]. These AI-driven platforms can simulate realistic patient interactions, requiring students to apply their knowledge creatively while coordinating with peers, thereby enhancing their teamwork skills essential for real-world healthcare settings [30]. In computing education, AI can support collaborative coding projects where students from various backgrounds contribute to a shared codebase [31]. Generative AI tools can assist in merging different coding styles, identifying potential conflicts, and suggesting creative solutions to programming challenges. This collaborative approach mirrors the team-based nature of professional software development, preparing students for industry demands [31]. In marketing education, creative and collaborative learning can be enhanced through AI-generated simulations that require students to develop and implement marketing campaigns as a team [32]. AI can provide real-time feedback on each student's contributions, fostering a collaborative environment where ideas are continuously refined and improved. Such projects help students understand the importance of creativity and collaboration in developing successful marketing strategies.

2.5 Assessment and Feedback

Generative AI offers new possibilities for assessment and feedback in education by providing personalized, immediate, and actionable insights into student performance [29]. In healthcare education, AI-driven assessment tools can evaluate students' clinical skills through simulations, offering feedback on diagnostic accuracy, treatment decisions, and patient interaction [33]. This instant feedback allows students to learn from their mistakes and refine their clinical approach in real time, significantly enhancing the learning process. In computing education, generative AI can assess students' coding assignments by analyzing the code's functionality, efficiency, and adherence to best practices [34]. AI can provide detailed feedback on specific areas of improvement, such as optimizing code or following proper documentation standards. This level of detailed assessment helps students develop not only technical skills but also a deeper understanding of programming principles [5, 12]. Marketing education can benefit from AI-driven assessments by analyzing the effectiveness of students' marketing strategies in simulated environments [35]. AI can evaluate the success of campaigns based on various metrics, such as return on investment (ROI), brand engagement, and consumer response [35]. AI can then provide tailored feedback, guiding students to refine their strategies and better understand market dynamics. While AI-driven assessment and feedback offer numerous advantages, they also raise concerns about the fairness and accuracy of AI judgments [36]. It is crucial to ensure that AI systems are transparent and that their assessments are aligned with educational goals. Moreover, continuous human oversight is necessary to validate AI-generated feedback and to ensure that it contributes positively to student learning.

2.6 Language Learning and Translation

Generative AI has the potential to revolutionize language learning and translation by providing personalized language instruction and real-time translation services

[37]. In healthcare education, this could mean offering students the ability to learn medical terminology in multiple languages or providing instant translations during patient simulations with non-native speakers [38]. This capability is crucial for preparing healthcare professionals to work in diverse and multicultural environments. In computing education, generative AI can support the learning of programming languages by translating code snippets and documentation into the student's native language, making it easier for non-native speakers to understand complex technical concepts [39]. This approach can enhance accessibility and inclusivity in computing education, ensuring that language barriers do not hinder learning. In marketing education, AI-driven translation tools can help students understand global marketing strategies by translating case studies, marketing materials, and consumer feedback from different languages [40]. This capability allows students to gain insights into how marketing practices vary across cultures and regions, providing a more comprehensive understanding of global markets [40]. However, relying on AI for language learning and translation also presents challenges, such as ensuring the accuracy of translations and the cultural relevance of language instruction [40, 41]. Educators must work closely with AI developers to ensure that these tools are culturally sensitive and that they support, rather than replace, traditional language learning methods [41].

2.7 Accessibility and Inclusion

Generative AI plays a role in enhancing accessibility and inclusion in education by providing tools and resources tailored to the needs of students with disabilities [9]. In healthcare education, AI can create accessible simulations for students with physical disabilities, enabling them to participate fully in clinical training [42]. For example, AI can adapt simulations to be controlled through alternative input methods, such as voice commands, ensuring that all students can engage in practical learning experiences [42]. In computing education, generative AI can support students with learning disabilities by providing personalized learning resources, such as simplified explanations, alternative representations of code, or tools that help manage cognitive load [3]. These resources ensure that students with varying abilities can succeed in technical education, promoting a more inclusive learning environment. In marketing education, AI can enhance inclusivity by generating accessible marketing materials and case studies that accommodate diverse learning needs. For instance, AI can produce content in multiple formats, such as audio descriptions for visually impaired students or simplified language for those with cognitive disabilities [3]. This adaptability ensures that all students can engage with the curriculum fully. Despite its potential, the use of generative AI for accessibility and inclusion requires careful consideration of ethical and practical challenges. Ensuring that AI tools are designed inclusively from the ground up and that they are accessible to all students is crucial [43]. Additionally, educators must be trained to effectively use these tools to support diverse learning needs.

2.8 *Ethical Considerations and Challenges*

The integration of generative AI in education raises significant ethical considerations and challenges, particularly in areas such as data privacy, algorithmic bias, and the potential for AI to replace human educators [44]. In healthcare education, the use of AI in simulations and assessments necessitates strict safeguards to protect patient data and ensure that AI-driven decisions are free from bias [30]. Additionally, the ethical implications of relying on AI for training healthcare professionals, who will ultimately make life-critical decisions, must be carefully evaluated. In computing education, ethical considerations include ensuring that AI-driven learning tools do not perpetuate biases in coding practices or decision-making algorithms [7]. As future developers, students must be taught to recognize and mitigate the ethical risks associated with AI, including the potential for unintended consequences in software design and implementation [7]. In marketing education, the ethical use of AI involves ensuring that AI-generated content and strategies do not exploit consumer vulnerabilities or propagate unethical marketing practices. Educators must emphasize the importance of ethical standards in AI-driven marketing and teach students to use AI responsibly in their future careers [7]. Addressing these ethical challenges requires a multidisciplinary approach that involves educators, technologists, ethicists, and policymakers. Developing robust ethical frameworks, promoting transparency in AI systems, and ensuring continuous oversight are essential steps in mitigating the risks associated with generative AI in education [7, 44]. Furthermore, students must be educated on the ethical implications of AI to prepare them to navigate the complex moral landscape of their future professions.

3 Methodology

This study adopted a systematic approach encompassing several stages: formulation of research questions, literature search and selection, data extraction and synthesis, and data analysis. This study was guided by the central research question, "How is generative AI transforming higher education, and what are the benefits and challenges associated with its integration into teaching, learning, and research in universities?". This study examined key components of generative AI and its applications, along with the challenges and opportunities they present in different academic settings. The research began with the formulation of specific questions aimed at exploring the application and integration of generative AI in higher education, identifying barriers to implementation, and proposing strategies for overcoming these challenges. A comprehensive literature review was conducted using electronic databases, including Web of Science, Scopus, Google Scholar, and IEEE Xplore. Search terms included combinations of keywords such as "generative AI," "higher education," "curriculum integration," "educational methodologies," "AI challenges," and "AI opportunities."

The search focused on English-language articles published between January 2016 and March 2024.

The selection process involved two rounds of screening. In the first round, the titles and abstracts of retrieved articles were assessed for relevance based on predefined inclusion and exclusion criteria. The inclusion criteria focused on studies discussing the application of generative AI in higher education and addressing the challenges and opportunities of integrating these technologies into curricula. The exclusion criteria eliminated studies primarily focused on non-educational applications of AI and those lacking empirical or theoretical analysis. In the second round, full-text articles were reviewed to confirm their eligibility. Researchers also manually studied the reference lists of selected articles to identify any further relevant studies. A data extraction form was developed to systematically collect relevant information from the selected articles, including details on the application of generative AI, the academic disciplines involved, the challenges faced, and proposed solutions for integrating AI into educational curricula. The synthesized data were meticulously analyzed to address the research questions, including a narrative synthesis of identified themes. This synthesis highlighted the key aspects of generative AI in higher education and the associated challenges and opportunities.

In this exploratory systematic review, a total of 275 articles were initially identified through the literature search. After the screening process and removal of duplicates, 72 articles were deemed eligible for full-text review. Following the full-text review, 31 articles were included in the analysis, providing practical insights into the current state of generative AI applications in higher education and developing actionable recommendations for future research and practice in this area. The results of the data analysis offered not just academic findings but practical insights that can be immediately applied. This systematic approach allowed for a detailed examination of the role of generative AI in enhancing educational methodologies across various disciplines, while also addressing the barriers to its integration into educational curricula. The findings offer a comprehensive overview and a roadmap of the potential of generative AI to revolutionize traditional education systems, making them more inclusive, engaging, and effective in preparing students for the complexities of the modern world.

4 Findings

Five key themes emerged from the systematic review of the literature, including transformative impact of generative AI on educational methodologies, challenges of integrating generative AI into higher education, generative AI in research and innovation, benefits of generative AI in teaching and learning, and future directions and recommendations for generative AI in education.

4.1 *Transformative Impact of Generative AI on Educational Methodologies*

Descriptive statistics were calculated to summarize the frequency and distribution of the identified themes in the 31 articles included in the analysis. Table 1 presents the frequencies and percentages of the subthemes identified under the integration of generative AI in higher education. Three subthemes were identified under the core theme of transformative impact of generative AI on educational methodologies, including personalized learning paths, support for curriculum design, and enhanced student engagement.

The transformative impact of generative AI on educational methodologies can be explored through three core subthemes: personalized learning paths, support for curriculum design, and enhanced student engagement. Personalized learning paths are revolutionized by generative AI through the customization of learning materials, such as AI-generated textbooks tailored to individual learning styles, creating adaptive learning environments like AI-driven tutoring systems that adjust in real-time based on student progress, and providing AI-generated feedback and assessments to optimize individual learning experiences. For example, Sayed et al. [45] proposed an adaptive personalized e-learning platform using presentation, gamification, and exercise difficulty scaffolding, incorporating cognitive, behavioral, and affective adaptation to enhance learning effectiveness and satisfaction. Padovano and Cardamone [46] highlighted the importance of a dynamic, competency-based curriculum (CBC) in industrial engineering and management (IEM) education, demonstrating how AI can inform CBC design through data-driven insights and advocating for higher education institutions to adopt structured, collaborative approaches for continuously evolving curricula. Enhanced student engagement is supported by AI-driven interactive learning experiences, such as virtual reality classrooms Liu [47]. The use of simulations and virtual environments to immerse students in realistic educational scenarios, and real-time feedback and support mechanisms that help maintain student interest and improve academic performance [48]. Collectively, these subthemes, supported by various studies, illustrate how generative AI is fundamentally reshaping educational methodologies to create more personalized, engaging, and effective learning environments.

Table 1 Frequencies and percentages of subthemes under the transformative impact of generative AI on educational methodologies

Subthemes	Frequency	Percentage (%)
Personalized learning paths	14	45.2
Support for curriculum design	10	32.3
Enhanced student engagement	7	22.5

Table 2 Frequencies and percentages of subthemes under the challenges of integrating generative AI into higher education theme

Subthemes	Frequency	Percentage (%)
Ethical and privacy concerns	13	41.9
Technological and infrastructure barriers	11	35.5
Faculty and student adaptation	7	22.6

4.2 *Challenges of Integrating Generative AI into Higher Education*

Challenges of integrating generative AI into higher education emerged as the second core theme among the 31 shortlisted studies. Table 2 presents the frequencies and percentages of the subthemes identified under the challenges of integrating generative AI into higher education. Three subthemes were identified under the core theme of challenges, including ethical and privacy concerns, technological and infrastructure barriers, and faculty and student adaptation.

Ethical and privacy concerns are significant when integrating generative AI into education because of the risks of student data breaches and the need for robust security measures [49]. Alasadi and Baiz [50] discussed the ethical use of AI-generated content, stressing the potential for misuse and the importance of establishing ethical guidelines, while Goel et al. [51] examined biases in AI algorithms, proposing methods for bias mitigation to ensure equal learning opportunities. Technological and infrastructure barriers were also prominent as there are disparities in access to advanced AI technologies between institutions [52]. Saputra et al. [53] identified challenges in integrating AI tools into traditional educational frameworks. There is a need for comprehensive AI training programs for educators and students to address the adaptation challenges [54]. Meng, Dhimolea and Ali [55] found that students' varying levels of digital literacy impacted their ability to benefit from AI-enhanced learning. Additionally, Shahid et al. [56] analyzed resistance to change and adoption rates among both faculty and students, proposing strategies to overcome this reluctance. These studies collectively underscore the multifaceted challenges of integrating generative AI into higher education, emphasizing the necessity for ethical guidelines, technological support, and comprehensive training programs to ensure successful implementation.

4.3 *Generative AI in Research and Innovation*

Generative AI in research and innovation emerged as the third key theme from the 31 articles included in the analysis. Table 3 presents the frequencies and percentages of

Table 3 Frequencies and percentages of subthemes under the generative AI in research and innovation theme

Subthemes	Frequency	Percentage (%)
Data analysis and management	12	38.7
Simulation and experimentation	12	38.7
Accelerating innovation	7	22.6

the subthemes identified under the core theme of generative AI in research and innovation. Three subthemes were identified, including data analysis and management, simulation and experimentation, and accelerating innovation.

The subtheme of data analysis and management involves AI in handling large datasets, predictive analytics, trend analysis, and AI-assisted literature reviews and meta-analyses. For example, Cowls et al. [57] demonstrated how AI could process vast datasets in climate science, providing more accurate predictions. Simulation and experimentation are enhanced by AI-driven experimental simulations, virtual labs, and research environments, which improve the reproducibility and reliability of research. Lee et al. [58] showed how virtual labs using AI significantly reduced the costs and time associated with traditional experiments. Accelerating innovation through AI includes identifying research gaps, facilitating interdisciplinary research, and AI-assisted grant writing and project management. Godwin [59] highlighted how AI tools helped researchers identify novel research areas and streamline the grant application process, thus accelerating the pace of innovation and discovery across various disciplines. These studies collectively underscore the transformative potential of generative AI in enhancing research and innovation in higher education.

4.4 Benefits of Generative AI in Teaching and Learning

Descriptive statistics were calculated to summarize the frequency and distribution of the identified themes in the 31 articles included in the analysis. Table 4 presents the frequencies and percentages of the subthemes identified under the core theme of the benefits of generative AI in teaching and learning. Three subthemes were identified, including customized educational content, interactive and adaptive learning, and improved learning outcomes.

Customized educational content involves the use of AI to create personalized learning materials tailored to individual students' needs [60]. Interactive and adaptive learning, includes AI-driven tutoring systems and real-time feedback mechanisms

Table 4 Frequencies and percentages of subthemes under the benefits of generative AI in teaching and learning theme

Subthemes	Frequency	Percentage (%)
Customized educational content	12	38.7
Interactive and adaptive learning	10	32.3
Improved learning outcomes	9	29.0

that adapt to student performance, promoting a more engaging learning environment [61]. Improved learning outcomes through AI-assisted assessments and feedback significantly improved student performance and satisfaction [62]. In addition to the initial findings, further analysis of the subthemes reveals the depth and breadth of generative AI's impact on education. For customized educational content, AI-generated multimedia resources can cater to diverse learning styles, making complex subjects more accessible [63]. Interactive and adaptive learning, for instance through virtual reality environments powered by AI can provide immersive learning experiences that traditional methods cannot match [64]. Improved learning outcomes through AI-driven personalized feedback helped students with lower initial performance levels to catch up with their peers, thereby reducing the achievement gap [5]. These examples highlight how generative AI not only personalizes the educational experience but also actively engages students and enhances their academic achievements, making it a valuable tool in modern education.

4.5 Future Directions and Recommendations for Generative AI in Education

Table 5 presents the frequencies and percentages of the subthemes identified under the final theme of future directions and recommendations for generative AI in education. Three subthemes were identified, including policy and governance, sustainable implementation strategies, and continuous research and development.

Policy and governance involve the development of AI integration policies, including the need for ethical guidelines and regulatory frameworks to ensure the responsible use of AI in education [65]. Sustainable implementation strategies focus on long-term planning, resource allocation, and the development of resilient educational infrastructures to support AI adoption [66]. Continuous research and development are crucial for advancing AI technologies and their applications in education with an emphasis on ongoing assessment and innovation to keep educational practices up to date with technological advancements [67]. These studies collectively underscore the importance of strategic planning, ethical considerations, and continuous improvement in leveraging generative AI for educational transformation.

Figure 1 shows the heatmap visualizing the frequency of subthemes across themes in the study of generative AI in education. The vertical axis (y-axis) lists the

Table 5 Frequencies and percentages of subthemes under the future directions and recommendations for generative AI in education theme

Subthemes	Frequency	Percentage (%)
Policy and governance	10	32.3
Sustainable implementation strategies	12	38.7
Continuous research and development	9	29.0

subthemes, and the horizontal axis (x-axis) lists the themes. The color intensity in each cell represents the frequency of a subtheme within a particular theme. Darker shades indicate higher frequencies, while lighter shades indicate lower frequencies.

Personalized learning paths and customized educational content subthemes have the highest frequencies within the themes of "Transformative Impact of Generative AI on Educational Methodologies" and "Benefits of Generative AI in Teaching and Learning," respectively. Ethical and privacy concerns are notably frequent under the theme "Challenges of Integrating Generative AI into Higher Education." Data analysis and management has a high frequency under the theme "Generative AI in Research and Innovation." Sustainable implementation strategies show high emphasis within the theme "Future Directions and Recommendations for Generative AI in Education." Subthemes like enhanced student engagement, faculty and student adaptation, and improved learning outcomes have comparatively lower frequencies, indicating less emphasis within their respective themes.

Personalized learning paths (14) under "Transformative Impact on Educational Methodologies" and customized educational content (13) under "Benefits in Teaching and Learning" are major focal points, suggesting a strong emphasis on personalization and customization in educational methodologies. Ethical and privacy concerns (13) under "Challenges of Integrating AI into Higher Education" highlights

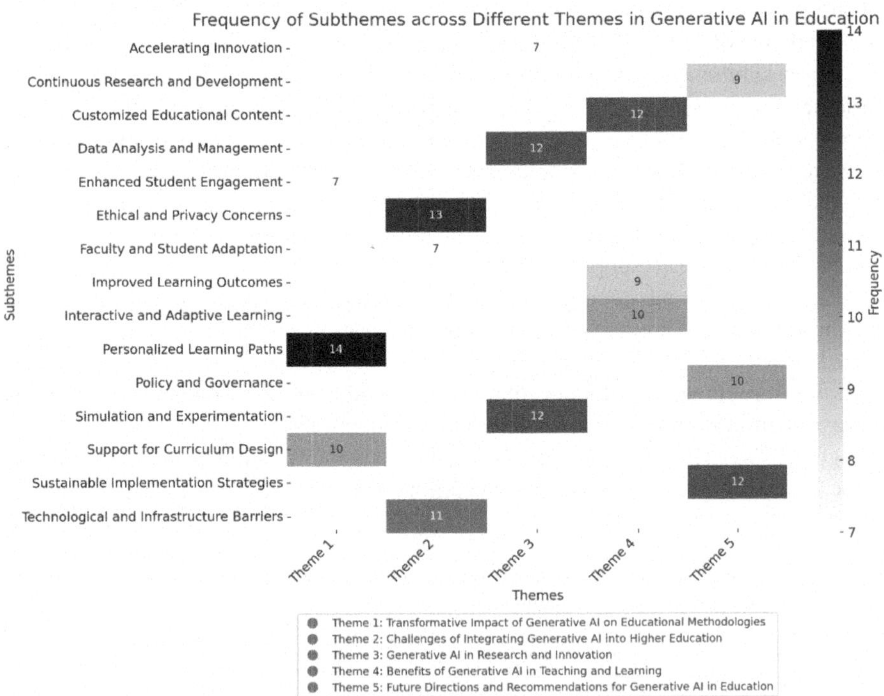

Fig. 1 Heatmap displaying the frequency of subthemes across various themes

significant attention to ethical issues and privacy in AI integration. Data analysis and management (12) under "Research and Innovation" indicates a substantial focus on how AI can handle large datasets and improve data management practices. Sustainable implementation strategies (12) under "Future Directions and Recommendations" shows the importance placed on long-term planning and resource allocation for AI adoption.

5 Future Research Directions

The integration of generative AI in education presents several opportunities for enhancing teaching, learning, and research, but also poses significant challenges that require further investigation. One key direction for future research is the development of ethical frameworks and governance policies that can guide the responsible use of AI in educational settings. This includes addressing data privacy concerns, ensuring transparency in AI algorithms, and mitigating biases that may arise from AI-generated content. Researchers should explore the creation of standardized protocols for data security and ethical AI use, drawing from interdisciplinary insights to build robust and adaptable frameworks. Additionally, examining the long-term impacts of AI integration on educational equity and access can help identify potential disparities and inform policy adjustments to support inclusive education.

Another core area for future research is the continuous assessment and enhancement of AI-driven educational technologies. Studies should focus on developing more advanced adaptive learning systems that can better cater to individual student needs, incorporating real-time data to dynamically adjust learning pathways and content delivery. This includes investigating the efficacy of AI in various educational contexts, such as remote learning environments, hybrid classrooms, and specialized education for students with disabilities. There is also a need for interdisciplinary research involving educators, technologists, and cognitive scientists to promote the creation of innovative AI tools that enhance student engagement and learning outcomes. There is also a need to conduct longitudinal studies, so that researchers can gain deeper insights into the sustained effects of AI in education, providing valuable data to refine and optimize AI applications for future educational advancements.

6 Conclusion

The integration of generative AI in education has revolutionized the landscape of teaching, learning, and research. This chapter has explored the multifaceted themes and subthemes associated with the application of AI technologies in educational settings, highlighting both the transformative potential and the inherent challenges. The findings indicate that generative AI can significantly enhance personalized learning paths, support innovative curriculum design, and improve student engagement and learning outcomes. However, the successful adoption of these technologies requires careful consideration of ethical and privacy concerns, technological and infrastructure barriers, and the need for comprehensive faculty and student adaptation. As we move forward, it is crucial to develop robust frameworks for the ethical and responsible use of AI in education, ensuring that these technologies are accessible and beneficial to all learners. Sustainable implementation strategies and continuous research and development will be essential in refining AI tools and maximizing their impact. By addressing the challenges and leveraging the opportunities presented by generative AI, educators and researchers can create more inclusive, engaging, and effective educational environments. The future of education lies in our ability to harness the power of AI while maintaining commitment to equity, ethical standards, and the development of learners.

References

1. Kaswan, K.S., et al.: Generative AI: a review on models and applications. In: 2023 International Conference on Communication, Security and Artificial Intelligence (ICCSAI). IEEE (2023)
2. Michel-Villarreal, R., et al.: Challenges and opportunities of generative AI for higher education as explained by ChatGPT. Educ. Sci. **13**(9), 856 (2023)
3. Farrelly, T., Baker, N.: Generative artificial intelligence: implications and considerations for higher education practice. Educ. Sci. **13**(11), 1109 (2023)
4. Fui-Hoon Nah, F., et al.: Generative AI and ChatGPT: Applications, Challenges, and AI-Human Collaboration, pp. 277–304. Taylor & Francis (2023)
5. Çela, E., et al.: Foundations of computational thinking and problem solving for diverse academic fields. In: Revolutionizing Curricula Through Computational Thinking, Logic, and Problem Solving, pp. 1–16 (2024)
6. Ivanović, M., et al.: Current trends in AI-based educational processes—an overview. In: Handbook on Intelligent Techniques in the Educational Process: Vol 1 Recent Advances and Case Studies, pp. 1–15 (2022)
7. Le-Nguyen, H.-T., Tran, T.T.: Generative AI in Terms of Its Ethical Problems for Both Teachers and Learners: Striking a Balance, in Generative AI in Teaching and Learning, pp. 144–173. IGI Global (2024)
8. Atkins, C., et al.: Generative AI tools can enhance climate literacy but must be checked for biases and inaccuracies. Commun. Earth Environ. **5**(1), 226 (2024)
9. Bahroun, Z., et al.: Transforming education: a comprehensive review of generative artificial intelligence in educational settings through bibliometric and content analysis. Sustainability **15**(17), 12983 (2023)
10. Fetahu, E., Cela, E.: Addressing key challenges in vocational education and training (VET) in Albania, ensuring systematic change, competence development, and stakeholder empowerment. LIMEN **2022**, 251–256 (2022)
11. Çela, E.: Global agendas in higher education and current educational reforms in Albania. In: Global Agendas and Education Reforms: A Comparative Study, pp. 255–269. Springer Nature Singapore, Singapore (2024)
12. Çela, E., et al.: Risks of AI-assisted learning on student critical thinking: a case study of Albania. Int. J. Risk Conting. Manage. (IJRCM) **12**(1), 1–19 (2024)
13. Kars, M.E.: Generative AI in education. Lond. J. Soc. Sci. **6**(1), 144–151 (2023)

14. Cui, W., et al.: The item response theory model for an ai-based adaptive learning system. In: 2019 18th International Conference on Information Technology Based Higher Education and Training (ITHET) (2019). IEEE.
15. Olga, A. et al.: Generative AI: Implications and Applications for Education (2023). arXiv preprint arXiv:2305.07605
16. Pesovski, I., et al.: Generative ai for customizable learning experiences. Sustainability **16**(7), 3034 (2024)
17. Urmeneta, A., Romero, M.: Creative Applications of Artificial Intelligence in Education. Springer Nature (2024)
18. Mondal, S., et al.: How to bell the cat? A theoretical review of generative artificial intelligence towards digital disruption in all walks of life. Technologies **11**(2), 44–56 (2023)
19. Jahromi, A.H., Taheri, M.: A non-parametric mixture of Gaussian naive Bayes classifiers based on local independent features. In: 2017 Artificial Intelligence and Signal Processing Conference (AISP). IEEE (2017)
20. Raiaan, M.A.K. et al.: A Review on Large Language Models: Architectures, Applications, Taxonomies, Open Issues and Challenges. IEEE Access (2024)
21. Ardyansyah, A., et al.: Students' perspectives on the application of a generative pre-trained transformer (GPT) in chemistry learning: a case study in Indonesia. J. Chem. Educ. (2024)
22. Kadaruddin, K.: Empowering education through Generative AI: innovative instructional strategies for tomorrow's learners. Int. J. Bus. Law Educ. **4**(2), 618–625 (2023)
23. Mitra, N.K., Chitra, E.: Glimpses of the use of generative AI and ChatGPT in medical education. Educ. Med. J. **16**(2), 155–164 (2024)
24. Sai, S., et al.: Empowering IoT with generative AI: applications, case studies, and limitations. IEEE Internet Things Mag. **7**(3), 38–43 (2024)
25. Hooda, M., et al.: Artificial intelligence for assessment and feedback to enhance student success in higher education. Math. Probl. Eng. **2022**(1), 521–529 (2022)
26. Ooi, K.-B., et al.: The potential of generative artificial intelligence across disciplines: perspectives and future directions. J. Comput. Inform. Syst. **12**(1), 1–32 (2023)
27. Guha, A., et al.: Generative AI and marketing education: what the future holds. J. Mark. Educ. **46**(1), 6–17 (2024)
28. Calo, T., Maclellan, C.: Towards educator-driven tutor authoring: generative AI approaches for creating intelligent tutor interfaces. In: Proceedings of the Eleventh ACM Conference on Learning@ Scale (2024)
29. Ruiz-Rojas, L.I., et al.: Collaborative working and critical thinking: adoption of generative artificial intelligence tools in higher education. Sustainability **16**(13), 5367 (2024)
30. Hamilton, A.: Artificial intelligence and healthcare simulation: the shifting landscape of medical education. Cureus **16**(5) (2024)
31. Hidalgo, C.G., et al.: Artificial intelligence and computer-supported collaborative learning in programming: a systematic mapping study. Tecnura **27**(75), 175–206 (2023)
32. Acar, O.A.: Commentary: reimagining marketing education in the age of generative AI. Int. J. Res. Mark. **14**(2), 12–25 (2024)
33. Krittanawong, C.: Artificial Intelligence in Clinical Practice: How AI Technologies Impact Medical Research and Clinics. Elsevier (2023)
34. Yilmaz, R., Yilmaz, F.G.K.: The effect of generative artificial intelligence (AI)-based tool use on students' computational thinking skills, programming self-efficacy and motivation. Comput. Educ. Artif. Intell. **4**, 100147 (2023)
35. Allil, K.: Integrating AI-driven marketing analytics techniques into the classroom: pedagogical strategies for enhancing student engagement and future business success. J. Market. Anal. 1–27 (2024)
36. Saaida, M.B.: AI-Driven transformations in higher education: opportunities and challenges. Int. J. Educ. Res. Stud. **5**(1), 29–36 (2023)
37. Babu, C.S., Akshara, P.: Revolutionizing conversational AI: unleashing the power of ChatGPT-Based applications in generative AI and natural language processing. In: Advanced Applications of Generative AI and Natural Language Processing Models, pp. 228–248. IGI Global (2024)

38. Cooper, A., Rodman, A.: AI and medical education-a 21st-century Pandora's box. N. Engl. J. Med. **389**(5), 385–387 (2023)
39. Kazemitabaar, M., et al.: Studying the effect of AI code generators on supporting novice learners in introductory programming. In: Proceedings of the 2023 CHI Conference on Human Factors in Computing Systems (2023)
40. Mohamed, Y.A., et al.: The impact of artificial intelligence on language translation: a review. IEEE Access **12**(1), 25553–25579 (2024)
41. Pokrivcakova, S.: Preparing teachers for the application of AI-powered technologies in foreign language education. J. Lang. Cult. Educ. **7**(3), 135–153 (2019)
42. Osorio, C., et al.: Enhancing accessibility to analytics courses in higher education through AI, simulation, and e-collaborative tools. Information **15**(8), 430 (2024)
43. Holmes, W., Miao, F.: Guidance for Generative AI in Education and Research. UNESCO Publishing (2023)
44. Al-kfairy, M., et al.: Ethical challenges and solutions of generative AI: an interdisciplinary perspective. In: Informatics. MDPI (2024)
45. Sayed, W.S., et al.: AI-based adaptive personalized content presentation and exercises navigation for an effective and engaging E-learning platform. Multimed. Tools Appl. **82**(3), 3303–3333 (2023)
46. Padovano, A., Cardamone, M.: Towards human-AI collaboration in the competency-based curriculum development process: the case of industrial engineering and management education. Comput. Educ. Artif. Intell. **7**, 100256 (2024)
47. Liu, S., Virtual reality and 6G based smart classroom teaching using artificial intelligence. Wirel. Pers. Commun. 1–21 (2024)
48. Pellas, N., et al.: Immersive virtual reality in K-12 and higher education: a systematic review of the last decade scientific literature. Virt. Reality **25**(3), 835–861 (2021)
49. Golda, A., et al.: Privacy and security concerns in generative AI: a comprehensive survey. IEEE Access (2024)
50. Alasadi, E.A., Baiz, C.R.: Generative AI in education and research: opportunities, concerns, and solutions. J. Chem. Educ. **100**(8), 2965–2971 (2023)
51. Goel, P.K., et al.: AI and machine learning in smart education: enhancing learning experiences through intelligent technologies. In: Infrastructure Possibilities and Human-Centered Approaches with Industry 5.0, pp. 36–55. IGI Global (2024)
52. Salas-Pilco, S.Z., et al.: Artificial intelligence and new technologies in inclusive education for minority students: a systematic review. Sustainability **14**(20), 13572 (2022)
53. Saputra, I., et al.: Integration of artificial intelligence in education: opportunities, challenges, threats and obstacles. A literature review. Indonesian J. Comput. Sci. **12**(4) (2023)
54. Chan, C.K.Y.: A comprehensive AI policy education framework for university teaching and learning. Int. J. Educ. Technol. High. Educ. **20**(1), 38 (2023)
55. Meng, N., et al.: AI-enhanced education: teaching and learning reimagined. In: Bridging Human Intelligence and Artificial Intelligence, pp. 107–124. Springer (2022)
56. Shahid, M.K., et al.: Exploring the relationship of psychological factors and adoption readiness in determining university teachers' attitude on AI-based assessment systems. Int. J. Manage. Educ. **22**(2), 100967 (2024)
57. Cowls, J., Tsamados, A., Taddeo, M., Floridi, L.: The AI gambit: leveraging artificial intelligence to combat climate change—opportunities, challenges, and recommendations. AI. Soc. 1–25 (2023)
58. Lee, I., Ali, S., Zhang, H., DiPaola, D., Breazeal, C.: Developing middle school students' AI literacy. In: Proceedings of the 52nd ACM technical symposium on computer science education, pp. 191–197 (2021)
59. Godwin, R.C., DeBerry, J.J., Wagener, B.M., Berkowitz, D.E., Melvin, R.L.: Grant drafting support with guided generative AI software. SoftwareX **27**, 101784 (2024)
60. Klašnja-Milićević, A., Ivanović, M.: E-Learning Personalization Systems and Sustainable Education, p. 6713. MDPI (2021)

61. Rizvi, M.: Investigating AI-powered tutoring systems that adapt to individual student needs, providing personalized guidance and assessments. Eurasia Proc. Educ. Soc. Sci. **31**, 67–73 (2023)
62. Yang, Y., Xia, N.: Enhancing students' metacognition via Ai-driven educational support systems. Int. J. Emerg. Technol. Learn. (Online) **18**(24), 133 (2023)
63. Pellas, N.: The influence of sociodemographic factors on students' attitudes toward AI-generated video content creation. Smart Learn. Environ. **10**(1), 57 (2023)
64. Kaswan, K.S., et al.: AI in personalized learning. In: Advances in Technological Innovations in Higher Education, pp. 103–117. CRC Press (2024)
65. Aler Tubella, A., et al.: How to teach responsible AI in higher education: challenges and opportunities. Ethics Inf. Technol. **26**(1), 3 (2024)
66. Wu, S., et al.: Evaluation of smart infrastructure systems and novel UV-oriented solution for integration, resilience, inclusiveness, and sustainability. In: 2020 5th International Conference on Universal Village (UV). IEEE (2020)
67. Adel, A.: The convergence of intelligent tutoring, robotics, and IoT in smart education for the transition from industry 4.0 to 5.0. Smart Cities **7**(1), 325–369 (2024)

The Impact of Generative AI on Healthcare

Atianashie A. Miracle and Chukwuma Chinaza Adaobi

Abstract The advent of generative artificial intelligence (AI) is revolutionizing the healthcare sector, bringing unprecedented advancements across various domains, including drug discovery and development, personalized treatment planning, medical imaging, and predictive health analytics. This chapter explores the multifaceted impact of generative AI in healthcare, highlighting its role in accelerating drug discovery processes, enhancing drug design and optimization, and improving drug safety and efficacy. The integration of generative AI in personalized treatment planning allows for the development of tailored therapeutic strategies based on a comprehensive analysis of patient-specific data, thereby improving treatment outcomes and patient satisfaction. In medical imaging, generative AI techniques such as generative adversarial networks (GANs) and variational autoencoders (VAEs) facilitate faster and more accurate image analysis and diagnosis, enabling early detection of diseases and abnormalities. Furthermore, generative AI's predictive health analytics and risk assessment capabilities empower healthcare professionals to predict disease outcomes, identify high-risk individuals, and implement timely preventive interventions. This chapter underscores the transformative potential of generative AI in enhancing healthcare delivery, reducing costs, and improving patient outcomes, paving the way for a more efficient and effective healthcare ecosystem. As technology continues to evolve, its integration into healthcare practices promises to drive further innovations and improvements in medical research and patient care.

Keywords Generative AI · Healthcare innovation · Personalized medicine · Medical imaging analysis · Predictive health analytics

A. A. Miracle (✉) · C. C. Adaobi
School of Research and Graduate Studies, Catholic University of Ghana, P.O. Box 363, Sunyani, Bono Region, Ghana
e-mail: miracleatianashie@cug.edu.gh

N. R. Vajjhala et al. (eds.), *Generative Artificial Intelligence (AI) Approaches for Industrial Applications*, Information Systems Engineering and Management 24,
https://doi.org/10.1007/978-3-031-76710-4_9

1 Introduction

The integration of generative artificial intelligence (AI) in healthcare marks a pivotal advancement in the medical field, promising transformative changes in how healthcare services are delivered and managed. In 2024, the adoption of generative AI has accelerated, driven by its profound capabilities in data analysis, predictive modeling, and personalized medicine. Generative AI encompasses sophisticated models like transformers and diffusion models, which have revolutionized diverse applications such as drug discovery, medical imaging, clinical decision support, and patient management. One of the most significant impacts of generative AI is seen in drug discovery and development. Traditional drug discovery processes are notoriously time-consuming and costly, often taking over a decade and billions of dollars to bring a new drug to market. Generative AI, however, has dramatically shortened this timeline by enabling the rapid generation and testing of novel drug candidates. AI-driven platforms, such as those developed by Exscientia and Pfizer, utilize machine learning algorithms to identify promising compounds and predict their efficacy, expediting the journey from lab to clinic [1].

In clinical settings, generative AI enhances diagnostic accuracy and efficiency. AI algorithms can analyze vast amounts of medical data, including imaging studies, genetic information, and electronic health records, to detect patterns and anomalies that may be imperceptible to human clinicians. This capability not only improves early disease detection but also facilitates personalized treatment plans tailored to individual patient profiles. Google's Med-PaLM 2, for instance, is a fine-tuned large language model designed to support clinical decision-making by providing contextually relevant insights during patient consultations [2]. Moreover, generative AI is transforming administrative and operational aspects of healthcare. By automating routine tasks such as documentation, scheduling, and patient follow-ups, AI reduces the administrative burden on healthcare professionals, allowing them to focus more on patient care. This automation also mitigates burnout among clinicians, enhances workflow efficiency, and improves overall patient satisfaction [3, 4].

Despite its numerous benefits, deploying generative AI in healthcare raises important ethical and practical considerations. Issues related to data privacy, algorithmic bias, and the need for rigorous validation and regulation of AI systems are critical to ensuring that AI-driven healthcare solutions are safe, equitable, and trustworthy. Continuous collaboration among healthcare providers, technologists, and policymakers is essential to address these challenges and fully harness the potential of generative AI [5]. As generative AI continues to evolve, its role in healthcare will likely expand, offering new opportunities for innovation and improvement in patient outcomes [6]. The ongoing research and development in this field promises a future where healthcare is more efficient, personalized, and accessible, fundamentally transforming the landscape of medical practice and patient care.

2 Drug Discovery and Development

Furthermore, the industry is tapping into generative AI methods to understand different diseases better and repurpose medicines typically used for different ailments. For instance, traditional pain management R&D initiatives have led to the discovery of potential solutions against Alzheimer's disease. Pharmaceutical enterprises and technology companies are leading these efforts, and many of them share a common goal: to fast-track the iterative process, which is currently expensive, time-consuming, and not as efficient as it potentially could be. The main difference between traditional drug development methods and generative AI-assisted methods lies in the sheer amount of information that scientists and IT specialists can feed to any given algorithm [7]. Given that self-learning algorithms improve their pattern-recognition abilities over time, this means that generative AI methods can accelerate the process of understanding the complicated biochemical processes behind diseases and possible cure options. On top of that, AI models can also help make predictions abouthe most likely candidates [8].

For instance, Exscientia, an AI-driven drug discovery company, has recently announced that they managed to identify a potential treatment for chronic kidney disease in a fraction of the time it would take a team relying on traditional drug discovery methods. Using several algorithms embedded in the Centaur Chemist platform, Exscientia pinpointed LRRK2 inhibitors, which may be suitable for repurposing as chronic kidney disease treatments. As a result, the second phase of the research—evaluating the most promising compounds with laboratory experiments—has effectively been reached within a year. In contrast, this interdisciplinary research effort could easily span six to twelve years using traditional drug discovery methods [8]. Finally, it is worth noting that AI and drug development field are still in the preliminary stages. Crisper screens, multi-model technology, and generative design approaches are processes that are often present in life science research trials but tend to be largely underrepresented in the development practice. As a result, the potential benefits of greater computational power and more robust self-learning algorithms can be dampened by a lack of investment and willingness to embrace generative AI [8]. However, dismissing AI's promise for the drug development industry is hard. The digitization and subsequent analysis of the vast amounts of patient health data promise to bring healthcare into the connected age, and hopefully, many more innovative solutions to unmet medical needs are yet to be discovered.

2.1 Accelerating Drug Discovery Process

Generative AI has significantly accelerated drug discovery, making it more efficient and cost-effective. Historically, developing a new drug is an expensive, laborious, and time-consuming process that may take up to 15 years from discovery to market, costing more than $2 billion. The drug discovery process typically involves several stages, including target identification and validation, lead discovery, lead optimization, and preclinical development. Generative AI expedites the lead discovery stage by generating novel lead molecules with desired therapeutic properties in a shorter period of time, utilizing machine learning and large datasets of molecule structures and their associated activities. Unlike traditional methods that require rigorous laboratory experiments and thorough testing on animals and human tissues, the unique advantage of generative AI in creating novel drug candidates is the ability to model and prioritize the most promising drug molecules for synthesis and testing based on in-silico, combined with minimal experimental validation [8, 8]. This approach saves significant time and resources in the initial stages, where manual work and testing procedures are repeated many times to identify lead molecules. For instance, the global biopharmaceutical company Pfizer has established a multi-year collaboration with IBM Watson Health to develop the world's first system for designing small molecule therapeutic drugs, utilizing advanced machine learning, AI, and robust computer systems. The collaboration has achieved remarkable progress on many diseases and biological targets, and the generated new molecules have novel structures that are not found in existing drugs, demonstrating the potential of AI-augmented research programs [8].

2.2 Enhancing Drug Design and Optimization

Generative AI significantly impacts healthcare in various areas, such as drug discovery and development, personalized treatment planning, medical imaging and diagnosis, and predictive health analytics and risk assessment. In drug discovery and development, generative AI accelerates the drug discovery process, enhances drug design and optimization, and improves drug safety and efficacy. First, let's understand the traditional drug design process. Drugs are essentially small molecules that control a process in a cell. Typically, drug molecules have a key-and-lock type structure, where the molecule is the key and a cell component, usually a protein, is the lock. According to the "lock and key" hypothesis, a drug will only activate a process in the body if it fits the protein exactly. Based on this structure–activity relationship understanding, drug designers would start with a small molecule that they knew was active in the body. Then, they would make small changes to the molecule by adding new bits or altering existing parts. This would generate many similar molecules called "analogs", which would all be tested for activity. However, making and testing analogs is actually a costly and time-consuming process. In addition, deciding which analog to make is also actually very difficult, and people cannot test all possible analogs because there are countless ways to change a molecule [9, 10]. This trial-and-error method to find the right structure that is a good fit for the protein target is one of the key reasons drug design is such a lengthy and expensive process. However, generative AI can analyze large amounts of data, learn from examples, and generate novel hypotheses. By using generative AI, researchers can investigate hundreds or thousands of potential designs by making and testing a small percentage of them and quickly identify new molecules with the right properties.

This approach of using AI to predict the probability of a sequence of molecules has been successfully used to generate molecules with specific chemical properties, such as a group of molecules known to control a certain process in the body. These generated molecules could be used in a completely new drug or found useful in understanding the importance of individual molecules in a given group. In addition, generative AI uses deep learning and recurrent neural networks, which are effective methods for modeling data that is composed of interdependent sub-units, such as the case in molecules, images, and language [1, 10]. By feeding the trained recurrent neural network with the simplified molecular-input line-entry system notation of a molecule, which represents a 2D map of the atoms in a molecule, generative AI can predict the likelihood of each atom forming a bond with another atom and ultimately generate a 2D visualization of a molecule [11]. At present, generative AI is becoming increasingly powerful in drug design, and several biotech companies and technology firms are pioneering a broad platform of generative chemistry using AI. AI could produce new molecules that are likely to be good drug candidates and be more advanced than humans in lead optimization. They could also be used in the initial drug discovery process to save time and money. Finally, it is important to note that human experts should always visualize and check computationally generated chemical structures. Generating and understanding potentially effective molecules is only one part of drug design and requires a huge amount of human insight, knowledge, wisdom, and interaction between computational methods and chemists.

2.3 Improving Drug Safety and Efficacy

At last, generative AI and machine learning play a great role in improving the safety and efficacy of drugs that are already on the market and under development. The conventional process of determining drug safety is to conduct experiments, mostly on animals, to study the drug's toxic effects and risks. This approach is time-consuming, expensive, and often leads to drug failures at a late stage of development. Besides, relying on animal data to predict human safety can be unreliable. Generative AI can learn from a combination of large and diverse chemical and biological data and predict molecular properties and biological activities of different chemical compounds. A machine learning model is often used to classify the safety profile of a compound by training the model to recognize the chemical features associated with toxic outcomes. This innovative strategy is called "predictive toxicology." With generative AI, designing drugs with minimized toxicity is now possible. Also, generative AI enables the synthesis of genetically tailored, safer, and more effective medicines [12, 13].

For example, Insilico Medicine, a bioinformatics company, has developed a machine-learning system to design and optimize small molecules interacting with a specific disease target [14]. This approach is highly efficient because the company only takes 18 months to identify and validate a new molecular target in disease and to develop a molecule that modulates that target, typically taking three to six years using the traditional methods. Besides, a unique feature of generative AI is that it can suggest new possibilities that human researchers may not consider, further driving drug discovery innovation. Due to the use of productivity and quality-enhancing AI technology and the accumulated evidence of improved success rates in drug research and development, more and more pharmaceutical companies are adapting and investing in AI. For example, in January 2020, a new strategic drug discovery collaboration was announced between GSK and China's leading AI and drug discovery company, WuXi AppTec, focusing on leveraging AI technology in support of the research and development of new medicines, the increased use of big data and the development of highly effective cognitive algorithms have turned drug development and many other fields into a much smarter and more efficient process. The role of AI will not just be another tool in the industry. It is set to completely reshape the drug research and development process, from identifying drug targets to designing and discovering candidate drugs, totally in line with the vision of precision medicine.

3 Personalized Treatment Planning

Personalized treatment planning, also known as precision medicine, is a medical model that separates people into different groups—with medical decisions, practices, interventions, and/or products being tailored to the individual patient based on their predicted response or disease risk. This branch of medicine has seen huge growth in recent years, in part thanks to the increasing availability of genetic testing. Generative AI is streamlining and enhancing personalized treatment planning by enabling the analysis of many factors and generating more optimized and effective treatment plans. By considering a wide range of patient-specific characteristics, including demographics, medical history, genetic information, and even lifestyle and environment data, generative AI can help identify the best course of action for a particular patient. Instead of relying on standardized treatment plans and making educated guesses, doctors can use generative AI to explore many potential options and receive data-driven recommendations tailored to individual patients. As a result, the treatment process can become much more personalized and effective, which may lead to improved patient outcomes and satisfaction [15, 16].

Generative AI makes it possible to generate a diverse set of possible treatment plans, each representing a different trade-off between, for example, potential side effects and risk of disease recurrence. These optimal plans can then be visualized and explored by physicians using advanced decision support tools, helping them understand and communicate the complex risk factors and uncertainties at the heart of personalized treatment planning. By involving the physician in exploring the possible solution space, generative AI supports a pragmatic rather than prescriptive approach to personalized treatment planning, recognizing and responding to the need for the physician to maintain autonomy in decision-making. In practice, generative AI enables physicians to develop a more sophisticated and nuanced understanding of the optimal treatment plans suggested by the data and to use this knowledge to tailor and refine them further [17, 18]. This explains the tremendous future of this treatment planning application in the healthcare ecosystem. As generative AI continues to evolve and data about the success of different plans and treatments is collected, it represents an opportunity to use cutting-edge technology to contribute to the national research strategy on personalized medicine and improve the evidence base for these innovative approaches. The Government's 2018 Life Sciences Sector Deal was committed to strengthening the UK's position in health research and translating pioneering research into healthcare innovation. Such a focus has the potential to advance technologies, including the use of generative AI, towards provision by the NHS, enhancing the delivery of precision medicine. By boosting patient treatment options and supporting healthcare professionals to refine and develop their practice, generative AI and personalized treatment planning could change the face of many medical fields, from cancer care to managing chronic conditions.

3.1 Tailoring Treatment Plans to Individual Patients

The goal, as outlined in the table of contents and expanded on in this, is to treat the patient as an individual. In a perfect world, the treatment you receive for whatever condition is ailing you would be tailored to you and nobody else. Of course, time and money are factors in this, and it is not practical. This manual analysis, however, is slowly changing. With generative AI, gathering and analyzing huge amounts of patient data ranging from medical histories and family genetics to lifestyle and environmental factors is now possible. This data creates 'virtual patients' or extensive computer simulations that can test various treatment combinations. By finding which combinations produce the best outcomes for virtual patients, doctors can use this information to pick the best treatments for the real patient. This sort of process is called personalized medicine and is a growing area of research and treatment worldwide. Not only is generative AI being used to find the most effective treatments for illness, but it is also being used to try and minimize the impact of any treatment on the patient's life, e.g., reducing the side effects while keeping the treatment's effectiveness [19]. For example, we can consider the application of this sort of technology in radiotherapy, a cancer treatment. By using generative AI to discover the most effective ways of protecting healthy tissue from radiation, doctors can now reduce the total number of doses a patient needs. This 'dose-sparing' effect of this discovery has led to a considerable increase in the quality of life of prostate cancer patients. Of course, there are many factors to consider when using a new treatment like this, when generating data about medication use, and whether the doctor's clinical judgment must be the determining factor in what course [20].

3.2 Optimizing Treatment Regimens

In this context, the term "generative" refers to the fact that these systems don't just look at the data and find trends or patterns—they use that information to create something new, in this case, a proposed optimal regimen for a particular patient. This can help to cut down the amount of trial and error needed to find the best treatment, meaning that patients could see the benefit of this technology through faster and more effective care. In recent years, researchers have begun to use generative AI to help optimize these complex treatment regimens. By modeling the interaction between different drugs at a molecular level and combining this with patient-specific information on things like genetics and medical history, it's possible to estimate how effective different drug combinations might be for an individual patient [21]. Taken together, this means that there are a vast number of possible drug regimens to consider—for example, in the case of epilepsy, it's been estimated that the number of potential combinations is larger than the number of atoms in the known universe!

Researchers have developed several techniques to try and solve this problem, many of which fall under "personalized medicine." The basic idea behind personalized medicine is to use as much information about a patient as possible to tailor their treatment to their unique genetic and lifestyle factors. However, this can be an incredibly complex process. There may be many potential drugs for many conditions, each of which may have several different dosages and possible combinations [22, 23]. Generative AI is currently used to optimize treatment regimens for patients with various conditions, including epilepsy. In epilepsy, a wide range of different drugs are available for treatment, each of which may be prescribed in multiple doses and combinations. Finding the best regimen for a patient—that is, the one that maximizes the time spent free of seizures while minimizing any side effects—can be a time-consuming and complex task. [24, 25].

4 Medical Imaging and Diagnosis

In the last decade, generative adversarial networks (GANs) and variational autoencoders (VAEs) have been applied to many different fields of medical imaging analysis, such as computer-aided detection, image reconstruction, data clustering, and synthesis. When GANs and VAEs are trained on large-scale medical image datasets, these networks can capture the data's inherent variability and underlying patterns and generate new samples that resemble the training data. When both labeled and unlabeled data are available, semi-supervised learning can be employed to obtain better performance [25, 26]. For example, a recent work proposed integrating unlabeled contextual information into a deep learning framework for lesion detection in breast ultrasound images and achieved better detection accuracy. Also, by using GANs, it's possible to transform images in one domain into another, e.g., reconstructing CT images from MRI images. Such image space translation has great potential to reduce

Table 1 Key variables and categories in medical imaging and diagnosis

Variable	Category
Generative Adversarial Networks (GANs)	Technique
Variational Auto-Encoders (VAEs)	Technique
Computer-aided detection	Field of application
Image reconstruction	Field of application
Large-scale medical image datasets	Data type
Inherent variability	Data characteristic
Semi-supervised learning	Learning method
Deep learning framework	Framework
Lesion detection in breast ultrasound images	Application example
Detection accuracy	Performance metric

repetitive radiation exposure in clinical routines where multiple scans with different imaging protocols for the same patients are commonly required [27, 28].

Another exciting application is the generation of realistic medical images with known pathology from the same cohort but without pathology presentations as shown in Table 1. Such techniques might help bridge the gap between the unlimited possibilities in medical image augmentation and the limited access to accurate patient data, which is crucial for image segmentation and disease classification. It would be exciting to see the solutions for improving the image quality and structural similarity between the generated images and the ground truth, as well as more successful stories about integrating generative models into the diagnostic and treatment workflow soon [29, 30]. There is still a lot to explore when it comes to developing novel generative models and pushing the boundaries of using them in medical imaging. There has recently been extensive research in creating 3D models based on GANs and VAEs, which provides another dimension of the variability in medical image data. We are also applying for big imaging genomics and clinical data and machine learning grants to develop new tools using VAEs for the personalized prediction of the treatment outcome for epilepsy [31, 32].

4.1 Automated Image Analysis for Faster Diagnosis

A related area of artificial intelligence known as computer-aided diagnosis is also showing great promise. These systems don't make a definitive diagnosis, but they alert radiologists to areas they should pay attention to, which can help speed up diagnosis times. One of the biggest areas of growth in medical imaging artificial intelligence is in automating some of what has, up until now, been the most time-consuming issue: image analysis. Medical images such as X-rays, CT scans, and MRI images comprise a series of dots, each with a different brightness or darkness value. A typical x-ray, for example, has about 2000 × 2000 columns of these little dots, which is about 4 million data points to study. This is why it currently takes experienced professionals a long time to decide what is or isn't visible on an image—and a long time to compare new images against ones already on file. Accurate, tight comparisons are crucial when diagnosing diseases, particularly those that rely on some kind of change showing up on consecutive scans. This is where generative or, rather, what's known as discriminative artificial intelligence comes in. For the last 20 years, AI analysis of medical scans has employed what is known as rule-based computer-aided diagnosis [33, 34]. That is to say, the systems in place compare the images against a set list of possible outcomes and flag up any areas of interest. However, human input is necessary to develop a set of standards for an algorithm to work.

As new diseases emerge or how a disease presents itself advances, the algorithms can quickly become outdated. Creative and generative AI is a way around this problem. By giving the AI some guidance—such as an initial starting point or a few human-identified areas of concern—generative AI can create its rulebook for that set of images. Because AI defines what's interesting in an image, it can then apply its findings to each study. This means that generative AI and the ability to quickly build a fully bespoke, case-specific standard of 'what's interesting' can give much faster and more informative results by allowing radiologists to move from trying to find a match in generic image libraries to fully understanding how a patient's scan both presents that disease and differs from it all in one go, diagnosis times can be massively reduced. As the AI continues to study incoming images and update its model of what represents exciting features, its diagnosis will become more detailed and accurate over time [34]. While this doesn't lessen the role of the professional in corroborating the results, a faster and more comprehensive computer-aided analysis of this level can only improve patient outcomes.

Figure 1 presents a comprehensive diagram illustrating the key variables and their interrelationships within the context of using artificial intelligence (AI) for automated image analysis to expedite medical diagnosis. The central focus is on how AI can influence and enhance various aspects of the diagnostic process.

Diagnosis Time is a critical variable influenced using AI. The diagram shows two pathways: one where AI assistance significantly reduces diagnosis time and another where traditional methods, without AI, take longer. The types of medical images analyzed by AI, such as X-rays, CT scans, and MRI, are depicted under Image Type. Each of these image types contributes to the data complexity, represented by Image Size, which includes the number of columns (e.g., 2000 × 2000) and the total data points (e.g., 4 million). The AI System Type differentiates between traditional Computer-Aided Diagnosis, which relies on rule-based systems, and Generative AI, which creates bespoke standards for image analysis. The necessity of Human Input is highlighted, showing how initial starting points and human-identified areas of concern guide AI analysis. The AI Analysis Method outlines the shift from Rule-Based Comparison to more sophisticated Generative/Discriminative AI approaches. This shift allows for more dynamic and accurate image analysis. Outcome Flagging shows how both AI and human radiologists identify areas of interest, enhancing diagnostic accuracy. Comparison Accuracy is crucial for diagnosis, with traditional

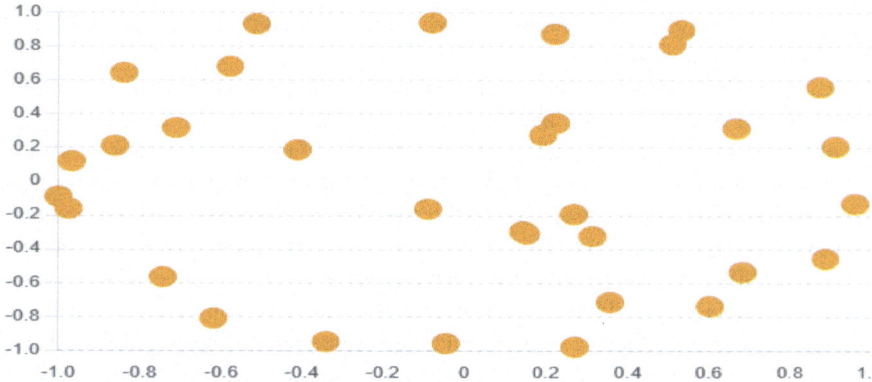

Fig. 1 Automated image analysis for faster diagnosis

methods relying on Existing Image Libraries and advanced methods using Case-Specific Standards created by AI. The Model Update Frequency variable emphasizes the need for regular updates to AI models to maintain their relevance and accuracy as new data is incorporated. Patient Outcome Improvement highlights the potential benefits of AI in enhancing patient care, showing improved outcomes with AI assistance compared to traditional methods. The diagram also considers the Emergence of New Diseases and their impact on the effectiveness of AI algorithms, demonstrating the need for continuous adaptation. Finally, Diagnosis Detail and Accuracy covers the granularity and precision of AI-assisted diagnoses, showing that, as AI systems learn from new data, their diagnostic accuracy improves over time.

4.2 Increasing Accuracy in Medical Imaging Interpretation

Medical image interpretation has been identified as a critical area in radiology that will benefit from generative AI. Traditional image interpretation relies on human visual inspection of images to look for specific patterns that might indicate a particular condition, illness or disease. However, due to the large numbers of images that are acquired in modern medical imaging and the complexity of human vision, an increasing number of images are going undiagnosed or not being diagnosed as early as they could be. Generative AI has the potential to completely revolutionize how medical images are interpreted [35, 36]. The concept is that generative models can take in all the key information from an image and then contextualize this information to search for the most optimal appearance of any damage, abnormality, or illness. By doing this, generative AI could help by providing significantly improved accuracy over traditional methods, thus leading to earlier diagnosis and better patient outcomes. This kind of technology will benefit more specialist areas of medicine, such as the reading and interpreting mammograms, where many images are taken, and a vast amount of data is acquired.

The hope with generative AI in this context is that it will be able to identify cancers sooner and reduce the number of false positives from the image analysis, thus reducing the number of patients called back for further investigations. These further tests would include things like biopsies, which are not only uncomfortable for the patient but also are associated with their own risks and can often result in unnecessary worry or distress. This impact on patient experience is important, and the unique ability of generative AI to increase accuracy and reduce false positive results will likely see it playing a key role in the future of patient care. However, it's important to note that generative AI is not intended to replace the skill and ability of the consultant or doctor reviewing the image. The real benefit of such technology is that it functions more as a supporting tool, allowing the user to have an insight and analysis of the image in a way that is independent and not at all influenced by human judgment [37, 38]. This might enable better imaging in cases where certain conditions, such as inflammation, might mask underlying causes of pain or debilitation. A generative model's ability to look past the obvious and identify smaller, less pronounced indications of problems will likely see more and more development of these technologies in the coming years. Given that we are in the early stages of implementing such technology in practice, research is ongoing to measure the impact of generative AI and validate its use as a supporting tool in imaging interpretation. However, initial feedback and studies suggest that users are optimistic about its potential, indicating that as technology advances and we can become sure of AI's safe and accurate functioning, it won't be long before generative AI is a commonly used tool for diagnostic imaging.

4.3 Early Detection of Diseases and Abnormalities

Early diagnoses brought about by such technology will open the door to early treatment and possible prevention. This will, in turn, alleviate some pressures faced by patients and healthcare systems worldwide, which are always under strain due to the overwhelming demand for medical services. The benefits of early detection also underscore the significance of the power of generative AI in amplifying human creativity in medical image analysis. With traditional statistical algorithms, it is important to provide a "hint" in the form of a general location in the images to indicate the possible area of interest. However, synthetic image iteration can be applied to all locations on the images, allowing for an algorithm that seeks to identify areas of abnormality without any prior indications. As early abnormality may not be visually recognizable, this represents a major step forward in our effort to achieve early detection of diseases and abnormalities [39].

After training, the generator can create realistic synthetic images of humans, which can be utilized to find abnormalities. When a patient's medical image is obtained, synthetic variations can be created by tweaking the noise input into the generator.

By feeding these synthetic images into a traditional statistical classifier in a trial-and-error manner and observing the response of the classifier, the researcher can identify the region in the images where the abnormality is indicated [38]. Generative adversarial networks (GANs) have the potential to provide an effective solution to this challenge. Comprising a generator and a discriminator, GANs work by training the generator to create synthetic images from computer noise and the discriminator to distinguish between real images and the synthetic images created by the generator. The generator and the discriminator will be improved iteratively through competition against each other in a zero-sum game [40]. Early detection of diseases and abnormalities is important in enabling timely intervention and effective treatment. Imaging tests serve as the front line in such detection. However, in many cases, the disease may not have progressed to the level where abnormalities can be visually detected or are too subtle to recognize.

5 Predictive Health Analytics and Risk Assessment

The application of modern data science research and technologies in the healthcare industry has given rise to a new field called predictive health. By generating predictive models from a wide range of data sources, including electronic health records, claims data, and genetic data, healthcare professionals can forecast and prevent the onset of chronic diseases for at-risk populations. As generative AI enables analytic models to be automatically developed, maintained, and applied, predictive health analytics can become more comprehensive and efficient. Modern machine learning technologies, such as generative adversarial networks (GAN) and reinforcement learning, result in more accurate predictive models, thus improving the effectiveness of preventive solutions and interventions [41, 42]. For example, researchers from Harvard Medical School have developed a GAN-based predictive model capable of simulating molecular and cellular activities over time, which can be applied to various medical research topics. The model, named "MedGAN," learns to generate large amounts of medical data which exhibit the same statistical dependencies as real-world observations. The predictions become significantly more accurate when the synthetic data is fed into predictive models. In another project, scientists at the Feinstein Institute for Medical Research have leveraged reinforcement learning to develop an adaptive sepsis treatment strategy using data from electronic health records. With an adaptive algorithm that optimizes treatment plans through learning from the model's predictions and real-patient outcomes, the system has shown a significant 21.5% reduction in mortality. Such cases demonstrate the potential of generative AI for predictive health research and manifest actual improvements in treatment and patient outcomes through machine learning capabilities [43, 44]. These technologies improve individuals' health, provider operations, payment strategies, and even governmental public health campaigns by providing valuable insights from a vast and complex data ecosystem.

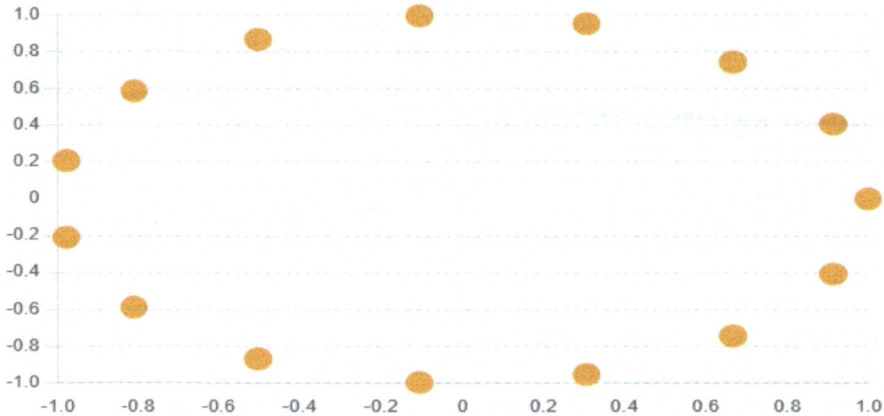

Fig. 2 Early detection of diseases and abnormalities

Figure 2 demonstrates the interconnected variables that play a crucial role in the early detection of diseases and abnormalities, emphasizing the relationships and flow of influence among them. The circular layout aids in visualizing how advancements in technology, particularly Generative AI and Generative Adversarial Networks (GANs), enhance the process of early detection and subsequent healthcare improvements.

At the core, Generative AI drives the development and application of GANs, showcasing its pivotal role in medical image analysis. GANs, comprising a generator and a discriminator, are trained to create and distinguish synthetic images. The generator produces synthetic variations, while the discriminator assesses their realism, iteratively improving both components. These components work together within the GANs framework to generate realistic synthetic images (generator) and validate them (discriminator). The generator produces synthetic variations, which are then fed into traditional statistical classifiers. This process is crucial for identifying subtle abnormalities in medical images. The traditional statistical classifier analyzes these synthetic variations, aiding in the detection of abnormalities that might not be visually recognizable at an early stage. Enhanced by synthetic image iterations and classifiers, the visual recognition of abnormalities is crucial for early detection, directly influencing the subsequent stages of early diagnoses and treatment.

Early detection of diseases and abnormalities is the primary goal of the entire process, facilitated by advanced AI technologies and thorough image analysis. This early detection leads to timely diagnoses, enabling prompt medical intervention. Prompt diagnoses facilitate early treatment, which can significantly improve patient outcomes and potentially prevent the progression of diseases. Early treatment contributes to prevention strategies, reducing the overall burden on healthcare systems. The alleviation of pressures on healthcare systems is a downstream effect of effective early detection, diagnoses, and treatment, showcasing the broader impact on public health infrastructure. Foundational elements such as traditional statistical algorithms and imaging tests support the entire process, providing the initial data and analytical frameworks necessary for advanced AI applications. Synthetic image iteration allows for comprehensive analysis without prior indications, representing a significant advancement in medical image analysis.

5.1 *Predicting Disease Outcomes and Progression*

As medical data grows in size and complexity, generative AI can help identify patterns and sound signals that would be impossible for individual doctors to recognize. Predictive models that use this data can take many forms, from those that forecast the future trajectory of an individual illness to those that process extensive data shared or referred to find as-yet unknown markers that can help define entirely new disease classifications. These models depend heavily on accessing large-scale global health data to develop and refine their accuracy over time, meaning that predictive health analytics will become increasingly international. This technology is still at a relatively early stage of development, but generative AI is proving to be highly effective in this area [45, 46]. For example, a recent study from the Francis Crick Institute in London applied deep learning techniques to define disease phenomena across various autoimmune and autoinflammatory disorders. The research found that distinct disease patterns in specific organs could be identified, allowing for more rapid and accurate diagnosis and potentially the move towards more targeted and successful treatment plans. Given that these techniques have demonstrated effectiveness in diseases that historically have been difficult to characterize and treat effectively, the application of generative AI in disease outcome prediction is a fascinating prospect for improving patient care worldwide. In the future, we can certainly hope to see more personalized and precise diagnoses and prognoses appraised from generative AI calculations, which have the potential to revolutionize treatment pathways [47].

5.2 *Identifying High-Risk Individuals*

In addition to predicting disease outcomes and progression, generative AI identifies high-risk individuals. AI algorithms can identify individuals more likely to develop a certain disease within a certain time frame. By analyzing patient medical records, family history, social determinants of health, and lifestyle data, generative AI calculates the probability and risk of everyone developing a disease [39]. As a result, it helps healthcare professionals prioritize their resources and preventive care for patients at higher risk, allowing for earlier and more targeted interventions. This undoubtedly has a positive and profound impact on both individual patients and public health. Early diagnosis and targeted, timely interventions may prevent the development of a disease, delay the progression of the disease, reduce complications, and improve the quality of life for the patients [48]. On the other hand, when

AI generates more accurate prevalence and incidence rates of certain diseases in the population, it helps public health agencies and policymakers to understand the systemic health needs better and allocate resources to the areas with the highest health risks. However, the effectiveness of using AI for identifying high-risk individuals depends not only on the accuracy of the AI algorithms but also on the availability of big data. It also requires a change of practice and workflow in the clinical setting [48, 49]. As generative AI provides more and more features that support clinical decision-making and patient risk assessment, it is essential to integrate new AI tools within the existing electronic health record system and the digitized workflow so that the potential of AI in transforming the capacity and efficiency of health delivery and the patient experience can be fully realized.

5.3 Informing Preventive Measures and Interventions

Generative AI will not only change healthcare for people who are not feeling well, but by changing and shaping therapies, it will also change healthcare for perfectly healthy people. Professor Ara Darzi, director of the Institute of Global Health Innovation at Imperial College London, has previously said, "The rising expectations of patient populations and more sophisticated technology, such as genomics and artificial intelligence, has the potential to truly shift focus within the healthcare field from merely treating patients when they are sick to a future whereby, we can predict and prevent illness—an empowering and exciting step." By using generative AI to look at lifestyle and genetic information, machines can understand more and more precisely what might lead to someone becoming ill [39]. Because of this, it may be possible to have that illness diagnosed years and even decades earlier than we would now. This can potentially change—and save—the lives of millions of people. Ts is the most inspiring part of working with generative AI in healthcare [50, 51]. Technology not only upscales the ability of researchers and doctors to provide medical support to people who are feeling unwell but also helps society rethink how we use modern technology to benefit the next generations.

5.4 Enhancing Population Health Management

While currently, most population health activities are retrospective—i.e., focus on the health of specific populations in the past—generative AI paves the way for prospective and more effective initiatives. Generative AI can analyze various data sources to identify population health needs, develop and evaluate the effectiveness of targeted interventions, and optimize available resources to improve the whole population's health [39]. For instance, a generative adversarial network model that uses deep learning technology to analyze electronic health records and claims data has been proposed to identify opportunities to improve population health in Tennessee in the

United States. As described in the proposal, the generative AI model will identify common and rare diseases that public health programs may not address by analyzing disease prevalences in different socio-demographic groups and geographical areas [52]. The proposal suggests that by evaluating disease progression pathways and the cost-effectiveness of recommended interventions using the generative AI model, the initiative will lead to more effective ways to improve the health of the underlying population. Such programs like the one proposed in Tennessee are essential not only to manage the health of the current population effectively but also to create an evidence base for the development and implementation of new practices and—over time—to create learning health systems that will continuously advance and reshape the way that we improve the health of the population through practical uses of data and technology [39].

6 Conclusion

Generative AI has emerged as a transformative force in the healthcare sector, significantly impacting various facets from drug discovery and development to personalized treatment planning, medical imaging, and predictive health analytics. This technology has shown immense potential in accelerating the drug discovery process, enhancing drug design, and improving drug safety and efficacy. By leveraging generative AI, pharmaceutical companies have been able to fast-track the identification of novel drug candidates and repurpose existing medications, thereby reducing the time and cost associated with traditional drug development methods. In personalized treatment planning, generative AI enables healthcare providers to tailor treatment plans to individual patients based on a comprehensive demographic, genetic, and lifestyle analysis. This personalized approach enhances the effectiveness of treatments and minimizes adverse effects, leading to improved patient outcomes and satisfaction. Generating diverse and optimized treatment plans helps physicians make informed decisions and adapt real-time strategies, ensuring patients receive the most appropriate care.

Medical imaging and diagnosis have also benefited from the integration of generative AI. Techniques such as GANs and VAEs have revolutionized the analysis and interpretation of medical images, enabling faster and more accurate diagnoses. AI-driven image analysis reduces the burden on radiologists, enhances the detection of diseases at early stages, and improves the accuracy of interpreting complex medical images. This leads to earlier interventions and better management of conditions, ultimately improving patient care. Predictive health analytics and risk assessment are other critical areas where generative AI has made significant strides. AI models can predict disease outcomes, identify high-risk individuals, and inform preventive measures by analyzing vast amounts of data from electronic health records, genetic information, and other sources. This proactive approach allows healthcare providers to implement timely interventions, reducing the incidence of chronic diseases and improving overall population health. The integration of generative AI in healthcare is paving the way for a new era of medical innovation and patient care. The technology's ability to process and analyze large datasets, generate novel insights, and support clinical decision-making is transforming how healthcare is delivered. As generative AI continues to evolve, it holds the promise of further advancements in medical research, personalized medicine, and healthcare delivery, ultimately leading to a healthier and more efficient future for patients and providers alike.

References

1. Anantrasirichai, N., Bull, D.: Artificial intelligence in the creative industries: a review. Artif. Intell. Rev. **55**(1), 589–656 (2022). https://doi.org/10.1007/s10462-021-10039-7
2. T. Hartung: Artificial intelligence as the new frontier in chemical risk assessment. Front. Artif. Intell. **6**, 2023. https://doi.org/10.3389/frai.2023.1269932
3. O.J. Erdélyi, J. Goldsmith: Regulating artificial intelligence proposal for a global solution. In: AIES 2018—Proceedings of the 2018 AAAI/ACM Conference on AI, Ethics, and Society, pp. 95–101 (2018). https://doi.org/10.1145/3278721.3278731
4. Thurzo, A. et al.: Where is the artificial intelligence applied in dentistry? Systematic review and literature analysis. Healthcare (Switzerland) **10**(7). https://doi.org/10.3390/HEALTHCARE10071269
5. Kleizen, B., Van Dooren, W., Verhoest, K., Tan, E.: Do citizens trust trustworthy artificial intelligence? Experimental evidence on the limits of ethical AI measures in government. Gov. Inf. Q. **40**(4) (2023). https://doi.org/10.1016/j.giq.2023.101834
6. Abulibdeh, A., Zaidan, E., Abulibdeh, R.: Navigating the confluence of artificial intelligence and education for sustainable development in the era of industry 4.0: challenges, opportunities, and ethical dimensions. J. Clean. Prod. **437** (2024). https://doi.org/10.1016/j.jclepro.2023.140527
7. Minopoulos, G.M., Memos, V.A., Stergiou, K.D., Stergiou, C.L., Psannis, K.E.: A medical image visualization technique assisted with AI-based haptic feedback for robotic surgery and healthcare. Appl. Sci. (Switzerland) **13**(6) (2023). https://doi.org/10.3390/app13063592
8. Wang, Y.-H., Lin, G.-Y.: Exploring AI-healthcare innovation: natural language processing-based patents analysis for technology-driven roadmapping. Kybernetes **52**(4), 1173–1189 (2023). https://doi.org/10.1108/K-03-2021-0170
9. De Lamotte, M.: Enlightenment, artificial intelligence and society. IFAC-PapersOnLine **53**(2), 17427–17432 (2020). https://doi.org/10.1016/j.ifacol.2020.12.2110
10. Haenlein, M., Kaplan, A.: A brief history of artificial intelligence: on the past, present, and future of artificial intelligence. Calif. Manage. Rev. **61**(4), 5–14 (2019). https://doi.org/10.1177/0008125619864925
11. He, A., et al.: A survey of artificial intelligence for cognitive radios. IEEE Trans. Veh. Technol. **59**(4), 1578–1592 (2010). https://doi.org/10.1109/TVT.2010.2043968
12. Mazzone, M., Elgammal, A.: Art, creativity, and the potential of artificial intelligence. Arts **8**(1), 26 (2019). https://doi.org/10.3390/arts8010026
13. Carrillo-Perez, F., et al.: Applications of artificial intelligence in dentistry: a comprehensive review. J. Esthet. Restor. Dent. **34**(1), 259–280 (2022). https://doi.org/10.1111/JERD.12844
14. Rodríguez-Hernández, C.F., Musso, M., Kyndt, E., Cascallar, E.: Artificial neural networks in academic performance prediction: systematic implementation and predictor evaluation. Comput. Educ. Artif. Intell. **2** (2021). https://doi.org/10.1016/j.caeai.2021.100018
15. White, G.R.T., et al.: Mapping the ethic-theoretical foundations of artificial intelligence research. Thunderbird Int. Bus. Rev. **66**(2), 171–183 (2024). https://doi.org/10.1002/TIE.22368
16. Brundage, M. et al.: The Malicious Use of Artificial Intelligence: Forecasting, Prevention, and Mitigation (2018). Accessed: 18 Apr. 2024 [Online]. Available: http://arxiv.org/abs/1802.07228

17. Nemitz, P.: Constitutional democracy and technology in the age of artificial intelligence. Philos. Trans. R. Soc. A: Math. Phys. Eng. Sci. **376**, 2133 (2018). https://doi.org/10.1098/RSTA.2018.0089
18. Belli, L., Zingales, N.: Data protection and artificial intelligence inequalities and regulations in Latin America. Comput. Law Secur. Rev. **47** (2022). https://doi.org/10.1016/j.clsr.2022.105761
19. Supovitz, J., Sirinides, P., May, H.: How principals and peers influence teaching and learning. Educ. Adm. Q. **46**(1), 31–56 (2010). https://doi.org/10.1177/1094670509353043
20. Brunetti, N.D., et al.: Telecardiology applied to a region-wide public emergency health-care service. J. Thromb. Thrombolysis **28**(1), 23–30 (2009). https://doi.org/10.1007/S11239-008-0241-Y
21. Duggal, R., Brindle, I., Bagenal, J.: Digital healthcare: regulating the revolution. BMJ (Online) **360** (2018). https://doi.org/10.1136/BMJ.K6
22. Iezzoni, L.I., Yu, J., Wint, A.J., Smeltzer, S.C., Ecker, J.L.: General health, health conditions, and current pregnancy among U.S. women with and without chronic physical disabilities. Disabil. Health J. **7**(2), 181–188 (2014). https://doi.org/10.1016/j.dhjo.2013.12.002
23. Grawitch, M.J., Trares, S., Kohler, J.M.: Healthy workplace practices and employee outcomes. Int. J. Stress. Manag. **14**(3), 275–293 (2007). https://doi.org/10.1037/1072-5245.14.3.275
24. Hackney, K.J., Daniels, S.R., Paustian-Underdahl, S.C., Perrewé, P.L., Mandeville, A., Eaton, A.A.: Examining the effects of perceived pregnancy discrimination on mother and baby health. J. Appl. Psychol. **106**(5), 774–783 (2021). https://doi.org/10.1037/apl0000788
25. Appiah, R., et al.: The causal link between anthropogenic activities, water pollution and health-related quality of life from residents' perspective: a review. Art. Int. J. Sci. Res. Sci. Technol. (2023). https://doi.org/10.32628/IJSRST52310242
26. Kuo, M.: How might contact with nature promote human health? Promising mechanisms and a possible central pathway. Front Psychol. **6** (2015). https://doi.org/10.3389/FPSYG.2015.01093
27. Frumkin, H. et al.: Nature contact and human health: a research agenda. Environ. Health Perspect. **125**(7) (2017). https://doi.org/10.1289/EHP1663
28. Choong, E.K.M., Shu, X., Leung, K.C.M., Lo, E.C.M.: Oral health-related quality of life (OHRQoL) after rehabilitation with removable partial dentures (RPDs): a systematic review and meta-analysis. J. Dent. **127** (2022). https://doi.org/10.1016/j.jdent.2022.104351
29. Bratman, G.N., et al.: Nature and mental health: an ecosystem service perspective. Sci. Adv. **5**(7) (2019). https://doi.org/10.1126/SCIADV.AAX0903.
30. Stuijfzand, S., et al.: Psychological impact of an epidemic/pandemic on the mental health of healthcare professionals: a rapid review. BMC Public Health **20**(1), 1–18 (2020)
31. Pang, H., Liu, Y.: Untangling the effect of cognitive trust and perceived value on health-related information seeking, sharing and psychological well-being: motivations sought perspective. Telemat. Inform. **79** (2023). https://doi.org/10.1016/j.tele.2023.101964
32. Srimaneepong, V., et al.: Fixed prosthetic restorations and periodontal health: a narrative review. J. Funct. Biomater. **13**(no. 1) (2022). https://doi.org/10.3390/JFB13010015
33. Timmons, S., Mann, C., Evans, C., Pearce, R., Overton, C., Hinsliff-Smith, K.: The advanced clinical practitioner (ACP) in UK healthcare: dichotomies in a new 'multi-professional' profession. In: SSM—Qualitative Research in Health, Vol. 3 (2023). https://doi.org/10.1016/j.ssmqr.2022.100211
34. Franzoi, I.G., et al.: Anxiety, post-traumatic stress, and burnout in health professionals during the covid-19 pandemic: Comparing mental health professionals and other healthcare workers. Healthcare (Switzerland) **9**(6) (2021). https://doi.org/10.3390/HEALTHCARE9060635
35. Nelson, B.W., Pettitt, A., Flannery, J.E., Allen, N.B.: Rapid assessment of psychological and epidemiological correlates of COVID-19 concern, financial strain, and health-related behavior change in a large online sample. PLoS One **15**(11) (2020). https://doi.org/10.1371/JOURNAL.PONE.0241990
36. Gonçalves, G.S.Y., de Magalhães, K.M.F., Rocha, E.P., dos Santos, P.H., Assunção, W.G.: Oral health-related quality of life and satisfaction in edentulous patients rehabilitated with implant-supported full dentures all-on-four concept: a systematic review. Clin. Oral Investig. **26**(1), 83–94 (2022). https://doi.org/10.1007/S00784-021-04213-Y

37. Lupu, E., Cernat, C., Petre, C.: Identifying the attitude of healthy individuals towards disabled children—a chance to be educated for all. Procedia Soc. Behav. Sci. **29**, 266–271 (2011). https://doi.org/10.1016/j.sbspro.2011.11.237
38. Shatte, A.B.R., Hutchinson, D.M., Teague, S.J.: Machine learning in mental health: a scoping review of methods and applications. Psychol. Med. **49**(9), 1426–1448 (2019). https://doi.org/10.1017/S0033291719000151
39. Miracle, A., Adaobi, C.C.: Advancement in the Healthcare Field of Wearable Technology and Future Perspective (2023) [Online]. Available: www.mkscienceset.com
40. Yousaf Gill, A., Saeed, A., Rasool, S., Husnain, A., Khawar Hussain, H.: Revolutionizing healthcare: how machine learning is transforming patient diagnoses—a comprehensive review of AI's impact on medical diagnosis. J. World Sci. **2**(10), 1638–1652 (2023). https://doi.org/10.58344/jws.v2i10.449
41. Braun, S.S., Kaihoi, C.A., McDaniel, H.L., Bradshaw, C.P.: Profiles of teachers' occupational health: associations with classroom management practices, gender, and race. Teach. Teach. Educ. **118** (2022). https://doi.org/10.1016/j.tate.2022.103819
42. Cesareo, et al. M. (2022). The effectiveness of nudging interventions to promote healthy eating choices: a systematic review and an intervention among Italian university students. Appetite **168** (2022). https://doi.org/10.1016/J.APPET.2021.105662
43. Penn, L., Goffe, L., Haste, A., Moffatt, S.: Management information systems for community based interventions to improve health: qualitative study of stakeholder perspectives. BMC Public Health **19**(1) (2019). https://doi.org/10.1186/S12889-018-6363-Z
44. Todua, N., Jashi, C., Todua, N.: The role of social media in healthcare marketing. In: Modern Healthcare Marketing in the Digital Era, pp. 26–41 (2023). https://doi.org/10.4018/979-8-3693-0679-6.CH002
45. Lu, W.: Adolescent depression: national trends, risk factors, and healthcare disparities. Am. J. Health Behav. **43**(1), 181–194 (2019). https://doi.org/10.5993/AJHB.43.1.15
46. Ashtari, S., Eydgahi, A., Lee, H.: Exploring Cloud Computing Implementation Issues in Healthcare Industry, p. 9 (2015)
47. Haque, A., Milstein, A., Fei-Fei, L.: Illuminating the dark spaces of healthcare with ambient intelligence. Nature **585**(7824), 193–202 (2020). https://doi.org/10.1038/S41586-020-2669-Y
48. Stahl, B.C., Coeckelbergh, M.: Ethics of healthcare robotics: towards responsible research and innovation. Rob Auton Syst **86**, 152–161 (2016). https://doi.org/10.1016/J.ROBOT.2016.08.018
49. Attenborough, J., Abbott, S., Brook, J., Knight, R.A.: Everywhere and nowhere: work-based learning in healthcare education. Nurse Educ. Pract. **36**, 132–138 (2019). https://doi.org/10.1016/J.NEPR.2019.03.004
50. Husnain, A., Rasool, S., Saeed, A., Yousaf Gill, A., Khawar Hussain, H.: AI'S healing touch: examining machine learning's transformative effects on healthcare. J. World Sci. **2**(10), 1681–1695 (2023). https://doi.org/10.58344/jws.v2i10.448
51. Ahuja, S.P., Mani, S., Zambrano, J.: A survey of the state of cloud computing in healthcare. Netw. Commun. Technol. **1**(2), 12–19 (2012). https://doi.org/10.5539/nct.v1n2p12
52. Jones, K.P., Brady, J.M., Lindsey, A.P., Cortina, L.M., Major, C.K.: The interactive effects of coworker and supervisor support on prenatal stress and postpartum health: a time-lagged investigation. J. Bus. Psychol. **37**(3), 469–490 (2022). https://doi.org/10.1007/s10869-021-09756-1

Revolutionizing Healthcare with Generative Artificial Intelligence Technologies

Fatemeh Rashidieranjbar, Amirfarhad Farhadi, and Azadeh Zamanifar

Abstract Generative Artificial Intelligence (GenAI) is a technology that can generate various types of content using a prompt, including text, images, videos, and code. This technology is well-known by OpenAI ChatGPT4 and Google DeepMind Gemini. This chapter assesses GenAI models and applications for assistance in various healthcare fields, such as medical imaging, pandemic prediction, synthetic data generation, clinical administration support, professional education, and patient engagement. The ability to tailor GenAI models to unique medical problems promises rapid progress, yet privacy, security, and ethical concerns demand robust strategies and legal frameworks to ensure that GenAI models are used more widely. Despite these obstacles, GenAI can adapt to the medical industry's specific needs. It is crucial to study GenAI in healthcare as it can revolutionize the healthcare profession by becoming an integral member of medical teams.

Keywords Generative artificial intelligence · Healthcare · Clinical administration support · Patient engagement · Syntactic data generation

1 Introduction

An incredible amount of money is poured into the field of healthcare every year in the hopes of enhancing results. However, expectations are not always met by reality. Approximately thirty percent of the $4.3 trillion spent yearly in the United States is not fruitful [1]. The healthcare industry is beset by several difficulties that lead to failure in avoiding preventable fatalities and subpar outcomes. Lives are tragically lost every year due to a variety of factors, including medical team tiredness, inaccurate patient information registration, imprecise medical imaging, and unequal access to care [2, 3]. Artificial Intelligence (AI) is a technology that has dramatically influenced

F. Rashidieranjbar · A. Farhadi · A. Zamanifar (✉)
Department of Computer Engineering, Science and Research Branch, Islamic Azad University, Tehran, Iran
e-mail: azamanifar@srbiau.ac.ir

N. R. Vajjhala et al. (eds.), *Generative Artificial Intelligence (AI) Approaches for Industrial Applications*, Information Systems Engineering and Management 24,
https://doi.org/10.1007/978-3-031-76710-4_10

various fields. A recent study [4] informs that integrating modern AI systems into healthcare could save up to 10% of spending in this field. Even though AI has made great contributions to healthcare, some challenges remain that Generative Artificial Intelligence (GenAI) must address. GenAI is a specialized field within AI focusing on designing algorithms and models that can generate several content types, like text, audio, images, and synthetic data. Generative algorithms, using existing content, learn and adapt to generate new and unique content. GenAI works as a subdomain of deep learning (see Fig. 1), enabling the neural networks to process labeled and unlabeled data. It usually encompasses the use of supervised, unsupervised, and semi-supervised methods of learning to tune and enhance the performance of the generating algorithms [5].

GenAI has immensely influenced various fields, such as education [6], marketing [7], art and music [8]. One area where GenAI holds great promise is healthcare. AI technologies generally need a large amount of data to be trained on for various tasks [9]. However, these data cannot be obtained easily especially in the field of healthcare for reasons such as disease rarity, clinic regulations, and costly data gathering processes. In this case, GenAI utilizes the acquired data to generate new data. The generated data can then be augmented to the initial dataset to enhance the usage of deep learning and machine learning models. In addition, since the data generated does not expose any patient information and is "artificially created", patient privacy can be maintained effectively.

Furthermore, GenAI can benefit the patient, administration workers, medical expertise, and medical students through various approaches. The GenAI can specially assist individuals who knows how to customize its models according to the task in their hands an also are aware of its applications. This chapter aims to provide

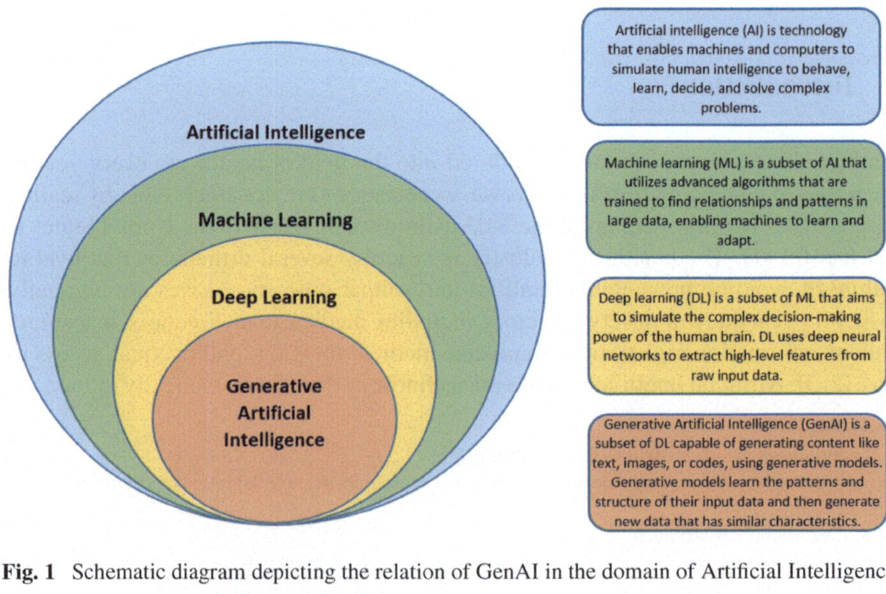

Fig. 1 Schematic diagram depicting the relation of GenAI in the domain of Artificial Intelligence

an overview of GenAI application in healthcare as well as illustrating a practical roadmap to customize and build GenAI model for specific medical tasks with practical example and case studies.

First, several GenAI applications, frameworks, and models on introduce in various filed of healthcare industry such as medical imaging, pandemic prediction, synthetic data generation, clinical administration support, professional education, and patient engagement. In the next section, a roadmap is briefly represented with a hypothetical scenario to deepen the understanding of how to build GenAI model for a specific medical task. Then two famous models, ChatGPT4 and Gemini are employed for them as well as a new scenario to evaluate their performance. Finally, ChatGPT4 and Gemini are assessed further by presenting real life case studies. Ultimately, we demonstrate the challenges that are on the way of GenAI revolutionizing the healthcare and whether its flourish will be successful despite the bumpy path ahead of it.

2 Generative Artificial Intelligence (GenAI) Applications in Healthcare

This section covers new and existing research projects where GenAI is being used in collaboration to advance medical imaging, prediction of pandemics, synthetic data creation, clinical administration, professional education, and patient engagement. The application of GenAI tools and platforms is targeted to help improve patient care, increase medical research, and reduce the administrative load on healthcare workers. By investigating the most recent breakthroughs in this field, better comprehension of the potential influence of cutting-edge GenAI models on the future of healthcare is provided, emphasizing the inclusive and collaborative nature of these groundbreaking efforts.

2.1 Generative Artificial Intelligence (GenAI) in Medical Imaging

Generative artificial intelligence in medical imaging has emerged as a potent way of revolutionizing medical imaging practices. Capable of synthesizing, translating, segmenting, and enhancing images, GenAI holds great promise for accelerating and enhancing many tasks within the field [10]. One of the most significant issues in medical imaging is having small datasets from only a few sources. This is primarily attributed to data acquisition barriers, such as the need to protect patients' privacy, the high cost of imaging procedures, and the prevalence of relatively rare diseases. This lack of diversity can lead to imbalanced datasets that do not represent the full range of conditions medical professionals may encounter [11].

GenAI offers a promising solution to these challenges by enabling researchers to create synthetic images that closely resemble authentic medical images [12]. Augmenting these datasets with generated images increases the size and diversity of the datasets, making a set of images more comprehensive for training and evaluation. It could assist with issues such as class imbalance and give machine learning algorithms a more representative sample to learn from in the medical image domain. For example, in [13], a medical image synthesis model called ANT-GAN is proposed to generate normal-looking images from abnormal-looking counterparts without needing paired training data. This model assists lesion segmentation and classification tasks by providing a "normal" counterpart to a medical image and generating realistic lesion-containing images for data augmentation. Experimental results on the BratS18 and LiTS challenge datasets validate the effectiveness of the ANT-GAN model in solving fundamental problems of medical image analysis.

One major utilization of GenAI in Medical imaging is image to image translation. Image-to-image translation is the act of converting an initial image into a final image while maintaining specific semantic characteristics [14]. This paper, [15] proposes a multi-domain medical image translation model, MI-GAN, based on the key transfer branch to address issues such as poor attention, insufficient key transfer, low-quality generated images, and lack of detailed features in existing translation models. The dataset of images of COVID-19 lungs is applied in training and testing the model to generate images of normal, viral pneumonia, and mild COVID-19. The obtained images can be much more real and diverse than those generated by other generative models and have a high accuracy rate in pneumonia diagnosis with enhanced sensitivity and specificity. The paper points out the importance of synthetic image data augmentation for lung image classification models based on generative adversarial networks to enhance the existing medical image dataset.

Another application of GenAI within the domain of image-to-image transformation is when synthetic CT images are generated from MRI data. The performance of GANs in generating CT images from MRI has been enhanced by deep learning; however, the outputs often need to be revised due to tissue distribution differences in current methods. The paper [16] proposes a new approach called MGDGAN, resulting in more accurate and realistic CT image generation by focusing on bone and soft tissue separately. This method, which reduces the need for expensive CT scans and lowers the risk of radiation exposure, offers potential for enhancing the caliber and effectiveness of MRI-only clinical imaging techniques.

Image enhancement is another way that generative models have improved medical imaging. In this realm, super-resolution and denoising techniques are two ways that GenAI models can enhance image quality and increase diagnostic accuracy. The accurate and effective identification of medical disorders made possible by generating high-quality images benefits both patients and healthcare professionals. For example, GAN-based models proved to be useful in clinical settings by improving image quality and helping radiologists identify lung nodules in low-dose CT (LDCT) scans, according to a study [17] that evaluated the efficacy of two denoising models using generative adversarial networks (GAN) on improving image quality and the ability to detect lung nodules in chest LDCT scans. Lowering radiation dosages while taking CT images can improve patient safety; however, this adversely influences CT image quality. A study [18] addresses this problem by proposing a GAN architecture that successfully improves CT image quality influenced by reduced radiation by minimizing noise while preserving structural details.

Medical image segmentation is essential in the healthcare field since it allows for accurately identifying and isolating interest regions within different types of medical images. Thus, image segmentation helps health professionals gain better insights into the anatomical data presented within medical images. Besides enhancing the accuracy of medical analysis, the segmentation process also helps evolve new medical procedures [19]. GenAI can significantly impact image segmentation, which may save lives. For example, cardiovascular diseases are among the top causes of death globally, which, among other reasons, are brought about by the formation of thrombus in blood vessels. The Left Atrial Appendage (LAA) is a crucial area for stopping the development of thrombus due to its high contractility; as such, the segmentation of LA is greatly needed. This paper [20] focuses on LA automatic segmentation in MRI images using the BLRSWSO and XSDGAN algorithms, with contrast enhancement using UHIBF. During experimentation, the LA was segmented well with the proposed approach compared to existing techniques.

Moreover, Redbrick AI has developed a potent tool, Fast Automated Segmentation Tool (F.A.S.T.) SAM, for segmenting visible objects in medical imaging, especially large organs, from CT and MRI scans. Physicians can see the mask computation live as they guide the model, making the segmentation process easier and more accurate. F.A.S.T. SAM automates much of the manual segmentation work, streamlining the classification of automatically generated segmentations for medical teams.

The IBM Watson [21] for Oncology model was only trained on oncology data and the desired results may not be easily obtained on other medical images. The purpose of this model is to help oncologists in choosing cancer treatment options. The positive points of this model include quick access to relevant medical information and evidence-based treatment recommendations. Meanwhile, these treatment recommendations lack widespread clinical validity. This model may not handle complex or rare cases and relies heavily on existing medical literature. The generative models can predict the further progress of diseases and the outcome by simulating the future condition of a patient based on the learned disease representations, thereby assisting clinicians in evaluating treatment choices. In this respect, for instance, a semi-supervised deep learning framework called Smile-GAN was developed to identify four distinct patterns of brain atrophy related to Alzheimer's disease using MRI data of more than 2800 participants [22]. These patterns range from mild to advanced atrophy, defining two progression pathways. The patterns bear clinical relevance by predicting cognitive performance, progression pathways, and progression to MCI and dementia. The patterns bear significant phenotypic information and could help in the targeted recruitment of clinical trial participants and offer insights into the disease process.

2.2 Generative Artificial Intelligence (GenAI) in Pandemic Prediction

Pandemic prediction in health is a basic principle for preparation, prevention, allocation of resources, and mitigation of effects during outbreaks. Precise predictions of pandemics will help health systems be adequately prepared against such eventualities, put in place measures to stop the spread of the disease and allocate their resources accordingly. Predicting pandemics will save lives, safeguard public health, and guarantee that health-care systems have the capacity to respond to and manage outbreaks. The COVID-19 pandemic has influenced global health significantly, and numerous approaches have been employed to study and predict its evolution. Traditional epidemiological models and machine learning models have been used in forecasting, but they have some limitations, either in generalization or data availability. In [23], T-SIRGAN has been introduced, which integrates epidemiological theories and deep learning models to predict the evolution of COVID-19. This method puts together the SIR model in simulation data generation with GAN in data augmentation and the Transformer in trend prediction. The work puts forward a T-SIRGAN approach that predicts the growth of COVID-19 spread and evaluates intervention effectiveness, such as vaccination. This work also [24] utilized CovidGAN to generate synthetic chest X-ray images to address the need for more training data. By adding these synthetic images to training set of CNN models in the study, the accuracy of COVID-19 detection improves from 85 to 95%. The proposed method enhances the efficiency of COVID-19 detection and strengthens radiology systems.

2.3 Generative Artificial Intelligence (GenAI) in Synthetic Data Generation

Synthetic data generation has revolutionized healthcare by using GenAI models that generate realistic synthetic patient records and images, maintaining patient privacy. These synthetic datasets have the same distributional properties as real healthcare data and give healthcare providers transparent access to information to analyze for research, education, and decision-making without disclosure. For instance, Syntegra Medical Mind employs transformer-based language models to understand the statistical distribution of various kinds of structured healthcare data (such as EHR, claims, and genomics), presented as a sequence of medical events over time. The model is trained to use this understanding to create new patient records that follow the same format and retain the statistical characteristics of the actual data, including rare cohorts and outliers. Additionally, Syntegra tackles data bias and encourages fair algorithms, aiding in creating unbiased and equitable treatment plans.

A model known as Correlation-Capturing Generative Adversarial Network (CorGAN) is introduced in [25]. Its purpose is to create realistic synthetic healthcare records with privacy protection. CorGAN incorporates Convolutional GANs and Convolutional Autoencoders (CAs) and can produce synthetic data that mirrors real data performance in classification tasks, as demonstrated by analysis and evaluations. Additionally, CorGAN preserves privacy by adjusting the ratio of generated synthetic data and available data to potential adversaries. On the other hand, a recent paper [26] has focused on investigating the applications of DALL-E 2 in the field of radiology. DALL-E 2 is one of the most significant OpenAI models for the task of text-to-image creation. It was built by training billions of text-image pairs, able to synthesize realistic synthetic images. Although, as the paper suggests, it may struggle with complex images such as CT, MRI, and ultrasound images, it can synthesize X-ray images almost indistinguishable from real ones. This demonstrates the possibility of fine-tuning DALL-E 2 with medical data and terminology to generate a specialized model for generating and augmenting radiological data.

Remote monitoring of medical devices, part of the "Internet of Medical Things" (IoMT), has made accessing patient information easier for healthcare providers. Despite potential challenges like connectivity issues that adversely affect Machine Learning models, GenAI can be employed to address this problem by generating synthetic datasets that can be augmented to the real dataset. In this work [27], generative adversarial networks (GANs) were utilized to generate synthesis data resembling the data collected from IoMT sensors for monitoring Chronic Obstructive Pulmonary Disease (COPD). The LLM algorithm was also implemented to validate the accuracy of the 1000 synthetic data by comparing it to actual data gathered from the sensor, demonstrating the effectiveness of using GAN to generate synthetic datasets in this case.

This paper [28] proposes a novel generative adversarial networks (GAN) model, named SynSigGAN, for automating the generation of four kinds of biomedical signals, including electrocardiogram (ECG), electroencephalogram (EEG), electromyography (EMG), photoplethysmography (PPG) with a high correlation coefficient. SynSigGAN can be applied on a small size of the original signal dataset to create new biomedical synthetic signals that align with the characteristics of the actual data, using bidirectional grid long short-term memory for the generator network and convolutional neural network for the discriminator network of the GAN model. The evaluation metrics used, MAE, RMSE, PRD, and FD scores, present that this model outperforms existing models in this field, promising a brighter future in generating accurate, realistic biomedical signals. The Unlearn AI platform leverages GenAI to build digital surrogates for the patients, popularly known as "Digital twins". They digitally model the clinical records of patients as their digital twin, simulating minute representations of health outcomes under different scenarios. All this enables the development of specialized treatment plans for patients by health providers.

2.4 Generative Artificial Intelligence (GenAI) in Clinical Administration Support

GenAI models in healthcare will revolutionize clinical administration and change the face of healthcare interaction in a manner that promises patient care improvement. Automation of clinical documentation makes the doctor devote less time to tasks like notetaking and focus on care progress. ChatGPT supports the rapid clinical note creation based on voice requests or patient information. Google Bard, on the other hand, advances healthcare interaction by offering 24/7 patient support; it assists physicians in answering patient queries, suggesting diagnoses, and supporting treatment plans. ChatGPT and Google Bard will also go a long way in improving patient understanding and engagement. They can increase the readability and reduce the complexity of language in clinical documentation, making it more accessible to patients and their families. This transforms specialized medical language into patient-friendly terms, reinforcing the treatment process and the communication between medical professionals and patients [29].

Another crucial part of clinic administration support is diagnosing patients' emergency conditions. In this case, results of [30] indicate that Google Bard is more promising than ChatGPT and Microsoft Bing Chat as it detected the truest emergency cases (87%) and true non-emergency cases (36%). Although all the GenAI tools require additional refining to identify emergencies reliably, these tools have effectively lowered ED pressure and improved emergency management. Microsoft Bing Chat, for example, can swiftly find relevant medical information, assisting in diagnosing and treating emergency situations. Several user-friendly GenAI platforms and tools have been created, providing their clients with many options to facilitate the clinic's administrative work. Although these tools use other GenAI tools that have been introduced before or are not as famous as them due to their focus on the health field, we will briefly introduce them in the following.

Corti, for instance, utilizes GenAI to deliver real-time transcription, guidance, and coding capabilities across a variety of channels of communication in healthcare. It automatically transcribes patient conversations and pulls out vital details like symptoms and medications, thus giving the healthcare practitioner so much more time to focus on care. Another example is Suki Assistant, an GenAI solution listening to doctor-patient interactions to automate clinical note creation, relieve administrative burdens, and offer diagnostic code suggestions. Ellen AI provides a tool for healthcare practitioners to improve patient communication and access. It is a means of converting written instructions into high-quality, spoken content so that patients with visual impairments have an easier time understanding their healthcare instructions.

Nuance, on the other hand, represents many variations targeted to a particular task. For example, Dragon Medical One offers high-precision speech recognition with features like automatic accent detection and voice calibration. It performs all functions, from dictation to note formatting and navigation, with built-in voice control and automatic punctuation. When paired with the PowerMic Mobile, clinicians can use their smartphones as secure wireless microphones. The AutoText feature automates also regular content entry. In addition, Naunce's other platform, DAX Copilot, captures multi-party conversations peripherally and automatically converts encounter conversations into specific clinical document summaries. The program has been fed millions of encounters to produce high-quality documentation proficiently. It scales to function in many medical environments across different healthcare organizations.

Regard acts as an assistant for healthcare providers, assisting them in the effective and efficient utilization of electronic medical records (EMRs). It simplifies the process by automating some of the administrative tasks related to EMRs, such as generating clinical notes or patient information from voice input. It presents a user-friendly interface with relevant EMR data, so healthcare professionals will not need to sift through many screens. It was found that Regard decreased physicians' documentation time by 20%.

Heidi is an online platform designed to improve productivity in the clinical system. By making use of GenAI, the tool captures important patient information after each medical visit and then generates a bullet point summarization. Heidi is able to create medical progress notes in a variety of formats, including but not limited to SOAP, CHEDDAR, or specialty-specific templates, and draft letters to the patient based on the summarized bullets. Heidi integrates with Epic, Cerner, and Athena, enabling prepared progress notes to be transferred into individual EHR systems. By inputting medical sessions' records into this multi-language platform, the AI model adapts to the user's approach with each interaction, addition, and modification, providing a personalized model.

The discussed platforms offer invaluable services in converting spoken content into written instructions and vice versa. At the same time, ChatGPT has huge potential in data analysis, decision support, and understanding and adherence to treatment plans. Integrating such tools with ChatGPT brings interesting possibilities to enhance patient and healthcare efficiency through novel voice-based interactions and smart text generation.

2.5 Generative Artificial Intelligence (GenAI) in Clinical Decision Support

Diagnosing and providing a treatment plan in health care is extremely important since it dramatically impacts the patient's health and life. With the increase in the number of patients, especially in special situations such as the sudden outbreak of COVID-19, doctors' fatigue can decrease concentration and cause errors in diagnosis. GenAI has made health decision-making easier in health care to deal with such predicaments by rapidly analyzing huge patient datasets to provide personalized treatment recommendations. Leveraging this power, health professionals can now make decisions more informedly with data-driven insights and reshape the way they treat patients. Several of these GenAI tools that have positively influenced medical decisions and diagnoses are noted below.

Glass AI is a niche tool designed for medical professionals; it helps doctors generate possible diagnoses and treatment plans. For instance, in the field of hematology, a hematologist presented a patient who complains of fatigue, shortness of breath, and pallor can input those findings into Glass AI. This system will then give a differential diagnosis that could be anemia, leukemia, or myelodysplastic syndrome. This technology could also help formulate a clinical plan and help the doctor decide what to do next regarding further tests or even treatments. In that way, Glass AI will be able to accelerate the diagnosis of diseases and expand the spectrum of diseases to consider beyond what could have been thought of if this technology had not been invented.

Another GenAI tool, the previously demonstrated Unlearn AI digital twins of patients, can be combined with real-world data to mimic the effects of various treatments on the patient's specific biology. Healthcare providers can tailor treatment programs, track patient progress, and make informed decisions to enhance patient outcomes. In hematology, for instance, a doctor can utilize the digital twin to simulate the potential disease progression of a patient affected by chronic lymphocytic leukemia under different kinds of treatment strategies—for example, chemotherapy or targeted therapy—before making a better-informed decision regarding the best treatment for this patient.

Likewise, Regard can provide a list of differential diagnoses that are feasible but less common based on the patient's data. The doctor can then sift through the alternatives to arrive at a diagnosis, either ruling it out or confirming it is in the best interests of good therapy. For instance, a general physician may have a patient with nonspecific symptoms of fatigue, weight loss, and night sweats. Regard will suggest possible diagnoses such as diabetes, thyroid disorder, tuberculosis, or lymphoma from the patient data. On the other hand, a hematologist-oncologist presented with a patient showing abnormal blood counts and lymphadenopathy may use Regard to consider possible diagnoses such as leukemia, lymphoma, myeloma, or infection in that patient. A dermatologist may also examine a patient for a skin lesion and can consider Regard for exploring possible diagnoses such as melanoma, basal cell carcinoma, squamous cell carcinoma, or seborrheic keratosis with the patient's data. Regard is of great contribution in clinical setups, as it helps increase the specificity of diagnosis since there was a 3.2% increase in the number of cases for which the diagnosis of congestive heart failure was pinpointed accurately.

RedBrick AI's annotation platform, previously noted, not only simplifies the segmentation process but is also suitable for medical professionals to utilize in clinical practice to optimize disease detection. For instance, in analyzing brain CT images that are crucial for tumor diagnosis, F.A.S.T. can quickly segment the tumor from the surrounding brain tissue, outperforming the traditional manual annotation techniques. This could also be extended to MRI scans of the spine to segment the discs for diagnoses of conditions such as Kyphosis, Degenerative disc disease, and Spondylolisthesis.

Furthermore, FullFocus can be utilized to analyze and detect numerous complex or subtle tissue patterns, broadening the profession's understanding and grasp of a wide range of challenging hematological and oncological illnesses and, as a result, improving diagnostic skills. For example, FullFocus can detect and quantify cancer cells in the colon's inner lining, as well as identify biomarkers that can assist doctors in determining the best treatment for their patients.

2.6 Generative Artificial Intelligence (GenAI) in Professional Education

Advancements in technology mean that the demand and population are increasing due to the possibility of better welfare services. This reiterates the role of medical systems in training professionals who can cover the growing medical needs of the population. GenAI tools, with their efficiency-boosting capabilities, can help medical professionals and interns enhance the speed and accuracy of their responses. GlassAI, mentioned earlier, is more than just a diagnostic aide. It is an environment that enables doctors to learn, organize, and monitor medical knowledge, hence promoting personal progress. It is an interactive learning platform that allows professionals to share and exchange their medical expertise. With such a wide community library, physicians can improve not only clinical skills but also gain and then maintain their medical knowledge throughout time, further progressing in their professional lives. One of the distinguishing features of GlassAI is the unique mechanism of clinical notetaking. This tool helps doctors to write and link their learning for the better understanding and retention of medical knowledge. The schema, scripts, and cases can be shared among clinicians for their peers and trainees to promote collaborative learning for better patient outcomes.

Med-PaLM [31] focuses on understanding medical questions, generating answers, and summarizing medical information. The strengths of this trained model include high accuracy in medical examinations, comprehensive medical knowledge, and the ability to produce reports and analyze medical images. Med-PaLM 2, an expanded version of Med-PaLM, was able to achieve a remarkable accuracy rate of 86.5% on United States Medical Licensing Examination (USMLE) questions, on par with "expert" test takers. The minimum score to pass this test is 60% and this trained model was able to win first place with a significant difference compared to GPT Neo, PubMedBert, Dragon, Galacitica and PubMed GPT [32]. Despite the high potential of this model, it should be noted that the need for careful monitoring for safe and ethical use and the potential to generate false information are some of the weaknesses of this model.

Although mainly trained in biomedical literature, BioGPT [33] has several challenges like ChatGPT, such as the generation of false information and the possibility of reinforcing biases in the training data. This pre-trained model focuses on the generation and analysis of biomedical texts for research and medical tasks, including answering biomedical questions, literature review, drug discovery, and protein modeling, and can greatly help speed up medical education for students. BioGPT-Large, a model of this family, achieved an accuracy of 81% on the PubMedQA dataset, which is higher than the accuracy typically achieved by human annotators and even well-known models such as Google's BERT. DeepHealth LLM greatly assists physicians in answering clinical questions, analyzing medical images, and supporting personalized medical plans. This model can provide faster access to medical information, improve image analysis and help plan treatment. However, it should be noted that this model requires a lot of training data, is prone to data bias and difficult to interpret.

Furthermore, the digital twins of Unlearn AI support physician education by bringing improvement in clinical research and decision-making of healthcare professionals and interns. This is applicable in clinical trials, whereby researchers can use digital twins to determine the possible effects of new treatments on several diseases without having to put up a large control group.

2.7 Generative Artificial Intelligence (GenAI) in Patient Engagement

GenAI is a key player in patient engagement, efficiently delivering personalized and interactive experiences for patients. It enables the development of chatbots and tools capable of answering patient questions, rendering real-time assistance, and support for chronic condition management. More than this, GenAI can be applied to send relevant health education content to patients based on their specific needs and preferences. This, in turn, enables structured education of the patients about conditions, options available for treatment, and lifestyle changes required while also giving much-needed assurance to stakeholders regarding the reliability of the information being delivered. Furthermore, according to [34], responses from the LLM-powered chatbot were preferred over physician responses and rated much higher for empathy. Two tools that contribute to patient engagement are illustrated below.

Clinicians can employ Midjourney Labs' AI-powered imaging generating platform to quickly generate high-quality, personalized visual aids for patient education and engagement. For example, in a hematology setting, a clinician could insert a textual description of a specific blood illness, such as sickle cell anemia, into the Midjourney prompt and obtain a generated image or series of images depicting how the disorder impacts red blood cells. As a result, the patient's understanding of their condition improves, and they become more engaged in the subject.

One of the most apparent issues patients have during their treatment process is forgetting the details discussed during their medical appointments. Abridge delivers patients a detailed text of the conversations through an application, which they can review later if desired. Abridge takes a step further by emphasizing crucial information from the conversation and simplifying complex medical terms to ensure patients thoroughly understand their diagnosis, treatment options, and follow-up plans. Eventually, this promotes patient adherence and improves health outcomes.

3 Building GenAI Models for Medical Applications

In the preceding section, a variety of projects were described in which GenAI was utilized as a tool to aid with healthcare applications. In this part, we will first demonstrate a systematic approach to developing a GenAI model for certain tasks. Then we examine a hypothetical scenario to improve the comprehension of the systematic approach. Following that, we discuss the outcome and how ChatGPT and Gemini will provide outputs within the mentioned scenario. Finally, we provide case studies to evaluate ChatGPT and Gemini in healthcare.

3.1 Introducing a Systematic Roadmap

Irrespective of the specific approach and the investigated domain, a systematic approach is imperative in achieving the desired GenAI model. These common steps to build GenAI models (see Fig. 2), problem definition, data collection and preprocessing, model selection, model training, model evaluation, model fine-tuning, deployment, and monitoring and maintenance, provide a clear pivotal roadmap [35].

Problem Definition: At the beginning, one must describe the problem that the GenAI model will handle in detail. This involves specifying the problem scope, desired output, required data, and limitations or restrictions. Knowing the problem well is fundamental as it will ensure the successful deployment of GenAI and permit effective data collection and model choice in later phases.

Data Collection and Preprocessing: During this phase, a large dataset should be collected, which represents the patterns and features that will be learned by the GenAI model. Collecting data is facilitated by a wide variety of devices or tools, including web scrapers [36–38], microphones, cameras [39], and sensors [23]. In the field of healthcare, proper medical equipment and tools for diagnosis purposes must be used accordingly, depending on the problem under investigation. For example, imaging, X-ray, MRI (Magnetic Resonance Imaging), CT (Computed Tomography), and PET (Positron Emission Tomography) [40–42] are among the devices utilized. In cases of medical data, EHR (Electronic Health Records) [43], ECG (Electrocardiogram recordings) [44], blood pressure readings, heart rate, body temperature, blood glucose, cholesterol levels, and Body Mass Index can be referenced. It is essential to emphasize the significance of ensuring the diversity and comprehensiveness of the data to identify underlying patterns and achieve the desired outcome effectively.

Model Selection: At this phase, the most suitable architecture or pre-trained model for the GenAI model should be chosen based on the problem being studied. A GenAI model can be developed from scratch by using various architectures according to the

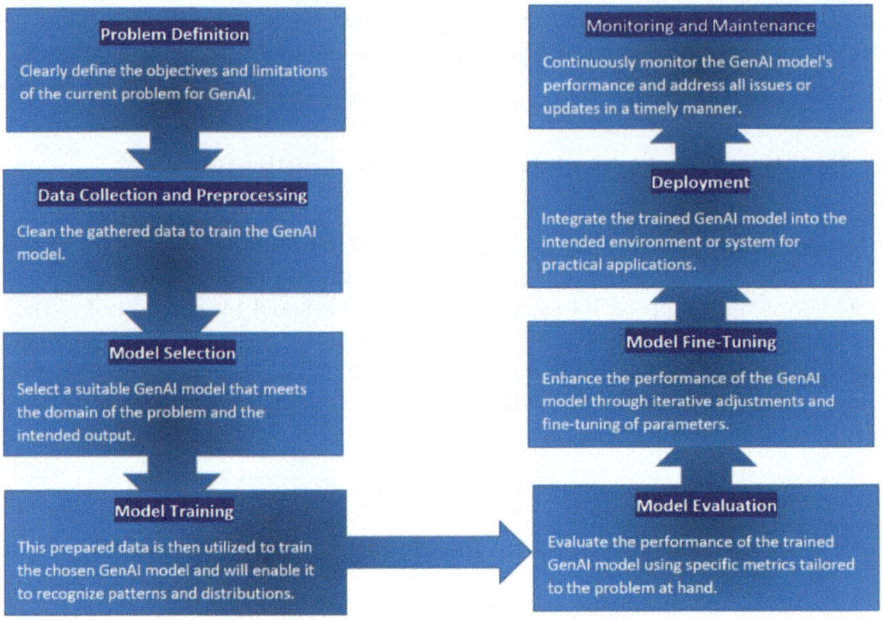

Fig. 2 Phases of building a GenAI model

requirements of the task at hand, including Variational Autoencoders (VAE), Generative Adversarial Networks (GANs), Transformers, and Diffusion models. However, this approach requires high expertise in coding skills and algorithm design. Another way is to leverage the pre-trained models to develop a customized GenAI model. Although many pre-trained models are reported in academic research, many face the issue of model file accessibility. In this case, the user needs to implement the model from the beginning, following the configuration described in the literature. On the other hand, some models provide model files and can be fine-tuned for a particular task without having to build the model from scratch.

There are two main ways to use a pre-trained GenAI model: web interfaces and programming libraries. Web interfaces facilitate user-friendly interaction with pre-trained generative models, whereby no installations or configurations of complex software environments are required from a user. It should, however, be noted that these web interfaces are not for developing new generative models. Additionally, even if some web interfaces allow limited possibilities for tuning the hyper-parameters of a generative model, fine-tuning is not normally supported. Despite requiring low technical knowledge, web interfaces make a great contribution to acquiring desired generated images, texts, or outputs by providing suitable prompts. In contrast, libraries such as TensorFlow and PyTorch provide access to a collection of pre-trained models and further allow customization or fine-tuning for particular tasks. While providing much more control and flexibility in modifying pre-trained generative models, utilizing programming libraries demands more technical skills and knowledge.

Model Training: In this phase, the selected model is trained on the collected data to learn the statistical relationships and underlying patterns within the dataset. Fine-tuning hyper-parameters, including the learning rate, batch size, network structure, and regularization methods, plays a key role in optimization of the GenAI model's behavior, convergence, and performance [45]. Considering the complexity of generative models and the large-scale datasets used in training, hardware such as graphics processing units (GPUs) [46] or tensor processing units (TPUs) [47], like Tesla V100 [48] or RTX 3090 [49], is typically used.

Model Evaluation: After training a model, it is critical to evaluate and validate it. Choosing the evaluation metrics depends on the application. For instance, visual inspection, Frechet inception distance (FID) [50], or inception score can be employed in image generation tasks to evaluate the quality and diversity of generated samples. On the other hand, evaluation metrics such as BLEU (Bilingual Evaluation Understudy) [51, 52] and ROUGE (Recall-Oriented Understudy for Gisting Evaluation) [53, 54] are commonly used in text or summary generation.

Model Fine-Tuning: In certain cases, it is necessary to adjust and optimize the hyper-parameters and weights of the pre-trained model to better fit a specific task or data set, comply with the specific limitations of the produced outputs, or increase the quality. Techniques such as smoothing the image or correcting the text can be applied based on the specific requirements of the domain.

Deployment: If the evaluation metrics show that the model's training was successful and that it can provide the expected outputs, the resulting generative model is ready to be deployed to generate new samples. In this phase, one can give an input to the model and expect an output that aligns with the data distribution. Now, the resulting GenAI model can generate multiple and diverse samples.

Monitoring and Maintenance: To effectively monitor and maintain the GenAI model, important tasks entail observing how effective, precise, and reliable the model performs, checking data quality for training and evaluation, as well as updating with new data while getting feedback from end users and those concerned. The continuity of the process requires implementing measures for monitoring these models alongside involvement with specific persons like data analysts and individuals utilizing these applications.

3.2 Bringing the Roadmap into Practice

For enhanced clarity regarding this roadmap, a hypothetical example is illustrated. In research [55], a computer vision approach was utilized to recommend a treatment plan for diseased teeth on panoramic radiography (OPGs). One challenge highlighted in this study is the low number of OPG images available for implant treatment, which causes an imbalance of classes and results in suboptimal outcomes concerning the formulation of treatment plans that include implant treatment. This example aims to generate OPG images that would encompass treatment plans involving implant treatment. The following lines will present how the roadmap is used to address the challenge.

Problem Definition: In this regard, the main purpose of developing an GenAI model is to generate artificial OPGs that will depict a series of dental treatments focusing on implant treatment as well as assisting the dental practitioner with treatment planning to get a clear view of what the outcome might look like.

Data Collection and Preprocessing: Initially, it is critical to collect a comprehensive and diversified OPG image dataset that depicts a wide range of treatment scenarios, particularly those involving implants, from reliable sources such as dental clinics. Afterward, the images must be labeled in accordance with a dentist's diagnostic and treatment plan. Using pre-trained model-based platforms and applications can be beneficial in this annotation process. For example, as previously noted, using Redbrick can speed up and enhance the accuracy of image annotations. This application's demo version is accessible on request through its website.

Prior to being fed into the generative model, data must go through a series of preparation steps. This involves normalizing the pixel size, resolution, and intensity values. Roboflow is a recommended tool for such tasks. OpenCV can also be used to normalize pixel dimensions and intensities; however, just like Roboflow it has limits in resolution control, particularly while enhancing it. In these cases, generative models designed for resolution enhancement, as previously mentioned, should be evaluated for optimal results.

Some modifications, including rotation, cropping, blurring, and contrast adjustment, must be performed to augment the data and reduce the possibility of overfitting. Aside from that, segmenting relevant zones in OPG images to increase emphasis on teeth and implantation sites is a significant strategy to strengthen the learning process. These segmentations can be easily formed utilizing generative or other models like the U-Net.

Model Selection: In this step, we choose the most suitable model depending on the unique needs of the problem at hand. As an illustration, if we employ a GAN architecture to generate synthetic images that mimic actual ones, the Generator will focus on making OPG images, while the Discriminator will distinguish between real and generated OPG images. GAN and VAEs may be implemented using frameworks such as TensorFlow or PyTorch. Figure 3 indicates an instance of GAN implementation.

As you can see, building a model from scratch using generative models' architectures may need technical skills and is time-consuming. Alternatively, rather than building a generative model from scratch, we might use pre-trained models. These models have previously been trained on a variety of datasets, improving the chance of generating better outcomes than untrained models. However, be alert that this may increase the possibility of overfitting. It should be noted that if the data distributions in the model's target domain differ from those in the source domain, domain adaptation is employed to improve the model's performance. The NF-ML technique developed in [56] manages complicated domain changes by combining meta-learning and a neuro-fuzzy system, which significantly improves the process.

For example, we may utilize the pre-trained StyleGAN2 model for imagine generation. Note that this model was not trained especially on medical images, but it is an

```
!pip install tensorflow
import tensorflow as tf
from tensorflow.keras import layers

# Generator model
def build_generator(latent_dim):
    model = tf.keras.Sequential([
        layers.Dense(256, activation='relu', input_dim=latent_dim),
        layers.BatchNormalization(),
        layers.Dense(512, activation='relu'),
        layers.BatchNormalization(),
        layers.Dense(1024, activation='relu'),
        layers.BatchNormalization(),
        layers.Dense(128 * 128 * 3, activation='tanh'),
        layers.Reshape((128, 128, 3))
    ])
    return model

# Discriminator model
def build_discriminator(image_shape):
    model = tf.keras.Sequential([
        layers.Flatten(input_shape=image_shape),
        layers.Dense(512, activation='relu'),
        layers.Dense(256, activation='relu'),
        layers.Dense(1, activation='sigmoid')
    ])
    return model

# GAN model combining generator and discriminator
def build_gan(generator, discriminator):
    discriminator.trainable = False
    model = tf.keras.Sequential([generator, discriminator])
    return model

    # Build models
    generator = build_generator(latent_dim)
    discriminator = build_discriminator(img_shape)
    gan = build_gan(generator, discriminator)
```

Fig. 3 Example pseudo-code for a GAN setup

accessible pre-trained model and can be loaded to use. The following Fig. 4 represents an approach on how we can load the model to use.

The StyleGAN2 repository can also be cloned from GitHub using the code (Fig. 5) below in window's terminal:

```
import torch
from torchvision import transforms
from stylegan2 import Generator

# Load pre-trained StyleGAN2 model
generator = Generator(style_dim=512, num_fp16_res=4, channel_multiplier=1, n_mlp=8)
generator.load_state_dict(torch.load('pretrained_stylegan2.pth'))
```

Fig. 4 Load pre-trained StyleGAN2 Model

```
git clone https://github.com/NVlabs/stylegan2.git
cd stylegan2
```

Fig. 5 Cloning StyleGan2 repository

Consider that seeking generative models trained on datasets with purposes like the task's goal at hand may yield better results.

Model Training: In this step, we train the model with the collected data. Since model training on massive data requires high processing hardware, we can use the Google Collab platform, which supports GPUs and TPUs. The input parameters of each model can be different according to the chosen model; for example, the training of the GAN model requires a training function with the following parameters and hyper-parameters: The generator, detector, and GAN were built in the previous step. These three hyper-parameters may be different based on different models. The dataset parameter is the address where the data is located. If Google Colab is used, the data should be uploaded to Google Drive. Epochs is the number of times the entire dataset is passed through the model during training. Batch size refers to the number of samples processed at each step during training. These two hyper-parameters plus the data address are needed to train any model. Latent-dim is the dimension of the random noise vector entered into the generator and directly influences the diversity of generated output.

Loss, optimizer, and metrics also should be specified before training the model at the time of compiling it. Loss is a measure of how well the model performs and is utilized to update model parameters during training. A suitable loss function, for GANs usually binary cross entropy is selected for both generator and detector, must be chosen. An optimizer is an algorithm that adjusts model parameters to minimize losses. Metrics such as accuracy for the detector are employed to evaluate the performance of the model. In the training phase, it is necessary to make sure that the model is not overfitting. This is the case where the model memorizes data instead of learning.

Model Evaluation: Use measures like initial score (IS) and initial Freisht distance (FID) to evaluate the model's performance in terms of image quality. Conducting qualitative assessments, such as having dental experts analyze the generated OPG images and apply them in practical scenarios, can also assist validate and evaluate the results. Consider using cross-validation approaches to ensure your model's reliability and efficacy.

Model Fine-Tuning: After the evaluation, we need to continuously readjust the hyper-parameters until we figure the hyper-parameters of optimal model.

Deployment: For example, at this stage, the final GenAI may be incorporated into dental software, allowing dentists to input OPG images, request the generation of an OPG with implant treatment, and teeth in need of such treatment being detected. Hosting the GenAI model on cloud services to ensure its scalability and availability may be another objective. A user interface could possibly be created, allowing dental practitioners to interact with the final trained model.

Monitoring and Maintenance: Following the launch of the GenAI model, feedback should be collected on an ongoing basis to develop and enhance the model. This might include expanding the dataset, changing the model architecture, retraining the model, or upgrading training approaches in response to new research.

3.3 Deploying ChatGPT and Gemini

In the preceding example, we explored how programming libraries are used to create GenAI models or personalize pre-trained GenAI models. Next, we're going to examine several common models that are readily intractable utilizing a user-friendly web interface. One of these models is Gemini, and the other is Chat GPT4.

ChatGPT4 (created by OpenAI) and Gemini (designed by Google DeepMind) are powerful artificial intelligence large language models which apply the Transformer architecture to comprehend and generate human-like prose. ChatGPT4 can aid with clinical decision support, patient diagnosis assistance, individualized medical treatment plans, and medical research and literature evaluation. Gemini can be used for a variety of assignments, including patient data management, disease outbreak prediction, personalized medicine and treatment, medical image analysis, patient engagement and education, and clinician decision support systems. It should be mentioned that the usage of these pre-trained models is risky because they are universal and have not been particularly trained for medical applications. They may possibly generate inaccurate data in some cases. Furthermore, if we utilize these models for specific medical problems, the model could fail to generate an acceptable result due to a lack of training on relevant special data. In the following, we will examine the scenario from the previous example to clarify.

After signing in to Gemini, we entered the prompt and obtained the response shown in Fig. 6. It should be noted that multiple prompts were executed to get the closest output to the intended outcome. One of the images generated by the model is a chest X-ray with no relation to oral OPG. Furthermore, detection is poor in one image and wrong in the other. The created images indicate that the simulated images are mostly concerned with dental implant treatment for the front teeth. This problem complicates the creation of a comprehensive dataset. As previously said, be cautious

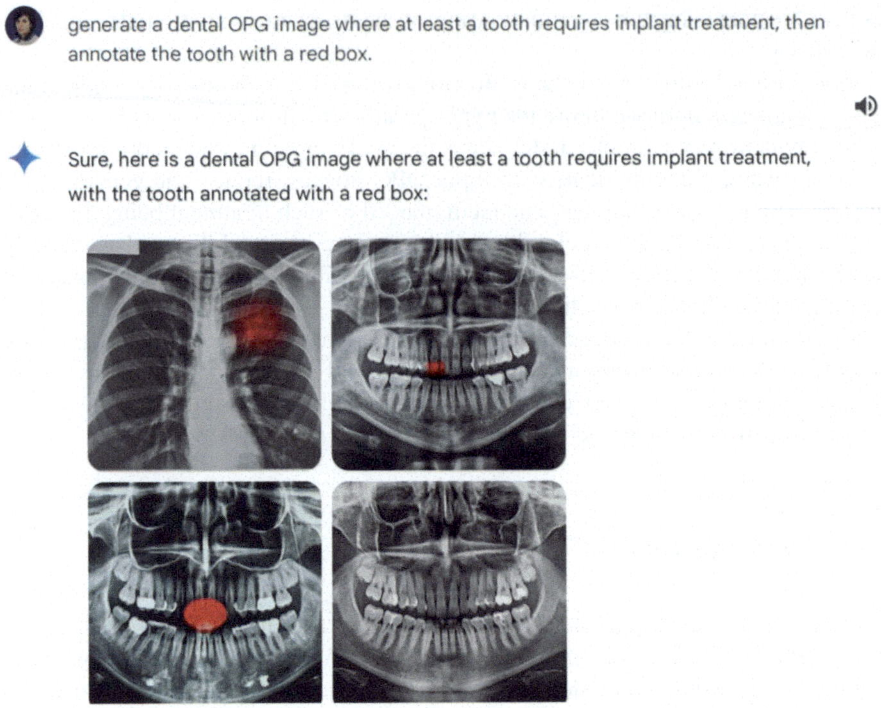

Fig. 6 Gemini generated image and response

when using this model for unique medical conditions. It should be noted that we did not utilize the Gemini advanced model and merely checked the free version.

Chat GPT4 does not directly generate images; instead, the DALL-E 2 and DALL-E 3 models must be utilized. To use OpenAI models, we first need to create an account. If you are an old-user of DALL-E 2, you may interact with the model via Web Inference; otherwise, you must run the following code in the Python environment (as pictured in Fig. 7).

The api_key parameter requires one's api_key, which is available on each individual's OpenAI account. The model parameter specifies which model, between DALL-E 2 and DALL-E 3, will be employed. In the prompt area, we type the desired prompt. In the prompt section, we input the desired prompt; in the size section, the desired size of the generated imagine; and in the quality section, the quality of the generated image, which can be standard or HD. The hyper-parameter n indicates how many images the model will generate. Keep in mind that DALL-E 3 requires n to equal 1.

To compare the two models, we asked both models to develop a treatment plan for a patient suffering from dengue illness.

Figure 8 presents Gemini's result while Fig. 9 depicts ChatGPT4's output. Chat GPT4 appears to have developed a more extensive treatment plan than Gemini did.

```
!pip install OpenAI

from openai import OpenAI

client = OpenAI(
  api_key= "OPENAI_API_KEY"
)

response = client.images.generate(
  model="dall-e-3",
  prompt="generate a dental OPG (orthopantomogram) image where the upper left canine tooth is severely decayed",
  size="1024x1024",
  quality="standard",
  n=1,
)

image_url = response.data[0].url
```

Fig. 7 Dall-E -3 model code to load in Google Colab

3.4 ChatGPT and Gemini Case Studies in Healthcare

The outcomes of generating a treatment plan for patients with dengue fever do not demonstrate that GPT is a superior model than Gemini, but they are largely dependent on the tasks at hand. Now, we'll investigate various case studies in which Gemini and ChatGPT are employed. Gemini and ChatGPT have been tested for many different tasks. For example, [57] evaluate how well the AI models GPT-4 and Gemini do, compared to emergency medicine professionals using the Emergency Severity Index (ESI) for patient triage in the emergency department. The study had 10 emergency medicine professionals, although their inexperience with the ESI model was acknowledged as one of the limitations.

The study was conducted in three phases: extracting 100 case examples, classifying replies to as correct triage, overtriage, or under-triage, and evaluating triage response accuracy. The results demonstrated substantial differences among the three groups. GPT-4 had the highest correct triage rate (70.60%), whereas Gemini had the highest rate of overtriage (35.2%). Emergency medicine experts had the greatest undertriage rate, at 32.90%. Gemini had the highest right triage rate related to ESI classes 1 and 2, at 87.77%, then GPT-4 follows by achieving 85.11%, and specialists at 49.33 percent.

The conclusion highlights that both GPT-4 and Gemini are very accurate in triaging critical and urgent patients, with GPT-4 outperforming Gemini overall. The results indicate that these AI models might successfully facilitate correct ESI triage in emergency situations. This study [58] looks on the efficacy of ChatGPT 4 and Gemini Advanced in delivering therapy recommendations for head and neck cancer. The study sought to test the value of these models in helping multidisciplinary assessments and decision-making by examining their responses to five hypothetical instances based on distinct anatomical sub-sites of head and neck cancer.

The findings indicated that both LLMs provided suitable treatment recommendations, with ChatGPT 4 outperforming Gemini Advanced in terms of adherence to guidelines and comprehensiveness of treatment planning. ChatGPT 4 had better AIPI

Treatment Plan for Dengue Fever

Disclaimer: This is a general treatment plan and should not replace professional medical advice. Always consult with a healthcare provider for any health concerns.

Goals of Treatment

- Reduce fever and pain
- Prevent dehydration
- Monitor for complications
- Prevent mosquito transmission

Treatment Plan

Supportive Care

- **Rest:** Encourage bed rest during the acute phase of illness.
- **Hydration:** Administer oral rehydration solutions (ORS) or intravenous fluids if necessary to prevent dehydration.
- **Pain Management:** Use acetaminophen (paracetamol) for fever and pain relief. Avoid aspirin, ibuprofen, and other nonsteroidal anti-inflammatory drugs (NSAIDs) as they can increase the risk of bleeding.
- **Diet:** Encourage consumption of easily digestible foods and fluids.

Monitoring

- **Vital Signs:** Monitor temperature, blood pressure, pulse, and respiratory rate regularly.
- **Platelet Count:** Closely monitor platelet count for signs of thrombocytopenia.
- **Fluid Balance:** Assess for signs of dehydration or fluid overload.
- **Organ Dysfunction:** Monitor liver and kidney function.
- **Warning Signs:** Educate the patient about warning signs of dengue hemorrhagic fever (DHF) or dengue shock syndrome (DSS), such as persistent vomiting, severe abdominal pain, rapid breathing, bleeding gums, nosebleeds, or blood in stool or urine.

Fig. 8 Gemini response

values, indicating improved overall performance; nonetheless, inconsistencies were identified in areas such as induction chemotherapy and surgical judgments relating to neck dissection. Finally, the study emphasizes that, while the inclusion of LLMs like as ChatGPT 4 and Gemini Advanced might improve healthcare decision-making,

Dengue Fever Treatment Plan

Patient Information:

- **Name**: [Patient Name]
- **Age**: [Patient Age]
- **Gender**: [Patient Gender]
- **Medical History**: [Relevant Medical History]

Diagnosis:

- Dengue Fever confirmed by clinical symptoms and laboratory tests (e.g., NS1 antigen, IgM/IgG antibodies, PCR).

Treatment Plan:

1. General Management:

- **Rest**: Advise the patient to take adequate rest.
- **Hydration**: Ensure the patient stays well-hydrated. Oral rehydration solution (ORS) is preferred. If the patient is unable to maintain oral intake, IV fluids may be necessary.
- **Fever Management**: Administer acetaminophen (paracetamol) for fever and pain relief. Avoid NSAIDs (e.g., ibuprofen, aspirin) due to the risk of bleeding.

2. Monitoring:

- **Vital Signs**: Monitor temperature, blood pressure, pulse, and respiratory rate regularly.
- **Hydration Status**: Monitor for signs of dehydration such as dry mucous membranes, decreased urine output, and skin turgor.
- **Laboratory Tests**: Regularly check complete blood count (CBC) to monitor hematocrit and platelet levels. Watch for signs of hemoconcentration and thrombocytopenia.

3. Warning Signs and Severe Dengue:

- Educate the patient and family about warning signs of severe dengue which include:
    - Severe abdominal pain
    - Persistent vomiting
    - Rapid breathing
    - Bleeding gums
    - Fatigue/restlessness
    - Blood in vomit or stool
    - Cold or clammy skin
    - Difficulty in breathing
- **Immediate Action**: If any warning signs develop, seek immediate medical attention.

4. Management of Severe Dengue:

- **Hospitalization**: Admit the patient to the hospital if any severe symptoms or complications arise.
- **Fluid Replacement**: Administer intravenous fluids (crystalloids or colloids) to manage shock and maintain adequate blood pressure.
- **Blood Transfusion**: Consider platelet transfusions or blood transfusions if there is significant bleeding or very low platelet counts.

Fig. 9 ChatGPT4 response

their usage must be approached with caution, ensuring that they support rather than replace traditional clinical knowledge.

This [59] examines the feasibility, inter-rater reliability, and accuracy of employing ChatGPT-4 and Gemini for quality assurance in emergency medical services (EMS), attempting to automate the assessment of patient care reports, thereby reducing medical directors' effort and improving patient outcomes. Two human reviewers as well as /gPT4 and Gemini evaluated 150 anonymized prehospital records from two metropolitan EMS organizations for compliance with cardiac care measures. The Kappa statistic was used to determine inter-rater reliability. Human reviewers indicated good dependability, with 91.2% agreement and a kappa score of 0.782. In contrast, ChatGPT-4 demonstrated significant agreement with humans for various criteria, including EKG recording and aspirin administration (76.2% agreement, kappa 0.401), while Gemini was abounded due its poor performance. The findings show that, while LLMs can help with quality assurance efforts by retrieving data, their ability to evaluate complicated and timely details is less reliable than that of human reviewers. As a result, even though these models provide modest benefits, they should serve as supplemental aids in the review process.

4 Discussion

The healthcare applications of GenAI include improvements in diagnostic decisions, medical image analysis, professional training, clinic administration, patient engagement, and clinical administrative support. However, according to recent research, privacy and security issues may negatively impact users' acceptance of GenAI systems. A study [60] claims that GANs' adversarial training feature, which includes training two neural networks to compete with one another, enables researchers to explore security and privacy-related problems without making assumptions about the adversaries' capabilities. This tool allows researchers to model a defense scenario or act as an attacker to simulate an attack. Examples of defense models include generative adversarial privacy [61], privacy-preserving adversarial networks [62], intensive adversarial privacy [63], and reconstructive adversarial networks [64]. They show high potential to effectively guarantee privacy and security in various applications, including medical image analysis. Nevertheless, using GANs to address privacy and security issues raises several new challenges. This section represents some GenAI privacy and security challenges and possible solutions.

Grounded in the principles of quality, efficiency, equity, and patient experience, GenAI holds great potential to make a substantial contribution to the field of healthcare. Everyone is enthusiastic about the opportunities that lie ahead due to the possibility of positive transformation. "The Paradox of Information Technology Productivity," however, suggests that various technologies have been introduced throughout history that initially seemed promising but sometimes—for decades—failed to deliver on their promise of improving productivity [65]. Now the question is, does GenAI in healthcare also face the productivity paradox? This section also aims to cite the responses to this question.

4.1 Protecting Patient Privacy and Security: Addressing Challenges in Generative Artificial Intelligence (GenAI) Healthcare Applications

One of the main challenges is the need for large and diverse training datasets for GenAI models to learn effectively and produce meaningful outputs. Obtaining quality and diverse data and managing such data sets can be laborious. On the other hand, in the field of healthcare, the collection and extensive use of medical data involve significant concerns due to the possible disclosure of patients' identities.

To address this problem in medical image data [66] proposes a deep convolutional framework based on GAN that uses encryption, binary classifier and a segmentation analysis network while maintaining the integrity and efficiency of medical images while preserving medical data privacy. In [67], medGAN is proposed to have taken a successful step [68] in maintaining patient identity privacy when using medical records by generating synthetic data from discrete EHR medical records. CorGAN [25] makes it possible to generate synthetic data with sequential EHR inputs instead of discrete inputs, generating more diverse data than [67]. Chen et al. [63] present a distributed AI paradigm, Federated Learning (FL), that preserves privacy for users by allowing models to be trained without access to participants' local data and by only sharing gradients during training. This approach uses techniques such as digital twins, and GANs, as well as IoMT networks. However, its implementation in IoMT networks has challenges, such as integrating FL with next-generation IoMT networks and creating sufficient security to prevent hacking in the network. It also involves the use of blockchain for decentralized and secure data storage. Note that using GANs is introduced as a possible solution, utilizing GANs and generally GenAI models causes privacy concerns and ethical considerations [69–71]. If GANs are used to generate synthetic data, they must be trained first, which requires extensive real data and may reveal identity and sensitive or private information. The lack of transparency and non-consensual access to data sources for training GenAI models is the source of a lawsuit in the United States of America [72]. A proposed solution is making GenAI models able to forget sensitive cases or people after training [73].

Additionally, GANs add noise to the data to prevent identifying people's confidential information, which can produce unrealistic data and disrupt intelligent diagnostic processes. Mode collapse, where a generative model cannot capture the full variety of training data and produces limited variation, is another challenge in GenAI. Techniques such as improving model architecture, optimizing loss functions, and using ensemble methods are explored to overcome this challenge and encourage the generation of a wider range of outputs. Developing reliable evaluation criteria for GenAI and incorporating meaningful feedback from users or experts to modify and improve the model is also necessary, but the problem of unpredictability of GenAI models and their output has not yet been solved.

Furthermore, GenAI faces regulatory and policy challenges, such as information asymmetry between tech giants and regulatory authorities, copyright issues for AI-generated content, data fragmentation and collection, and the lack of human control over AI behavior. For example, many GenAI models are considered black boxes, making their decision-making processes challenging to interpret and out of control. Another major judicial challenge of GenAI is who should be liable in case of dissatisfaction that can be caused by the disclosure of confidential information of a person or the damage that arises from the production of inaccurate information by artificial intelligence? Who has the right to publish the content of productive artificial intelligence, and who is responsible for the use of GenAI, damages, and losses caused [74–76]?

Although companies such as Microsoft have added "your content generated using our AI services" regarding AI services, there is a need for strict rules and regulations. While, like other emerging technologies, it will take time for relevant laws and regulations to be enacted, this does not mean that it is impossible to establish laws and regulations for this technology. As we can see from the European Parliament and the UK Cabin case, regulating artificial intelligence laws is proceeding thoughtfully, and countries have announced a series of restrictions. These efforts include preventing potential misuse of GenAI technologies for malicious purposes, promoting transparency and accountability, and addressing biases in the outputs produced. These are essential steps to ensure the responsible and ethical use of generative models.

4.2 *Protecting Patient Privacy and Security: Addressing Challenges in GenAI Healthcare Applications*

Considering that AI has had previous setbacks in healthcare and that the first decade of the Electronic Health Record (EHR) era was problematic, the future of GenAI comes into question. Will GenAI also face challenges that slow down its flourishing in healthcare? There are characteristics of GenAI that can aid in swift progress:

- The GenAI-related tools are very user-friendly. While effective prompts can enhance the output, using the tools does not require specific skills [77].
- Unlike the implementation of Electronic Health Records (EHRs), which require significant investment in hardware and a complete restructuring of healthcare workflows, GenAI can be smoothly delivered to users via software [78].
- Advancements in APIs and plug-in technologies make integrating GenAI applications with EHR systems easier [79].
- GenAI has the advantage of being able to improve over time with minimal human intervention [80].

Initial studies on GenAI in fields outside of healthcare indicate that these tools have the potential to significantly improve productivity and quality at a faster rate than previous technologies [81]. For example, although early large language models suffered from issues like 'hallucinations,' racial and ethnic biases, and unsuitable outputs, GenAI has made tremendous progress in minimizing these concerns in a relatively short time [82, 83]. Recent research has demonstrated that GenAI can result in significant productivity gains for software engineers [84], management consultants [85], and authors [86]. One research even suggested that up to 80% of occupations, including many in healthcare, could see significant productivity enhancements through GenAI tools [87]. However, because of the inherent hazards associated with GenAI, integrating it into the healthcare industry presents greater obstacles than in other fields. For instance, the GenAI feature of rapid transformation can cause concerned changes in patient outcomes to often go unreported, as even developers struggle to completely grasp how their technologies generate certain results [88].

GenAI is expected to first target healthcare delivery systems in order to reduce inefficiencies and enhance administrative processes before directly engaging in patient care activities such as diagnostic and treatment suggestions [4]. This could include automation of processes in the creation of doctor notes, scheduling of visits, and processing of billing and authorization requests. Furthermore, on the productivity paradox factors, leaders in the healthcare sector have increased the implementation of advances in technology within the business, learned from previous failures, and adapted to prepare better for the future. Companies in the digital health area have also improved their grasp of the industry's intricacies and problems, with an emphasis on complementing innovations and the value of humility and patience. These changes in the healthcare marketplace, based on limited reimbursement and increasing competition, are going to further accelerate the investment in and deployment of GenAI within the sector, possibly for problems like administrative pressure and waste within the healthcare delivery systems. While GenAI may confront issues in the future, it is better prepared to handle and overcome them, resulting in accelerated development and improvement.

5 Conclusion

In this chapter, we reviewed a selection of practical examples of GenAI applications in healthcare. Moreover, we briefly mentioned how a GenAI model can be personalized for specific medical tasks. As it has not been a long time since GenAI was introduced and considering its evolution in recent years, we believe that GenAI will revolutionize the healthcare industry shortly and at a quicker rate than EHRs have emerged in clinics. It should be noted that through leveraging models such as Gemini and ChatGPT, plus the case studies and presented research, it is evident that, just as [89, 90] pointed out, GenAI is not going to replace doctors but is becoming so commonplace in hospitals that clinics prefer replacing other doctors with doctors proficient in using GenAI. Even though challenges such as protecting patient privacy and security, regulatory and policy issues, and ethical considerations must be addressed, given the effort countries have put into publishing regulations, a smoother path for GenAI advancement is anticipated. Despite the failures of previous technology implementations because of the unique attributes of healthcare, the future of GenAI looks promising due to its special capabilities. Additionally, emerging AI tools have taught technology companies many lessons, preparing the conditions for GenAI to quickly improve efficiency, quality, and patient outcomes.

References

1. Berwick, D.M., Hackbarth, A.D.: Eliminating waste in US health care. JAMA **307**(14), 1513–1516 (2012)
2. Kavanagh, K.T., Saman, D.M., Bartel, R., Westerman, K.: Estimating Hospital-Related Deaths Due to Medical Error. J. Patient Saf. **13**(1), 1–5 (2017). https://doi.org/10.1097/pts.0000000000000364
3. Zirpe, K., Seta, B., Gholap, S., Aurangabadi, K., Gurav, S.K., Deshmukh, A.M., Wankhede, P., Suryawanshi, P., Vasanth, S., Kurian, M., Philip, E., Jagtap, N., Pandit, E.: Incidence of Medication Error in Critical Care Unit of a Tertiary Care Hospital: Where Do We Stand? Ind. J. Crit. Care Med. **24**(9), 799–803 (2020). https://doi.org/10.5005/jp-journals-10071-23556
4. Sahni, N.R., Stein, G., Zemmel, R., Cutler, D.: The Potential Impact of Artificial Intelligence on Healthcare Spending. National Bureau of Economic Research (2023). https://www.nber.org/system/files/chapters/c14760/c14760.pdf
5. Zhuhadar, L.P., Lytras, M.D.: The application of AutoML techniques in diabetes diagnosis: current approaches, performance, and future directions. Sustainability **15**(18), 13484 (2023). https://doi.org/10.3390/su151813484
6. Qadir, J.: Engineering education in the era of ChatGPT: promise and pitfalls of generative AI for education. In: IEEE Global Engineering Education Conference (EDUCON), pp. 1–9 (2023). https://doi.org/10.1109/educon54358.2023.10125121
7. Chaisatitkul, A., Luangngamkhum, K., Noulpum, K., Kerdvibulvech, C.: The power of AI in marketing: enhancing efficiency and improving customer perception through AI-generated storyboards. Int. J. Inform. Technol. **16**(1), 137–144 (2024). https://doi.org/10.1007/s41870-023-01661-5
8. Taylor, J., Ardeliya, V.E., Wolfson, J.: Exploration of artificial intelligence in creative fields: generative art, music, and design. Int. J. Cyber IT Serv. Manage. **4**(1), 39–45 (2024). https://doi.org/10.34306/ijcitsm.v4i1.149
9. Farhadi, A., Mirzarezaee, M., Sharifi, A., Teshnehlab, M.: Domain adaptation in reinforcement learning: a comprehensive and systematic study. Front. Inform. Technol. Electron. Eng. (2024). https://doi.org/10.1631/FITEE.2300668
10. Koohi-Moghadam, M., Bae, K.T.: GenAI in medical imaging: applications, challenges, and ethics. J. Med. Syst. **47**(1) (2023). https://doi.org/10.1007/s10916-023-01987-4
11. Zhuang, J.-X., Cai, J., Zhang, J., Zheng, W., Wang, R.: Class attention to regions of lesion for imbalanced medical image recognition. Neurocomputing **555**, 126577–126577 (2023). https://doi.org/10.1016/j.neucom.2023.126577
12. AlAmir, M., AlGhamdi, M.: The Role of generative adversarial network in medical image analysis: an in-depth survey. ACM Comput. Surv. **55**, 1–36 (2022)
13. Sun, L., Wang, J., Huang, Y., Ding, X., Greenspan, H., Paisley, J.: An adversarial learning approach to medical image synthesis for lesion detection. IEEE J. Biomed. Health Inform. (2020). https://doi.org/10.1109/JBHI.2020.2964016
14. Pang, Y., Lin, J., Qin, T., Chen, Z.: Image-to-Image Translation: Methods and Applications. ArXiv (2021). https://doi.org/10.48550/arXiv.2101.08629

15. Jiang, Y., Chen, H., Loew, M., Ko, H.: COVID-19 CT image synthesis with a conditional generative adversarial network. IEEE J. Biomed. Health Inform. **25**(2), 441–452 (2021). https://doi.org/10.1109/JBHI.2020.3042523
16. Luo, Y., Zhang, S., Ling, J., Lin, Z., Wang, Z., Yao, S.: Mask-guided generative adversarial network for MRI-based CT synthesis. Knowl.-Based Syst. 111799–111799 (2024). https://doi.org/10.1016/j.knosys.2024.111799
17. Cao, Q., Mao, Y., Qin, L., Quan, G., Yan, F., Yang, W.: Improving image quality and lung nodule detection for low-dose chest CT by using generative adversarial network reconstruction. Br. J. Radiol. **95**(1138) (2022). https://doi.org/10.1259/bjr.20210125
18. Marcos, L., Babyn, P., Javad Alirezaie.: GenAI in Medical Imaging and Its Application in Low Dose Computed Tomography (CT) Image Denoising, pp. 387–401. Springer EBooks (2024). https://doi.org/10.1007/978-3-031-46238-2_19
19. Chowdhury, S., Acharjya, D.P.: Segmentation and feature extraction in medical imaging: a systematic review. Procedia Comput. Sci. **167**, 26–36 (2020). https://doi.org/10.1016/j.procs.2020.03.179
20. Bhan, A., Mangipudi, P.S., Goyal, A.: Left atrium MRI image segmentation using efficient Xception stochastic depth based generative adversarial network. Int. J. Healthc. Manage. 1–12 (2023). https://doi.org/10.1080/20479700.2023.2166206
21. Jie, Z., Zhiying, Z., Li, L.: A meta-analysis of Watson for oncology in clinical application. Sci. Rep. **11**(1), 5792 (2021)
22. Yang, Z., Nasrallah, I.M., Shou, H., Wen, J., Doshi, J., Habes, M., Erus, G., Abdulkadir, A., Resnick, S.M., Albert, M.S., Maruff, P., Fripp, J., Morris, J.C., Wolk, D.A., Davatzikos, C., Fan, Y., Bashyam, V., Mamouiran, E., Melhem, R., Pomponio, R.: A deep learning framework identifies dimensional representations of Alzheimer's Disease from brain structure. Nat. Commun. **12**(1) (2021). https://doi.org/10.1038/s41467-021-26703-z
23. Wang, H., Tao, G., Ma, J., Jia, S., Chi, L., Yang, H., Zhao, Z., Tao, J.: Predicting the epidemics trend of COVID-19 using epidemiological-based generative adversarial networks. IEEE J. Select. Top. Signal Process. **16**(2), 276–288 (2022). https://doi.org/10.1109/jstsp.2022.3152375
24. Waheed, A., Goyal, M., Gupta, D., Khanna, A., Al-Turjman, F., Pinheiro, P.R.: CovidGAN: data augmentation using auxiliary classifier GAN for improved Covid-19 detection. IEEE Access (2020). https://doi.org/10.1109/ACCESS.2020.2994762
25. Torfi, A., Fox, E.A.: CorGAN: Correlation-Capturing Convolutional Generative Adversarial Networks for Generating Synthetic Healthcare Records. ArXiv (2020). https://doi.org/10.48550/arXiv.2001.09346
26. Adams, L.C., Busch, F., Truhn, D., Makowski, M.R., Aerts, H.J.W.L., Bressem, K.K.: What does DALL-E 2 know about radiology? J. Med. Internet Res. (2023)
27. Vaccari, I., Orani, V., Paglialonga, A., Cambiaso, E., Mongelli, M.: A Generative Adversarial Network (GAN) technique for internet of medical things data. Sensors **21**(11), 3726 (2021). https://doi.org/10.3390/s21113726
28. Hazra, D., Byun, Y.-C.: SynSigGAN: generative adversarial networks for synthetic biomedical signal generation. Biology **9**(12), 441 (2020). https://doi.org/10.3390/biology9120441
29. Moons, P., Van Bulck, L.: Using ChatGPT and Google bard to improve the readability of written patient information: a proof of concept. Eur. J. Cardiovasc. Nurs. (2023). https://doi.org/10.1093/eurjcn/zvad087
30. Juana, G., Zúñiga, D., Vindel, C. L., Yoong, A. M., Sofía Hincapié, Zúñiga, A., Zúñiga, P., Salazar, E., Zúñiga, B.: Efficacy of AI chats to determine an emergency: a comparison between OpenAI's ChatGPT, Google Bard, and Microsoft Bing AI. Chat. Cureus. (2023). https://doi.org/10.7759/cureus.45473
31. Singhal, K., Tu, T., Gottweis, J., Sayres, R., Wulczyn, E., Hou, L., Clark, K., Pfohl, S., Cole-Lewis, H., Neal, D., Schaekermann, M.: Towards Expert-Level Medical Question Answering with Large Language Models. arXiv, 2305.09617 (2023)
32. Sai, S., Gaur, A., Sai, R., Chamola, V., Guizani, M., Rodrigues, J. J.: Generative AI for transformative healthcare: a comprehensive study of emerging models, applications, case studies and limitations. IEEE Access (2024)

33. Luo, R., Sun, L., Xia, Y., Qin, T., Zhang, S., Poon, H., Liu, T.Y.: BioGPT: generative pre-trained transformer for biomedical text generation and mining. Brief. Bioinform. **23**(6) (2022)
34. Ayers, J.W. Poliak, A. Dredze, M. Leas, E.C. Zhu, Z. Kelley, J.B. Faix, D.J. Goodman, A.M. Longhurst, C.A. Hogarth, M. et al.: Comparing physician and artificial intelligence Chatbot responses to patient questions to a public social media forum. JAMA Internal Med. **183**(6), 589–596 (2023). https://doi.org/10.1001/jamainternmed.2023.1838
35. Bandi, A., Adapa, P.V.S.R., Kuchi, Y.E.V.P.K.: The power of GenAI: a review of requirements, models, Input–output formats, evaluation metrics, and challenges. Fut. Internet **15**(8), 260 (2023). https://doi.org/10.3390/fi15080260
36. Thoppilan, R., De Freitas, D., Hall, J., Shazeer, N., Kulshreshtha, A., Cheng, H.T., Jin, A., Bos, T., Baker, L., Du, Y., et al.: Lamda: Language Models for Dialog Applications. arXiv:2201.08239 (2022)
37. Schick, T., Dwivedi-Yu, J., Jiang, Z., Petroni, F., Lewis, P., Izacard, G., You, Q., Nalmpantis, C., Grave, E., Riedel, S. PEER.: A Collaborative Language Model (2022). arXiv:2208.11663
38. Li, Y., Choi, D., Chung, J., Kushman, N., Schrittwieser, J., Leblond, R., Eccles, T., Keeling, J., Gimeno, F., Dal Lago, A., et al.: Competition-level code generation with alphacode. Science 1092–1097 (2022)
39. Fang, W., Ding, Y., Zhang, F., Sheng, J.: Gesture recognition based on CNN and DCGAN for calculation and text output. IEEE Access 28230–28237 (2019)
40. Jain, G., Mittal, D., Thakur, D., Mittal, M.K.: A deep learning approach to detect Covid-19 coronavirus with X-Ray images. Biocybern. Biomed. Eng. **40**(4), 1391–1405 (2020). https://doi.org/10.1016/j.bbe.2020.08.008
41. Domingues, I., Pereira, G., Martins, P., Duarte, H., Santos, J., Abreu, P.H.: Using deep learning techniques in medical imaging: a systematic review of applications on CT and PET. Artif. Intell. Rev. **53**(6), 4093–4160 (2019). https://doi.org/10.1007/s10462-019-09788-3
42. Naser, M.A., Deen, M.J.: Brain tumor segmentation and grading of lower-grade glioma using deep learning in MRI images. Comput. Biol. Med. **121**, 103758 (2020). https://doi.org/10.1016/j.compbiomed.2020.103758
43. Yuan, Q., Cai, T., Hong, C., Du, M., Johnson, B.E., Lanuti, M., Cai, T., Christiani, D.C.: Performance of a machine learning algorithm using electronic health record data to identify and estimate survival in a longitudinal cohort of patients with lung cancer. JAMA Netw. Open **4**(7), e2114723 (2021). https://doi.org/10.1001/jamanetworkopen.2021.14723
44. Machine learning for predicting epileptic seizures using EEG signals: A review. IEEE J. Mag. | IEEE Xplore (2021). https://ieeexplore.ieee.org/abstract/document/9139257/
45. Thoppilan, R., De Freitas, D., Hall, J., Shazeer, N., Kulshreshtha, A., Cheng, H.T., Jin, A., Bos, T., Baker, L., Du, Y., et al.: Lamda: Language Models for Dialog Applications (2022). arXiv:2201.08239
46. Ren, Y., Ruan, Y., Tan, X., Qin, T., Zhao, S., Zhao, Z., Liu, T.Y.: Fastspeech: Fast, Robust and Controllable Text to Speech (2019). arXiv:1905.09263
47. Brock, A., Donahue, J., Simonyan, K.: Large Scale GAN Training for High Fidelity Natural Image Synthesis (2018). arXiv:1809.11096
48. Elizalde, B., Deshmukh, S., Al Ismail, M., Wang, H.: Clap learning audio concepts from natural language supervision. In: Proceedings of the ICASSP 2023–2023 IEEE International Conference on Acoustics, Speech and Signal Processing (ICASSP), Rhodes Island, Greece, 4–10.2023, pp. 1–5. IEEE (2023)
49. Manco, I., Benetos, E., Quinton, E., Fazekas, G.: MusCaps: Generating Captions for Music Audio. In Proceedings of: the 2021 International Joint Conference on Neural Networks (IJCNN), Shenzhen, China, 18–22. pp. 1–8 (2021)
50. Salimans, T., Goodfellow, I., Zaremba, W., Cheung, V., Radford, A., Chen, X.: Improved Techniques for Training Gans (2016). arXiv:1606.03498
51. Oniani, D., Sreekumar, S., DeAlmeida, R., DeAlmeida, D., Hui, V., Lee, Y. J., Zhang, Y., Zhou, L., Wang, Y.: Toward improving health literacy in patient education materials with neural machine translation models. In: AMIA Joint Summits on Translational Science Proceedings. AMIA Joint Summits on Translational Science, pp. 418–426 (2023). https://www.ncbi.nlm.nih.gov/pmc/articles/PMC10283125/

52. Birkenbeuel, J., Joyce, H., Sahyouni, R., Cheung, D., Maducdoc, M., Mostaghni, N., Sahyouni, S., Djalilian, H., Chen, J., Lin, H.: Google translate in healthcare: preliminary evaluation of transcription, translation and speech synthesis accuracy. BMJ Innovations **7**(2), 422–429 (2021). https://doi.org/10.1136/bmjinnov-2019-000347
53. Ibrahim, A., Alfonse, M., Aref, M.: A systematic review on text summarization of medical research articles. Int. J. Intell. Comput. Inform. Sci. **23**(2), 50–61 (2023). https://doi.org/10.21608/ijicis.2023.190004.1252
54. Chaves, A., Kesiku, C., Garcia-Zapirain, B.: Automatic text summarization of biomedical text data: a systematic review. Information **13**(8), 393 (2022). https://doi.org/10.3390/info13080393
55. Rashidi Ranjbar, F., Zamanifar, A.: Autonomous dental treatment planning on panoramic x-ray using deep learning based object detection algorithm. Multimed. Tools Appl. **83**(14), 42999–43033 (2024)
56. Farhadi, A., Sharifi, A.: Leveraging meta-learning to improve unsupervised domain adaptation. Comput. J. **67**(5), 1838–1850 (2024). https://doi.org/10.1093/comjnl/bxad104
57. Meral, G., Ateş, S., Günay, S., Öztürk, A., Kuşdoğan, M.: Comparative analysis of ChatGPT, Gemini and emergency medicine specialist in ESI triage assessment. Am. J. Emerg. Med. **81**, 146–150 (2024)
58. Lorenzi, A., Pugliese, G., Maniaci, A., Lechien, J.R., Allevi, F., Boscolo-Rizzo, P., Vaira, L.A., Saibene, A.M.: Reliability of large language models for advanced head and neck malignancies management: a comparison between ChatGPT 4 and Gemini Advanced. In: European Archives of Oto-Rhino-Laryngology, pp. 1–6 (2024)
59. Brant-Zawadzki, G., Klapthor, B., Ryba, C., Youngquist, D.C., Burton, B., Palatinus, H., Youngquist, S.T.: The performance of ChatGPT-4 and Gemini Ultra 1.0 for quality assurance review in emergency medical services chest pain calls. In: Prehospital Emergency Care, pp. 1–12 (2024)
60. Cai, Z., Xiong, Z., Xu, H., Wang, P., Li, W., Pan, Y.: Generative adversarial networks: a survey toward private and secure applications. ACM Comput. Surv. **54**(6), 1–38 (2021). https://doi.org/10.1145/3459992
61. Huang, C., Kairouz, P., Chen, X., Sankar, L., Rajagopal, R.: Context-aware generative adversarial privacy. Entropy **19**(12), 656 (2017). https://doi.org/10.3390/e19120656
62. Tripathy A., Wang Y., Ishwar, P.: Privacy-preserving adversarial networks. In: Proceedings of the 57th Annual Allerton Conference on Communication, Control, and Computing, pp. 495–505 (2019). https://doi.org/10.1109/allerton.2019.8919758
63. Chen, C.S., Chang, S.F., Liu, C.H.: Understanding knowledge-sharing motivation, incentive mechanisms, and satisfaction in virtual communities. Soc. Behav. Pers. **40**(4), 639–647 (2012). https://doi.org/10.2224/sbp.2012.40.4.639
64. Liu, S., Shrivastava, A., Du, J., Zhong, L.: Better accuracy with quantified privacy: representations learned via reconstructive adversarial network. ArXiv (2019). https://doi.org/10.1090/mbk/121/79
65. Brynjolfsson, E., Rock, D., Syverson, C.: The productivity J-curve: how intangibles complement general purpose technologies. Natl Bureau Econ. Res. (2023). https://www.nber.org/papers/w25148
66. Kim, B.N., Dolz, J., Jodoin, P.M., Desrosiers, C.: Privacy-net: an adversarial approach for identity-obfuscated segmentation of medical images. IEEE Trans. Med. Imaging **40**(7), 1737–1749 (2021). https://doi.org/10.1109/TMI.2021.3065727
67. Choi, E., Biswal, S., Malin, B., Duke, J., Stewart, W.F., Sun, J.: Generating Multi-Label Discrete Patient Records using Generative Adversarial Networks (2018). https://arxiv.org/abs/1703.06490
68. Yale, A., Dash, S., Dutta, R., Guyon, I., Pavao, A., Bennett, K.P.: Generation and evaluation of privacy preserving synthetic health data. Neurocomputer **416**, 244–255 (2020). https://doi.org/10.1016/j.neucom.2019.12.136
69. Rane, N.: ChatGPT and similar generative artificial intelligence (AI) for smart industry: role, challenges and opportunities for industry 4.0, industry 5.0 and society 5.0. Soc. Sci. Res. Netw. (2023). https://doi.org/10.2139/ssrn.4603234

70. Bale, A.S., Dhumale, R., Beri, N., Lourens, M., Varma, R.A., Kumar, V., et al.: The impact of generative content on individuals privacy and ethical concerns. Int. J. Intell. Syst. Appl. Eng. **12**(1), 697–703 (2023). https://ijisae.org/index.php/IJISAE/article/view/3503
71. Nova K.: GenAI in healthcare: advancements in electronic health records, facilitating medical languages, and personalized patient care. J. Adv. Anal. Healthc. Manage. **7**(1), 115–131 (2023). https://research.tensorgate.org/index.php/JAAHM/article/view/43
72. Nolan, M.: Llama and ChatGPT Are Not Open-Source—Few Ostensibly Open-Source LLMs Live up to the Openness Claim. IEEE Spectrum (2023). https://spectrum.ieee.org/open-source-llm-not-open
73. Zhang, D. Finckenberg-Broman, P. Hoang, T. Pan, S.; Xing, Z. Staples, M. Xu, X.: Right to be Forgotten in the Era of Large Language Models: Implications, Challenges, and Solutions (2023)
74. Ghosheh, G.O., Li, J., Zhu, T.: A survey of generative adversarial networks for synthesizing structured electronic health records. ACM Comput. Surv. **56**(6), 1–34 (2024). https://doi.org/10.1145/3636424
75. Hernandez, M., Epelde, G., Alberdi, A., Cilla, R., Rankin, D.: Synthetic data generation for tabular health records: a systematic review. Neurocomputer **493**, 28–45 (2022). https://doi.org/10.1016/j.neucom.2022.04.053
76. Fui-Hoon Nah, F., Zheng, R., Cai, J., Siau, K., Chen, L.: GenAI and ChatGPT: applications, challenges, and AI-human collaboration. J. Inform. Technol. Case Appl. Res. **25**(3), 277–304 (2023). https://doi.org/10.1080/15228053.2023.2233814
77. Holmgren, A.J., Apathy, N.C.: Trends in US hospital electronic health record vendor market concentration, 2012–2021. J. Gen. Intern. Med. **38**(7), 1765–1767 (2023). https://doi.org/10.1007/s11606-022-07917-3
78. Baily, M., Brynjolfsson, E., Korinek, A.: Machines of Mind: The Case for an AI-Powered Productivity Boom (2023)
79. Glaser, J., Gardener, E.: Standardized APIs could finally make it easy to exchange health records. Harvard Bus. Rev. (2022)
80. Ajay Agrawal, Gans, J., Avi Goldfarb.: The Economics of Artificial Intelligence: An Agenda. The University of Chicago Press (2019). https://press.uchicago.edu/ucp/books/book/chicago/E/bo35780726.html
81. Brynjolfsson, E., Li, D., Raymond, L.R.: GenAI at Work. Natl Bureau Econ. Res. (2023). https://www.nber.org/papers/w31161
82. Alkaissi, H., McFarlane, S.I.: Artificial Hallucinations in ChatGPT: implications in scientific writing. Cureus **15**(2), e35179 (2023). https://doi.org/10.7759/cureus.35179
83. Hanna, J.J., Wakene, A.D., Lehmann, C.U., Medford, R.J.: Assessing Racial and Ethnic Bias in Text Generation for Healthcare-Related Tasks by ChatGPT. MedRxiv (2023). https://www.medrxiv.org/content/https://doi.org/10.1101/2023.08.28.23294730v1
84. Peng, S., Kalliamvakou, E., Cihon, P., Demirer, M.: The Impact of AI on Developer Productivity: Evidence from Github Copilot. ArXiv (2023). https://arxiv.org/abs/2302.06590
85. Dell'Acqua, F., McFowland, E., Mollick, E.R., Lifshitz-Assaf, H., Kellogg, K., Rajendran, S., Krayer, L., Candelon, F., Lakhani, K.R.: Navigating the jagged technological frontier: Field experimental evidence of the effects of AI on knowledge worker productivity and quality. In: Harvard Business School Technology & Operations Management Unit Working Paper (24-013) (2023)
86. Noy, S., Zhang, W.: Experimental evidence on the productivity effects of generative artificial intelligence. Science **381**(6654), 187–192 (2023)
87. Eloundou, T., Manning, S., Mishkin, P., Rock, D.: Gpts are gpts: an early look at the labor market impact potential of large language models. ArXiv **2303**, 10130 (2023)

88. Meskó, B., Topol, E.J.: The imperative for regulatory oversight of large language models (or GenAI) in healthcare. NPJ Digit. Med. **6**(1), 120 (2023). https://doi.org/10.1038/s41746-023-00873-0
89. Brito, S.D.P.: In the future will artificial intelligence be able to replace doctors?—narrative review. Mexican J. Med. Res. ICSA (2024). https://repository.uaeh.edu.mx/revistas/index.php/MJMR/article/view/12397
90. Shuaib, A., Arian, H., Shuaib, A.: The increasing role of artificial intelligence in health care: will robots replace doctors in the future? Int. J. Gener. Med. 891–896 (2020). https://doi.org/10.2147/ijgm.s268093

Identification of Face Age Progression and Rejuvenation Using Generative Adversarial Networks

Abhishek Kulkarni and Amit Joshi

Abstract Face age synthesis aims to alter an individual's face to predict the facial image's future or past appearance. Face age synthesis has applications in many areas, including law enforcement and tracking victims of human trafficking. This work uses conditional variational autoencoder generative adversarial networks. The encoder network converts a facial image into a latent representation. The decoder network is used to generate images of the desired age by using this latent representation. The Discriminator network is used to check the authenticity of the image. The proposed system is evaluated on a publicly available Cross-Age Celebrity Dataset. This chapter focuses on facial images from people in their late teens to older adults. Different tests are conducted to measure and analyze the outcomes. Experimental results obtained using metrics such as age estimation error, cosine similarity, and identity preservation show the feasibility of the framework. The proposed solution provides a viable methodology for solving the problem of face-age synthesis.

Keywords Face age synthesis · Generative adversarial networks · Conditional variational autoencoder generative adversarial networks · Cross age celebrity dataset · Face++

1 Introduction

The human face exhibits distinctive features such as eyes, nose, ears, cheeks, forehead, and more. Each person has individual variations in the size and shape of these features which makes them distinguishable from one another. From the moment of birth till the end of life, these features continuously evolve at an unpredictable pace,

A. Kulkarni (✉) · A. Joshi
Department of Computer Science and Engineering, School of Computational Sciences, COEP Technological University, Pune, India
e-mail: kulkarniar20.comp@coeptech.ac.in

A. Joshi
e-mail: adj.comp@coeptech.ac.in

N. R. Vajjhala et al. (eds.), *Generative Artificial Intelligence (AI) Approaches for Industrial Applications*, Information Systems Engineering and Management 24,
https://doi.org/10.1007/978-3-031-76710-4_11

resulting in a noticeable contrast between the current facial appearance and the aged or younger face of the individual. Researchers have done a considerable amount of work trying to unravel how an aged or a younger face would look based on the present-day face. This process of modeling the transformations that occur as the age of an individual changes is known as face age synthesis.

The application of face age synthesis extends across various areas, such as aiding people who have lost their loved ones in human trafficking. This area is receiving increased attention from researchers to determine if technology can help solve the problem. Montasari and Jahankhani [1] proposed various methods that can be used to combat human trafficking. One of the most useful methods involves the use of face recognition technology. However, the usefulness of this technology decreases as the victim ages. Sometimes, people get lost in simply unfortunate circumstances, such as abduction, runaways, or senior citizens suffering from dementia. Law enforcement authorities may need to know the present-day appearance of people in their databases. Taister et al. [2] published a detailed study about the aging of individuals, describing the changes in facial appearance associated with aging, as well as the characteristic changes influenced by environmental factors, alcohol and drug consumption, diseases, and dental, medical, or cosmetic treatments.

The conventional method of manually checking each individual and determining potential matches is inefficient and unrealistic. Face age synthesis offers a more streamlined and efficient approach, allowing authorities to project the current facial characteristics of individuals and facilitating a more targeted and effective search process. Usually, seekers have an image of the target's face. However, factors such as the age at which the image was taken, the number of years between today and the day the image was taken as face age recognition performance decreases as the number of years increase [3], plastic surgery, etc., make it difficult to identify the individual even if the person is standing in front. As the face ages, various changes occur, such as changes in skin texture, weight, hair [4], etc. To tackle this problem, various classical and deep learning methods have been proposed. This work proposes a deep learning-based model to tackle the problem.

This work proposes the use of Conditional Variational Autoencoder Generative Adversarial Networks (CVAE-GAN), which is a combination of two deep models: Conditional Variational Autoencoders (CVAE) and Generative Adversarial Networks (GANs). This work uses the Cross-Age Celebrity Dataset (CACD) for model training and evaluates its accuracy using standard evaluation methods and performance metrics. It also compares this new approach to face age synthesis with already established GAN based models.

1.1 Generative Adversarial Networks

GAN based models are semi-supervised or unsupervised deep learning models. GANs have two different networks: A Generator and a Discriminator. The Generator generates data in the target domain and the Discriminator's role is to label it as real or synthetic. A latent vector of random values, denoted as z is fed to the Generator (G) which generates fake data, represented as G(z). The Discriminator (D) trains to increase the probability of correctly labeling both real data and generated data from G. Essentially, D and G are engaging in the following two-player minimax game with the value function V(D, G).

$$\text{Min max V(D, G)} = [\log(D(x))] + [\log(1 - D(G(z)))] \quad (1)$$

As per Eq. 1, initially, the Generator is slightly less trained than the Discriminator and thus the Discriminator can reject samples with ease. It punishes the Generator by increasing the loss so that Generator generates better fake data. The training continues until the Discriminator cannot distinguish between training and generated data [5].

GANs have found applications in areas other than age synthesis. Peng et al. [6] used GANs for beautification of face. It employs a dual Encoder, one is used to learn the facial features of the given image and the other learns the mask of geometrically beautiful faces. The Generator generates a new image based on these two inputs. A neural network called Segmentor helps the Generator in the learning process.

GANs are also used to increase stability in text to image synthesis. An Encoder is used to encode a text into a semantic vector before passing it to the triple GAN architecture. First GAN is used to generate an image based on the text input while the other two GANs change the resolution of the generated image to improve its quality [7]. Attempts to detect deepfakes by using GANs have also been made. A variant of GAN called DCGAN has been used to achieve this purpose [8].

Areas other than image processing include information extraction in E-commerce where the Generator takes queries that are formed through user searches and corresponding condition vectors as input. Conditions are introduced using a continuous bag of words method to help the Generator. Discriminator uses LSTM layers to differentiate between real documents in the collection and the sequence of words formed by the Generator [9].

GANs have also found application in the field of medicine by synthesizing CT images from brain MRI. A Generator takes an MRI image as input and outputs a CT image. The DeepResNet network is employed to take input images which are smoothed randomly and outputs fake CT images. It also helps in regularizing the Discriminator (ADvNet) which takes fake and corresponding real CT images as input and helps networks in their training [10].

1.2 Varitional Autoencoder

Variational Autoencoder (VAE) is a popular neural network used for reconstruction and generation. It consists of an Encoder that converts the input into a latent representation using probabilistic encoding and The Decoder network converts this latent representation into the intended output. If a label is added to the encoded latent representation, then it is called a Conditional Variational Autoencoder (CVAE). It is often done to generate new samples.

The loss function for VAE is described in Eq. 2. Minimizing this loss function is equivalent to optimizing an upper bound of an expected Kullback-Liebler (KL) divergence [11].

$$\begin{aligned} L(\theta, \phi;\ x^{(i)}) = & -D_{KL}\big(q_{\emptyset}(z|x^{(i)})||p_{\emptyset}(z)\big) \\ & + Eq_{\emptyset}\big(z|x^{(i)}\big)\big[\log p_{\emptyset}\big(x^{(i)}|z\big)\big] \end{aligned} \tag{2}$$

VAE finds applications in a variety of areas like query expansion in ad-hoc information retrieval where the Encoder is given the query which is represented by an embedded vector. Based on this input, it creates a latent vector. The Decoder samples from the distribution and estimates a language model for the query [12].

CVAEs can also be used on time series datasets such as using it for sensing anomalies of photovoltaic systems. Inference and recurrent neural networks along with the usual Encoder and Decoder networks are used to achieve this purpose [13]. This study introduces an approach combining the above mentioned Conditional Variational Autoencoders (CVAE) and Generative Adversarial Networks (GANs), referred to as Conditional Variational Autoencoder Generative Adversarial Networks (CVAE-GAN). The model is trained using the Cross-Age Celebrity Dataset (CACD) and its performance is assessed through standard evaluation techniques and metrics. Additionally, this new method is compared against the latest GAN-based models in the context of face age synthesis. In the literature review, this work will introduce the latest research in face age synthesis. The proposed methodology section will explain the architecture of the model and the algorithm in detail. The results and discussions section is split into four parts: experimental details will describe the setup of training the model and various hyperparameters. The dataset part describes the Cross-Age Celebrity Dataset (CACD) and data preprocessing methods in detail. Three performance metrics used in this work are explained in the next part. The Results part talks about results which are achieved using this methodology using the above-mentioned setting. Lastly, the work is concluded, and the future scope is presented.

2 Literature Review

There are three popular approaches to solve the problem of face age synthesis. The first approach deals with physical modeling. Bando et al. [14] proposed a process where the target face systematically develops wrinkles. Suo et al. [15] introduced a face aging model based on probabilistic Markov chains. In this work, an AND-OR graph consisting of multiple layers is used to represent the human face. It focuses on three main aspects: hairstyle and its aspects, deformations and aging, and wrinkles.

However, physical models tend to focus only on certain aspects of face aging and often fail to consider intricate details. These approaches are computationally expensive and less accurate. Physical methods were popular from the late 1990s to the early 2010s until the development of sophisticated neural networks.

The second approach employs prototype-based methods that rely on a non-parametric model. Age categorization is done on the faces, and prototype is defined as average face of each group. The variation between these prototypes signifies the model. To synthesize age, this variation is applied to the input face. Kemelmacher-Shilzerman et al. [16] employed this approach to calculate the average image subspace with image illumination awareness. However, these methods are often simple in nature and thus unable to replicate complex personalized features that change differently from person to person.

The last approach utilizes deep learning models, with the most widely used models being GANs [17]. These models augment certain neural networks to the base GANs and conditional GANs. In a paper by Chen et al. [18], the relationship between latent space and age attribute was modeled. A residual channel attention module, which learns about specific feature emphasis and suppression, was also trained. This enabled the Encoder to only learn features that are most fitting for age translations. The model also employed a pair of Discriminators, with one determining the authenticity of the image and the other providing extra response to the Generator which helps the network generate features within the center of the face that are more photo-realistic.

AW-GAN contains a block attention module comprising of a channel attention module which focuses on meaningful features in the given image and a spatial attention module which focuses on the locations of those meaningful features. It also added wavelet transformation to the Discriminator of GAN. This increased the consistency of the model and captured the local textures of the images [19]. Shi et al. [20] proposed CAN-GAN which uses different attention factors for different facial aging regions. Zhu et al. [21] proposed UGAN for multi-domain translations. The model is trained to not only perform aging of the face but also change expressions and makeup. This is done by training the Generator such that it masks the source domain of the image and retains characteristics related to the target domain only. Li et al. [22] brought a fresh approach with a framework that learns continuous aging. It uses an Encoder to obtain the encoding of an image. The Obtained encoding is given as input to an age estimator which determines the probability distribution of the age. Considering the target age and the estimated age distribution, the age transformation is applied to the said encoding. This transformed encoding is then fed to the Generator which generates fake images.

There are also approaches like FaceGAN which did both progression and regression of the images and unlike previously mentioned methods, it used unpaired training data and could be extended to facial expression synthesis also. It employs two deep recognition models in addition to the usual GAN model. The Generator generates two images rather than one belonging to the actual class and target class. The Discriminator is in turn divided into two parts. One part predicts the reality of the image while the other outputs the probability distribution over label classes. In addition to the adversarial loss employed by most works, it also employs age classification loss to train both models about age and identity preservation loss which ensures personalized characteristics belonging to each face are preserved [23]. Tatikonda et al. [24] proposed two different models for face aging and facial attribute manipulation. A similar approach was also seen in [25]. ROUTINGGAN added a dropout layer between the Encoder and Decoder. The age invariant features were disentangled from the input face. It also employs an age classifier along with the Discriminator which attempts to improve age accuracy [26].

Sharma et al. [27] proposed a dual GAN model of attention, and a custom GAN model called SRGAN which was specifically used to increase the resolution of the image and thus made the quality of the generated image better. One major drawback of all GAN based models is that they operate on different datasets and use different techniques to measure the accuracy of their models. This work uses one of the most used dataset and accuracy measurement techniques.

The above models used datasets that span a person's lifetime. These approaches generally have a lower accuracy for child and teenage image synthesis. Some have focused on the synthesis of these images only like ChildGAN and PatchGAN. ChildGAN has used an encoder-based approach along with CGANs. The Encoder is aided by a self-attention block which helps to model dependencies across face regions. The Discriminator is also paired with a self-attention block [28]. PatchGAN used three Discriminators instead of one. These Discriminators worked on three different resolutions of the image. It used additional blocks for learning geometric and structural patterns. This block helps the Generator generate better images [29]. These models are trained on datasets that focus on child and early teen images. These datasets are different from the adult images' datasets in the sense that the transformation of the face from that of a newborn baby to that of a late teen is more drastic than the transformation of a late teen face to an old age face [30].

GANs have found applications in numerous image related tasks, such as enhancing resolution, Image to image conversion, creating better quality images and manipulating visual content, etc. Various GAN variants are used for these purposes like StyleGANs, SRGANs etc. Kokate et al. [31] has given an analysis of various types of GANs. One of the more popular approaches in GAN based models involves use of a separate Autoencoder network to encode the data about the image into a latent representation (z) which is then given as input to the GAN network. To generate new samples, a label is concatenated to this latent representation. Huang et al. [32] uses an inverse Generator network to down sample image into latent representation. Then, the interconvertible conditional translational module was used to perform transformations on the latent representation based on the target age category. This was done by using the information from the prior distributions. A Generator then takes this transformed latent space as input and creates an image in the target category. A multi-layer perceptron is used as a Discriminator.

Liu et al. [33] used two VAE based Autoencoders for encoding images. One learns age specific features while the other one learns age irrelevant features. It also makes use of cycle consistency learning that is the image converted to the target domain should revert to the source image upon applying inverse transformation. Barve and Joshi [34] used two separate linear networks. First is a convolutional

Autoencoder which has pair of Encoder-Decoder networks. The Encoder is trained to produce a deterministic representation of latent space. This trained Encoder then gives the input to a custom conditional GAN model. Zhang et al. [35] proposed a Conditional Adversarial Autoencoder (CAAE) in which a Decoder network is replaced by a Discriminator which imposes latent uniform distribution on z. L2 loss between input and output images is used to update the encoder and the generator. A separate Discriminator helps the output to be photorealistic and plausible for the target label.

Recent works prove that GANs are single handedly the most widely used in the face age synthesis. All these methodologies revolve around enhancing a base GAN by incorporating different custom networks. Encoder-Decoder networks are by the most popular additions to the GANs. VAEs are Encoder-Decoder type networks which are widely used for generation of new data. Thus, this work uses the combination of these two networks to solve the problem of face age synthesis.

3 Proposed Methodology

The proposed methodology uses a single deep learning architecture—Conditional Variational Autoencoder Generative Adversarial Networks. The architecture comprises of three independent neural networks, an Encoder, a Decoder, and a Discriminator. The Encoder is used to convert the input image into a latent representation. The dimension of the latent vector is 128. A hot coded target age vector having a dimension of 5 is concatenated to this encoded vector and is given as an input to the Decoder. The hot coded target age vector is filled with zeroes except for one dimension where it is filled with 1. The person's in the input image determines this particular dimension. The Decoder converts the modified latent representation into an image. The reality of the image is determined by the discriminator. The Discriminator takes three inputs to calculate its loss, the real image which was given as an input to the Encoder, The image which was generated by the Decoder after taking the latent representation and hot coded target age vector and the third is a Decoder generated image after taking noise as an input. The overall loss for the model is given in Eq. 3.

$$\mathrm{L} = \mathrm{L}_{\mathrm{prior}} + L_{llike}^{Disl} + \mathrm{L}_{\mathrm{GAN}} \tag{3}$$

The first term refers to loss imposed on z from prior distribution. The second term of the loss is imposed from the real image and fake image from the Generator. The third term is GAN loss which can be further defined as per Eq. 4.

$$\begin{aligned}\mathrm{L}_{\mathrm{GAN}} &= \log(\mathrm{Dis}(\mathrm{x})) + \log(1 - \mathrm{Dis}(\mathrm{Dec}(\mathrm{z}))) \\ &+ \log(1 - \mathrm{Dis}(\mathrm{Dec}(\mathrm{Enc}(\mathrm{x}))))\end{aligned} \tag{4}$$

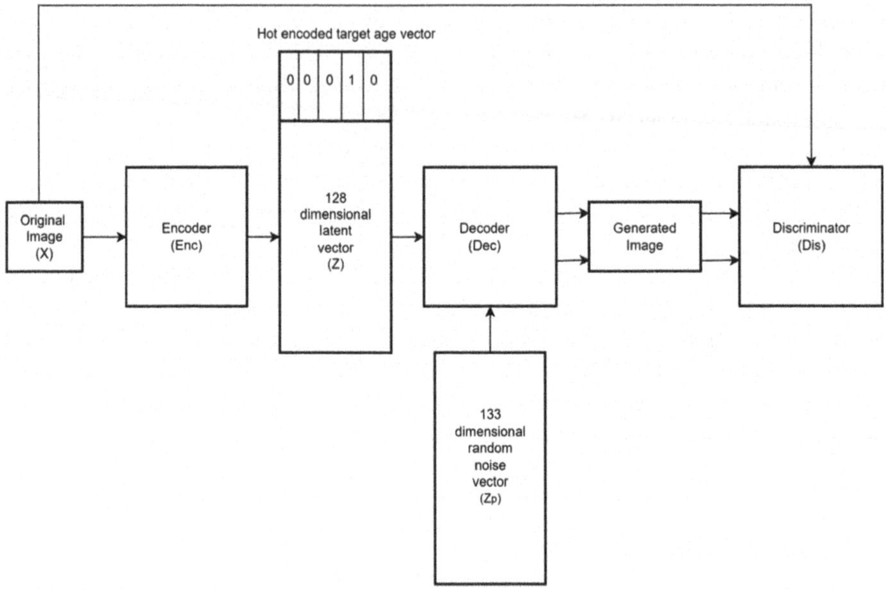

Fig. 1 Architecture of the model

where the real image is denoted by x, The image generated by the Decoder from the random noise is denoted by Dec(z) and the latent representation of the real image concatenated by the hot coded target age vector is denotes by Enc(x) and Dec(Enc(x)) is the image generated from the input received from the Encoder. Figure 1 shows the architecture of the model.

3.1 Architecture of Encoder

The Encoder network takes a 64 * 64 * 3 image as an input. It has 3 convolutional blocks and a fully connected block at last. Each convolutional block contains a convolutional layer, a batch normalization layer, followed by a ReLU layer. A kernel of size of 5 * 5 and stride 2 are used in each convolutional layer. In the first block, the 3 input channels are increased to 64. In the 2nd block, the channels are further increased to 128. In each block, both width and height of the image are halved. In the 3rd block, the channels further increased to 256. Then in the fully connected block, the input of 8 * 8 * 256 is converted 128. After a layer of batch normalization and ReLU, the mean and standard deviation are calculated based on the output.

3.2 Architecture of Decoder

The Decoder network takes a latent vector (z) of 133 dimensions which is either a concatenation of the latent vector which came as an output from the Decoder network and hot coded target age vector or a random noise. The first block in the Decoder network is a fully connected block that converts z into 8 * 8 * 256. It is followed by batch normalization which is then followed by a ReLU layer. Following this block, there are 3 convolutional blocks which consist of a Transposed convolutional layer, a batch normalization layer, followed by a ReLU layer. The channels are reduced by a factor of 2 in each convolutional block while height and width are doubled. The kernel size and stride are the same as that of the Encoder. The last convolutional block has only a transposed convolutional layer. The final output has a Tanh activation layer.

3.3 Architecture of Discriminator

The Discriminator network has 4 convolutional blocks which have a convolutional layer, a batch normalization layer, and a ReLU layer except for the first one which does not have a batch normalization layer. The channels are expanded from 3 to 32 in the first block, from 32 to 128 in the second, and from 128 to 256 in the third. The input and output channels in the fourth block are the same. After these blocks, there is a fully connected block which has a linear layer having 512 output channels, batch normalization layer, and a ReLU layer. There is a final single layer having only a single channel which is then followed by a sigmoid activation layer. The sigmoid layer gives an output in the range of [0,1]. If the output is 0 then the Discriminator thinks that the image is fake while if it is 1 then the image is real. The loss function is calculated by using Binary Cross Entropy [36].

3.4 Algorithm

Repeat the steps below till the entire dataset is processed.

(1) Take a sample X from the dataset.
(2) Let Z be Enc(X) i.e. latent representation generated by the decoder.
(3) L_{prior} is calculated as $D_{KL}(q(Z|X) \| p(Z))$.
(4) Modify Z by appending hot coded target age vector.
(5) Let χ be Dec(Z) i.e. image generated by the decoder from the modified latent representation.
(6) L_{llike}^{Disl} is calculated as $-\ E_{q(Z|X)}\ [p(Dis_l(X)|Z)]$.
(7) Let Z^* be random sample from η (0,1).
(8) Let X^* be $Dec(Z^*)$.

(9) L_{GAN} is calculated as $\log(Dis(X)) + \log(1 - Dis(\chi)) + \log(1 - Dis(X^*))$.
(10) Update the encoder gradient according to $L_{prior} + L_{llike}^{Disl}$.
(11) Update the decoder gradient according to $15 * L_{llike}^{Disl} - L_{GAN}$.
(12) Update the discriminator gradient according to L_{GAN}.

This methodology combines the use of two powerful generative models VAE and GANs. Thus, this methodology can perform on par with GAN based methodologies. On the downside, the probabilistic nature of encoding causes facial attributes to be different for generated images which may slightly hinder the performance.

4 Results and Discussions

This section discusses the details about the experimental setup, the dataset, performance metrics and results that are obtained from the proposed methodology.

4.1 Experimental Setup

For training the model, the Google Colab platform is used. It has access to a single NVIDIA T4 GPU card with 16 GB RAM and a 100 GB disk space. The model is trained for around 350 epochs. It took around 12 h to train the model. All three networks are trained with the RMSprop [37] optimizer with the Encoder and the Decoder network having a learning rate of 0.0003 and the Discriminator network having a learning rate of 0.00003. Google Colab is used as it removes hardware dependencies while training the model. The images are processed in a batch of 64. The entire experimental setup is summarized in Table 1.

Table 1 Summarized experimental setup

Hyperparameter	Name/value
Training platform	Google Colab
GPU	NVIDIA T4
Epochs	350
Optimizer	RMSProp
Encoder and Decoder learning rates	0.0003
Discriminator learning rate	0.00003
Batch size	64

Table 2 Categorization based on age

Categories	Age span
1	0–20
2	21–30
3	31–40
4	41–50
5	50+

4.2 Dataset

For training the system, the Cross Age Celebrity Dataset (CACD) which has 163,446 images is used [38]. In the CACD dataset, the information about the person in the picture is contained in the image name itself. The information given in the image name is the age, the real name, and the picture number of that person at that age. Only the age of the person is extracted for training purposes.

After the necessary preprocessing which includes checking the existence of the image and whether it can be opened or not, 155,156 images from the CACD dataset remain. Out of those images, 8000 images are randomly chosen. Table 1 specifies all age categories. Unnecessary background is removed from the images using the OpenCV face detection tool [39]. Further, the images that are not cropped correctly are manually removed from the dataset. Lastly, all the images are converted to 64 * 64 * 3 size. Finally, Facial images are grouped into appropriate categories based on the person's age as per the Table 2.

The dataset is divided the training and testing set into an approximate 4:1 ratio. The training set consists of 100 such batches while the testing set consists of 25 batches.

4.3 Performance Metrics

Age Estimation Error

This work utilizes the Face++ [40] tool to evaluate age estimation error. This involves analyzing the differences in age distributions between generated and real images within specific age categories.

The age estimation error is calculated when the model generates an image in the category of the original image. The image is then sent to the Face++ tool as a JSON request. The request returns the value of the age predicted by the Face++ tool. The absolute difference between the age values of the real and generated image is added to that age category. Lastly, this difference is averaged across all categories and is defined as age estimation error. For a precise prediction, the value of age estimation error should be low [25].

Identity Preservation

Face comparison is done with the help of Face++. The methodology includes submitting two images of an individual as input to Face++. The tool generates the probability of correctly identifying the same person within the given image pair. The pairs consist of the image generated by the model and its corresponding real image for all the categories. Images are generated for each category by concatenating the hot coded target age vector for that category. The real image and generated image for each age category are sent as a JSON request to the Face++ tool. It returns the probability that the faces in the two images are of the same person. For two images associated with an individual, the probability is computed. This process is iterated across various image pairs and the outcomes are subsequently averaged for each distinct age category. The ultimate result is presented as the average percentage of successful persons matching within each age category [41].

Cosine Similarity

Cosine similarity is a measure which is used to quantitatively find similarity between two images. The value of cosine similarity varies between -1 to 1 with 1 indicating two images are completely similar, 0 indicating the images are completely dissimilar while -1 indicates that the images are diametrically opposed.

A pre trained vision transformer model is used in calculation of cosine similarity [42]. Real image and generated image in the age category of the real image are converted to tensor followed by normalization and resizing. These image tensors are individually sent to the model which outputs representation features. This pair of representation features is used to calculate cosine similarity [43].

4.4 Results

Figure 2 displays the generated faces corresponding to all the categories for an image sourced from the testing set. Each row represents the image in a distinct category, while columns correspond to one of the five age categories listed in Table 2.

Figure 2 demonstrates that with increasing age, the shape of the input image's face undergoes transformation. The CVAEGAN adds some wrinkles on the forehead for male face images. For female face images the wrinkles are rarer due to the rarity of wrinkles on female face images on 50+ age category. The noticeable difference in age-related changes for both genders is particularly evident around the jaw area. The system also addresses beard presence, reducing the amount of beard in the 0–20 and 50+ age categories. The beard also becomes grayer as age increases. 0–20 category is also categorized by presence of bangs for the females. CVAEGAN also changes expressions at random as the images have people either smiling or a neutral facial expression.

Fig. 2 Generated faces corresponding to all the categories for an image sourced from the testing set

Age Estimation Error

Age estimation error is calculated by using The Face++ API. This performance metric performs best for 21–30 age category which is then closely followed by 0–20 and 31–40 category. The 41–50 and 50+ have abnormally high values for this performance metric. The reason for such high values can be attributed to the faces having seemingly younger features especially for women. As aged features are added while reconstruction, the ages are overestimated by a huge margin.

Identity Preservation

Identity preservation is more uniform than age estimation error. The highest identity preservation is for the 0–20 age category. The lowest is for 50+ age category. The 50 + category has slightly lower results which might be due to the addition of few wrinkles and sunken cheeks.

Cosine Similarity

The cosine similarity is also uniform like identity preservation. It is the lowest for 41–50 category followed by 50+ age category. The results in these categories are slightly lower which can be due to random changes in facial attributes of the generated images. Table 2 shows category wise values of both performance metrics.

Comparative Analysis

The results are compared with two other significant works. The first one would be Cross Age Face Generator which is used in [34]. This work will refer to the model as CAFG. CAFG uses two separate linear networks, one is Encoder—Decoder network while the other is a custom CGAN model. The second one is CAAE [35] which uses two Discriminators to regulate the generated images and the latent representation. In this proposed work uses a single model consisting of an Encoder, a Decoder/Generator and a Discriminator.

Results are compared on two of our performance metrics, age estimation error and identity preservation. Both of the mentioned works use age categorization like this work, but the number of categories and age span of each category is different. For simplicity's sake, this work will directly compare the mean results across all categories with our mean result. The comparison is shown in Table 3.

Our proposed method beats CAAE in Identity Preservation. It does not perform as well against CAFG which can be attributed to the factor of limited size of the dataset

Table 3 Quantitative analysis

Age category	Age estimation error cosine similarity	Identity preservation (%)
0–20	8.82 0.7498	61.44
21–30	8.69 0.7525	60.75
31–40	8.82 0.7482	61.05
41–50	11.07 0.7339	60.94
50 +	14.45 0.7475	60.72

Table 4 Comparative analysis

Methods	Age estimation error	Identity preservation (%)
CAAE	5.16	3.59
CAFG	2.75	96.82
Proposed method	**9.68**	**60.89**

which is kept small due to the limitations on computing power. But even with such small dataset, the proposed method performs very well. Except for the 50+ and 40–50 age categories in Age Estimation Error, the results are consistent. Due to the low percentage of images in 50+ age category, the performance metrics perform poorly in this category. With well diversified and a greater number of images, The proposed method will be able to provide highly consistent results across all age groups in all performance metrics which is not the case for above methods where the result seem to vary a bit more across age categories than proposed method.

CVAEGAN has a lot of potential as it is a combination of two widely used generative networks. The simultaneous training of all three sub networks and the hyper-parameters used help to reduce training time of the architecture. Custom neural networks can be added to enhance the performance of the existing architecture. However, appending a hot-coded target age vector may not be the best method. Learned similarity metric or similar method as suggested in [36] might give better results. The methodology might underperform if more diverse facial data is used.

5 Conclusion and Future Scope

This chapter proposes the use of Conditional Variational Autoencoder Generative Adversarial Networks to address the issue of face age synthesis. The system consists of an Encoder, a Decoder, and a Discriminator. The Encoder is used to convert the given image into a 128-dimensional vector. The Decoder is used to generate images in the required age category from the input dimensional vector. The Discriminator checks the reality of the image. The model is evaluated on the CACD dataset and results are measured quantitatively. The system gives consistent results of 60.89% for identity verification, 0.7457 for cosine similarity and 9.68 for age estimation error. The work can be extended to include more images of the 0 to 20 and 50+ age categories. Synthetic images can be generated in this category. This model can further be trained to perform face age synthesis from infancy to old age. This will expand the use case of the model. VAE based models may produce blurry images. Models like SRGANs can be appended to the existing model which will improve quality of images generated by the model. Custom neural networks can be incorporated in this work which will help in generation of distinct age dependent features.

References

1. Montasari, R., Jahankhani, H.: The application of technology in combating human trafficking. In Cybersecurity, Privacy and Freedom Protection in the Connected World: Proceedings of the 13th International Conference on Global Security, Safety and Sustainability, London, January 2021, pp. 149–156. Springer International Publishing, Cham (2021)
2. Taister, M.A., Holliday, S.D., Borrmman, H.I.M.: Comments on facial aging in law enforcement investigation. Forens. Sci. Commun. **2**(2) (2000)

3. Klare, B., Jain, A.K.: Face recognition across time lapse: on learning feature subspaces. In 2011 International Joint Conference on Biometrics (IJCB), pp. 1–8. IEEE (2011)
4. Coleman, S.R., Grover, R.: The anatomy of the aging face: volume loss and changes in 3-dimensional topography. Aesthet. Surg. J. **26**(1_Supplement), S4–S9 (2006). https://doi.org/10.1016/j.asj.2005.09.012
5. Goodfellow, I., Pouget-Abadie, J., Mirza, M., Xu, B., Warde-Farley, D., Ozair, S., Bengio, Y., et al.: Generative adversarial nets. In: Advances in Neural Information Processing Systems, p. 27 (2014)
6. Peng, T., Li, M., Chen, F., Xu, Y., Xie, Y., Sun, Y., Zhang, D.: ISFB-GAN: interpretable semantic face beautification with generative adversarial network. Expert Syst. Appl. **236**, 121131 (2024). https://doi.org/10.1016/j.eswa.2023.121131
7. Luo, X., He, X., Chen, X., Qing, L., Zhang, J.: Dualg-gan, a dual-channel generator based generative adversarial network for text-to-face synthesis. Neural Netw. **155**(155–167), 155 (2022). https://doi.org/10.1016/j.neunet2022.08.016
8. Kumar, M., Sharma, H.K.: A GAN-based model of deepfake detection in social media. Procedia Comput. Sci. **218**, 2153–2162 (2023). https://doi.org/10.1016/j.procs.2023.01.191
9. Cakir, A., Gurkan, M.: Modified query expansion through generative adversarial networks for information extraction in e-commerce. Mach. Learn. Appl. **14**, 100509 (2023). https://doi.org/10.1016/j.mlwa.2023.100509
10. Peng, Y., Sun, J., Ren, Y., Li, D., Guo, Y.: A histogram-driven generative adversarial network for brain MRI to CT synthesis. Knowl.-Based Syst. **277**, 110802 (2023). https://doi.org/10.1016/j.knosys.2023.110802
11. Kingma, D.P., Welling, M.: Auto-Encoding Variational Bayes (2013). arXiv preprint arXiv:1312.6114
12. Ou, W., Huynh, V.N.: Conditional variational autoencoder for query expansion in ad-hoc information retrieval. Inf. Sci. **652**, 119764 (2024). https://doi.org/10.1016/j.ins.2023.119764
13. Li, D., Zhang, Y., Yang, Z., Jin, Y., Xu, Y.: Sensing anomaly of photovoltaic systems with sequential conditional variational autoencoder. Appl. Energy **353**, 122124 (2024). https://doi.org/10.1016/j.apenergy.2023.122124
14. Bando, Y., Kuratate, T., Nishita, T.: A simple method for modeling wrinkles on human skin. In: 10th Pacific Conference on Computer Graphics and Applications, 2002. Proceedings, pp. 166–175. IEEE (2002). https://doi.org/10.1109/PCCGA.2002.1167852
15. Suo, J., Min, F., Zhu, S., Shan, S., Chen, X. (2007). A multi-resolution dynamic model for face aging simulation. In: 2007 IEEE Conference on Computer Vision and Pattern Recognition, pp. 1–8. IEEE. https://doi.org/10.1109/CVPR.2007.383055
16. Kemelmacher-Shlizerman, I., Suwajanakorn, S., Seitz, S.M. Illumination-aware age progression. In: Proceedings of the IEEE Conference on Computer Vision and Pattern Recognition, pp. 3334–3341 (2014). https://doi.org/10.1109/CVPR.2014.426
17. Kale, A., Altun, O.: Face age synthesis: a review on datasets, methods, and open research areas. Pattern Recogn. **109791**,(2023). https://doi.org/10.1016/j.patcog.2023.109791
18. Chen, X., Sun, Y., Shu, X., Li, Q.: Attention-aware conditional generative adversarial networks for facial age synthesis. Neurocomputing 451, 167–180 (2021). https://doi.org/10.1016/j.neucom.2021.04.068
19. Chandaliya, P.K., Nain, N.: AW-GAN: face aging and rejuvenation using attention with wavelet GAN. Neural Comput. Appl. **35**(3), 2811–2825 (2023). https://doi.org/10.1007/s00521-022-07721-4
20. Shi, C., Zhang, J., Yao, Y., Sun, Y., Rao, H., Shu, X.: CAN-GAN: conditioned-attention normalized GAN for face age synthesis. Pattern Recogn. Lett. **138**, 520–526 (2020). https://doi.org/10.1016/j.patrec.2020.08.021
21. Zhu, D., Liu, S., Jiang, W., Gao, C., Wu, T., Wang, Q., Guo, G.: Ugan: Untraceable Gan for Multi-Domain Face Translation (2019). arXiv preprint arXiv:1907.11418
22. Li, Z., Jiang, R., Aarabi, P.: Continuous face aging via self-estimated residual age embedding. In Proceedings of the IEEE/CVF Conference on Computer Vision and Pattern Recognition, pp. 15008–15017 (2021)

23. Zeng, J., Ma, X., Zhou, K.: Photo-realistic face age progression/regression using a single generative adversarial network. Neurocomputing **366**, 295–304 (2019). https://doi.org/10.1016/j.neucom.2019.07.085
24. Tatikonda, S., Nambiar, A., Mittal, A.: Face age progression with attribute manipulation. In: International Conference on Pattern Recognition and Artificial Intelligence, pp. 639–652. Springer International Publishing (2022, June)
25. Yang, H., Huang, D., Wang, Y., Jain, A.K.: Learning face age progression: a pyramid architecture of gans. In: Proceedings of the IEEE Conference on Computer Vision and Pattern Recognition, pp. 31–39 (2018)
26. Huang, Z., Zhang, J., Shan, H.: RoutingGAN: Routing age progression and regression with disentangled learning. In: ICASSP 2021–2021 IEEE International Conference on Acoustics, Speech and Signal Processing (ICASSP), pp. 2145–2149. IEEE (2021, June). https://doi.org/10.1109/ICASSP39728.2021.9414735
27. Sharma, N., Sharma, R., Jindal, N.: Prediction of face age progression with generative adversarial networks. Multimed. Tools Appl. **80**(25), 33911–33935 (2021). https://doi.org/10.1007/s11042-021-11252-w
28. Chandaliya, P.K., Nain, N.: ChildGAN: Face aging and rejuvenation to find missing children. Pattern Recogn. **129**, 108761 (2022). https://doi.org/10.1016/j.patcog.2022.108761
29. Chandaliya, P.K., Nain, N.: Child face age progression and regression using self-attention multi-scale patch gan. In 2021 IEEE International Joint Conference on Biometrics (IJCB), pp. 1–8. IEEE (2021, August). https://doi.org/10.1109/IJCB52358.2021.9484329.
30. Best-Rowden, L., Jain, A.K.: Longitudinal study of automatic face recognition. IEEE Trans. Pattern Anal. Mach. Intell. **40**(1), 148–162 (2017). https://doi.org/10.1109/TPAMI.2017.2652466
31. Kokate, P., Joshi, A.D., Tamizharasan, P.S.: An empirical comparison of generative adversarial network (gan) measures. In: Advances in Communication and Computational Technology: Select Proceedings of ICACCT 2019, pp. 1383–1396 (2021). Springer Singapore.
32. Huang, Z., Chen, S., Zhang, J., Shan, H. (2021). AgeFlow: Conditional Age Progression and Regression with Normalizing Flows. arXiv preprint arXiv:2105.07239. https://doi.org/10.24963/ijcai.2021/103
33. Liu, L., Wang, S., Wan, L., Yu, H.: Multimodal face aging framework via learning disentangled representation. J. Vis. Commun. Image Represent. **83**, 103452 (2022). https://doi.org/10.1016/j.jvcir.2022.103452
34. Barve, P.V., Joshi, A.D.: Cross Age Face Generator: A Generative Adversarial Networks (GANs) Based Approach. In: Advances in VLSI, Communication, and Signal Processing: Select Proceedings of VCAS 2021, pp. 39–53. Springer Nature Singapore, Singapore (2022)
35. Zhang, Z., Song, Y., Qi, H.: Age progression/regression by conditional adversarial autoencoder. In: Proceedings of the IEEE Conference on Computer Vision and Pattern Recognition, pp. 5810–5818 (2017)
36. Larsen, A.B.L., Sønderby, S.K., Larochelle, H., Winther, O.: Autoencoding beyond pixels using a learned similarity metric. In: International Conference on Machine Learning (pp. 1558–1566). PMLR (2016)
37. Ruder, S.: An Overview of Gradient Descent Optimization Algorithms (2016). arXiv preprint arXiv:1609.04747
38. Chen, B.C., Chen, C.S., Hsu, W.H.: Cross-age reference coding for age-invariant face recognition and retrieval. In: Computer Vision–ECCV 2014: 13th European Conference, Zurich, Switzerland, 6–12 Sept. 2014, Proceedings, Part VI 13, pp. 768–783. Springer International Publishing (2014)
39. Culjak, I., Abram, D., Pribanic, T., Dzapo, H., Cifrek, M.: A brief introduction to OpenCV. In: 2012 Proceedings of the 35th International Convention MIPRO, pp. 1725–1730. IEEE (2012)
40. Face++ homepage. https://www.faceplusplus.com/
41. Liu, Y., Li, Q., Sun, Z.: Attribute-aware face aging with wavelet-based generative adversarial networks. In: Proceedings of the IEEE/CVF Conference on Computer Vision and Pattern Recognition, pp. 11877–11886 (2019). https://doi.org/10.1109/CVPR.2019.01215

42. Dosovitskiy, A., Beyer, L., Kolesnikov, A., Weissenborn, D., Zhai, X., Unterthiner, T., Houlsby, N., et al.: An Image is Worth 16×16 Words: Transformers for Image Recognition at Scale (2020). arXiv preprint arXiv:2010.11929
43. Lahitani, A.R., Permanasari, A.E., Setiawan, N.A.: Cosine similarity to determine similarity measure: Study case in online essay assessment. In: 2016 4th International Conference on Cyber and IT Service Management, pp. 1–6. IEEE (2016)

Empowering Clinical Decision-Making with Generative AI in Intelligent Decision Support Systems

Mohsen Ghorbian and Saeid Ghorbian

Abstract Wilm's tumor (WT) is a type of cancer that primarily affects children. In medical centers, inadequate equipment and a shortage of experienced staff make it difficult for families suffering from this disease to receive post-treatment care after the treatment period is over. This chapter proposes an intelligent mechanism combining Machine Learning (ML), the Internet of Things (IoT), and blockchain technology to support families of patients with WT, which primarily affects children. The hybrid system enables continuous child health monitoring using IoT devices, with ML algorithms analyzing data to provide real-time health models. The system categorizes the child's condition into acute, requiring further investigation, and normal. By incorporating blockchain technology, the system ensures secure data exchange, addressing privacy concerns. This intelligent mechanism aims to improve early detection of disease recurrence, thereby enhancing recovery rates and survival chances. The proposed system demonstrates significant potential in reducing diagnosis time and providing timely medical intervention.

Keywords Machine learning · Internet of things · Blockchain · Wilm's tumor

1 Introduction

The WT is a type of childhood cancer that develops in the colon. Most cases occur in children between the ages of two and five. A Wilms tumor usually occurs only in one clone and may arise in several tissues within the clone. However, genetic factors and the mother's age during pregnancy may contribute to the development of this type of cancer, which has not yet been determined. This disease can recur after

M. Ghorbian
Department of Computer Engineering, Qom Branch, Islamic Azad University, Qom, Iran
e-mail: mohsen.ghorbian@iau.ir

S. Ghorbian (✉)
Department of Molecular Genetics, Ahar Branch, Islamic Azad University, Ahar, Iran
e-mail: saeid.ghorbian@iau.ac.ir

N. R. Vajjhala et al. (eds.), *Generative Artificial Intelligence (AI) Approaches for Industrial Applications*, Information Systems Engineering and Management 24,
https://doi.org/10.1007/978-3-031-76710-4_12

treatment, as with all cancers. Identifying the recurrence (return) of Wilms tumor after treatment is extremely important [1]. The patient must be followed carefully and provided with continuous care after the initial treatment of Wilms tumor. Several imaging methods, potentially enhanced by AI Generative technologies, can be used to detect recurrence. These include abdominal ultrasound, radiography, CT scan, and MRI. Blood and urine tests are conducted during the follow-up process to identify changes indicative of tumor recurrence. Upon identifying suspicious symptoms or changes in tests, additional tests, potentially guided by AI Generative insights, may be ordered, including a biopsy, to confirm the presence of recurrence of the tumor. By using these methods, doctors can provide appropriate and timely treatment for Wilms tumor recurrence if necessary [2]. IoT refers to a set of electronic devices connected to the Internet that can exchange data with each other. These devices include sensors, medical devices, and other electronic devices. Using sensors and wearable devices, IoT can monitor and control cancer patients and collect information about their daily activities, such as sleep, activity, and food consumption. In turn, physicians can use this information, along with new methods and techniques derived from AI Generative analytics methods, to better understand the patient's condition and provide appropriate treatment or, if necessary, modify the treatment process. With the use of IoT, complemented by AI Generative models, cancer patients can improve control, monitoring, and quality of treatment. It is essential, however, to address data privacy and security issues and protect sensitive patient data from unauthorized access to utilize IoT effectively [3]. ML is a subset of artificial intelligence that enables computers to predict and learn from new data. They provide highly accurate diagnoses and treatments for a wide range of diseases. ML in cancer diagnosis allows us to identify patterns in the data collected from cancer patients to assist in cancer diagnosis and predict the likelihood of disease recurrence. AI Generative tools can enhance this process by generating synthetic data or predictive models. As a result, doctors can better plan their treatment and follow-up. It is important to note that ML and generative artificial intelligence are related fields within the field of artificial intelligence. Each field has its focus and methodology. As a subset of artificial intelligence, ML involves training algorithms to learn patterns from data and make predictions or decisions without explicit programming. It includes techniques such as supervised learning and reinforcement learning. By contrast, generative artificial intelligence generates original and original images, texts, and audio files. Generative models, such as generative adversarial networks (GANs), create synthetic data similar to real data using ML techniques. Thus, generative artificial intelligence is considered a subset of ML in generating new and creative data, as it utilizes ML techniques [4]. Blockchain technology is a digital technology that provides high levels of security and transparency for data exchange. Blockchain data is stored in blocks containing a list of transactions and a security code (hash) linked to previous blocks. This technology's high level of security and transparency makes it an ideal technology for healthcare use. Blockchains can provide adequate security for sensitive patient data when controlling and monitoring patients via IoT, potentially in conjunction with AI Generative methods [5]. In addition to protecting patient privacy and preventing unauthorized access to patient data, blockchain identifies and tracks changes or modifications to patient data by storing it securely in interconnected blocks. As a result, patients' data can be securely stored in interconnected blocks using blockchain, and AI Generative learning machines can be trained based on this information to diagnose and predict cancer recurrence. Therefore, this increases the accuracy and efficiency of the cancer diagnosis and the prediction of its recurrence. As a result, blockchain technology can address issues related to the privacy and security of sensitive patient data and problems associated with IoT and ML [6]. This chapter presents a new Intelligent mechanism based on IoT, ML, and blockchain technologies. This intelligent mechanism attempts to address the challenges and concerns associated with the use of new technology in the field of healthcare through its presentation, including maintaining data security, protecting a person's identity, preventing unauthorized access to data, preventing doctors from promptly checking patients' conditions, and ensuring that patients can reach their doctors when necessary.

This chapter is structured in the following manner: The first part examines the necessary prerequisites for cancer WT, IoT, ML, and Blockchain technologies, and the capabilities, capabilities, and features of IoT, ML, and blockchain technology for diagnosing WT in healthcare. In the second part of this chapter, we explore the use of WT, ML, Blockchain, and IoT technologies in healthcare individually. The third part examines the application of blockchain and IoT in healthcare by examining subsets, such as an overview of HLF blockchain technology, identification of WT using various techniques, and classification approaches and methods in ML. In the fourth section of this chapter, ML and IoT are comprehensively examined through blockchain technology. In the fifth section, the proposed intelligent mechanism is presented. As a final step, the sixth section concludes this explanation.

2 Background

This section provides information and terminology related to WT and technologies such as blockchain, ML, and the IoT. These technologies have also been discussed as solutions for detecting diseases, monitoring and controlling affected patients, protecting patients' privacy, and demonstrating that they can enhance treatment effectiveness and reduce the time required to identify recurrences.

2.1 WT Cancer

WT is a rare form of childhood cancer that can be treated. Generally, this type of cancer is found in children between the ages of two and five and is often detected incidentally, meaning without any specific symptoms. It remains unclear why WT occurs. This type of cancer can occur in both sexes, and there has been no evidence that it is more prevalent in one gender than the other. Thus, WT incidence is similar between men and women, and both sexes may be affected. However, evidence suggests that genetic factors and family disorders play a significant role in its occurrence. WT symptoms include a mass in the abdomen, swelling, loss of appetite, fever, urine bleeding, and increased kidney volume. A biopsy of the tumor mass is performed to confirm the diagnosis. An abdominal ultrasound, CT scan, and MRI are commonly used to diagnose WT [7]. The treatment for Wilms tumors generally consists of surgery to remove the tumor along with the affected kidney, as well as chemotherapy as well as radiation therapy in some cases. Surgery to remove the tumor mass and kidney is performed in order to destroy the remaining cells and reduce the risk of recurrence. In addition to chemotherapy, radiation therapy may be used if necessary. Following the treatment, a thorough follow-up and postoperative care program is conducted to prevent tumor recurrence. Hence, this includes regular monitoring, frequent imaging, and blood tests. In the event of suspicious symptoms or signs of recurrence, the attending physician must be informed immediately so that the necessary steps can be taken for retreatment [8].

2.2 Blockchain Technology

The blockchain technology is used to store and manage data. Blockchain is a distributed system that stores data in blocks and links each block to the previous one. Thus, it is protected from fraud and data changes and provides a high-security level. There are two types of blockchains in general: permission-less and permission-based. The permission-less Blockchain is a distributed system that enables storing and verifying financial and non-financial transactions without central management. The Blockchain facilitates the review, approval, and rejection of recorded data and transactions by anyone accessing them and using such blockchains for e-commerce, supply chains, and internet payments. Bitcoin is one of the most widely recognized permission-less blockchains. Bitcoin uses blockchain technology, and all transactions are recorded and verified publicly without needing a management center. Ethereum is another example of a permission-less blockchain, a platform for developing and managing Ethereum applications. It uses Solidity as a programming language to develop Ethereum smart contracts and applications. Due to their high levels of security, transparency, and lack of management centers, permissionless blockchains are widely used in many different sectors [9]. In light of their lack of dependence on companies and organizations, these blockchains are a novel method of establishing trust and transparency in digital environments. In addition to permission-based blockchains, permission-based systems are generally designed for use in organizations and businesses. A limited number of individuals, often including the organization's members, have the right to access the Blockchain and verify transactions in this type of Blockchain. Blockchain users are required to have access permissions granted by blockchain management. The Linux Foundation (Linux Foundation) has developed HLF as a permission-based blockchain designed for organizations and corporations. This Blockchain has the characteristics of transparency, security, reliability, and repeatability and is used as a development platform for developing applications in various languages, including Go, Java, and Node.js. It is more secure to use permission-based blockchains than permission-less blockchains, as they are only authenticated by those who have access to them. These blockchains are also typically used for specific applications, such as supply chains, insurance, banking, etc. Blockchain can serve numerous purposes in healthcare and cancer diagnosis as a new technology [10]. A permission-based blockchain offers increased security and greater control over data access. Blockchain can store and share medical data about cancer patients in this field. The medical data of many patients is contained in several different systems, most of which need to communicate with one another, resulting in difficulty in diagnosing and treating patients. Blockchain technology can store medical data in a distributed and integrated system. As a result, doctors and health professionals can diagnose and treat patients accurately by having full access to their patient's medical data [11]. Blockchain technology can also be utilized for cancer diagnosis and treatment research. It is possible to store and share data automatically without requiring intermediaries, and these data can contribute to the diagnosis and treatment of cancer. Figures 1 and 2 illustrate the structure of the permission-less and permission-based blockchain, respectively.

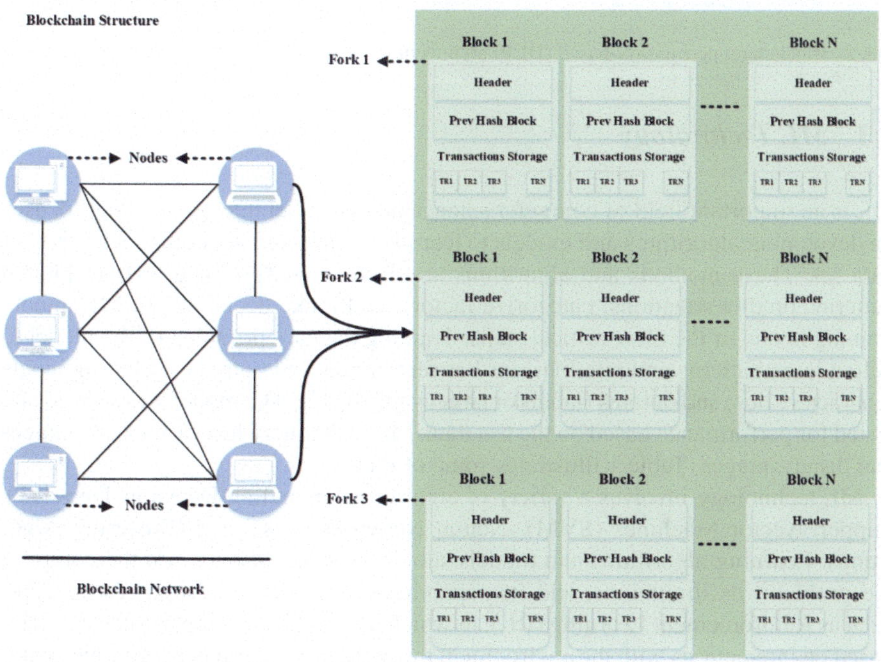

Fig. 1 Blockchain permission-less architecture

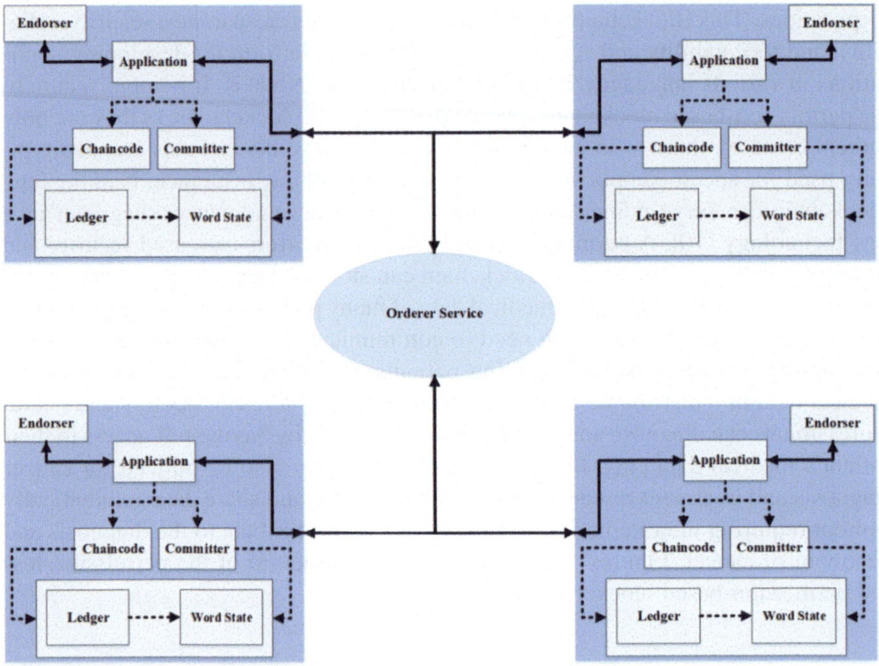

Fig. 2 Blockchain permission-based (HLF) architecture

2.3 ML Technology

ML is an important field of computer science and artificial intelligence that focuses on developing algorithms and models to learn from data and generalize it to another data set. These methods and algorithms are used to analyze data, extract hidden patterns, predict situations, categorize factors, and make decisions. In ML, model training is one of the most critical steps. Training data are provided to the model to help it learn patterns and relationships between inputs and outputs, including inputs (characteristics) and outputs (labels or target values) [12]. The model is evaluated and tested for performance based on the test data. ML technology has various advantages and disadvantages. Table 1 illustrates some of these.

ML technology involves a variety of algorithms, including Decision Tree (DT), Support Vector Machines (SVM), Neural Networks (NN), and Clustering Algorithms. Learning algorithms and models should be selected based on the nature of the problem, its limitations, and its capabilities. ML also encompasses concepts such as Reinforcement Learning (RL), in which a model learns appropriate behavior based on interactions with the environment through rewards and penalties, and semi-supervised learning, in which labeled and unlabeled data is used to train the model. ML is widely utilized in many fields, including natural language processing, pattern recognition, imaging, audio processing, extensive data analysis, robotics, etc. Due

Table 1 A comparison of the advantages and disadvantages of ML

Advantage	Disadvantage
Analysis of big data for learning	Data sets must be complete and of high quality
Analyzing data to detect hidden patterns	Availability and quality of data are important factors
Predictive and decision-making abilities based on data	Interpretation and justification of model performance are complex
Utilization of semi-structured or heterogeneous data	Processing power and large resources are required
Utilization of new data for improving and updating models	Parameter setting and model selection require specialist knowledge
Generalizable and flexible compared to new data	Making decisions may be subject to bias and instability
Applications in various fields, including medicine, business, automobiles, etc.	Issues related to privacy and security in the use of data

to the ever-growing amount of data and technological advancements, ML has gained much attention as a powerful tool for solving complex problems and providing accurate predictions [13].

2.4 *IoT Technology*

The IoT is a network of physical objects equipped with sensors, communication devices, and Internet connections that enable communication and data exchange between them. In this technology, objects and smart devices can communicate, creating a network of connected objects. As a result, these objects can collect and send data to various systems for analysis and processing, increasing efficiency, optimizing processes, and developing various applications in fields such as smart cities, smart buildings, and Industry 4.0 [14]. It is important to understand that, like all technologies, the IoT has advantages and disadvantages. Knowing these advantages and disadvantages will help users gain a more comprehensive understanding of this technology. Table 2 shows several advantages and disadvantages of the IoT.

By connecting various objects to Internet networks and utilizing the IoT, creating a dynamic, intelligent world has become possible. Hence, IoT is widely used in several sectors, including smart homes, industries, healthcare, transportation, healthcare, agriculture, and smart cities. Some of the applications include:

- **Smart Home**: The IoT enables different objects in the home to be connected and provides capabilities such as temperature control, lighting, security systems, and other amenities.

Table 2 A comparison of the advantages and disadvantages of IoT

Advantage	Disadvantage
Interactions between objects and their relationships	Privacy protection concerns
Enhancements to the functionality and services provided	Implementation and management complexity
Optimizing processes and increasing productivity	Communication problems and wireless interference
Collecting and analyzing large amounts of data	Increasing security risks and cyber threats
Increasing productivity and reducing costs	Issues relating to the legal protection of data and property
Development of smart cities and intelligent resource management	Heterogeneous standardization
Improving the quality of life and the intelligence of individuals	The need for a strong and stable infrastructure

- **Smart Industries**: IoT can improve the performance of several industries, including equipment forecasting and maintenance, supply chain management, and quality control.
- **Healthcare**: The IoT allows medical equipment and monitoring systems to be connected to a network, improving patient monitoring and treatment [15].
- **Smart Transportation**: Connecting cars to the IoT enables traffic monitoring and management, tracking and navigation, smart car services, and enhanced driver safety.
- **Smart Agriculture**: The IoT can monitor and control agricultural products, improve irrigation systems, and control greenhouse environments.
- **Smart City**: IoT can connect and manage various objects in cities, including smart lighting systems, parking management systems, air pollution monitoring systems, and energy resource management systems [16].

Therefore, the applications of IoT are extensive. Interaction between objects enables intelligent facilities and services to be provided for users. Figure 3 illustrates the structure of IoT technology.

3 Incorporating Technologies into Healthcare

Several new technologies, including the IoT, ML, and blockchain, have been applied to healthcare. In addition to creating a context for presented efficient services in healthcare, these technologies can contribute significantly to ensuring that accurate and reliable data is used to provide health and medical services. By integrating IoT technology, ML, and blockchain into the healthcare field, facilities such as early disease detection, better treatment management, preventing complications, and

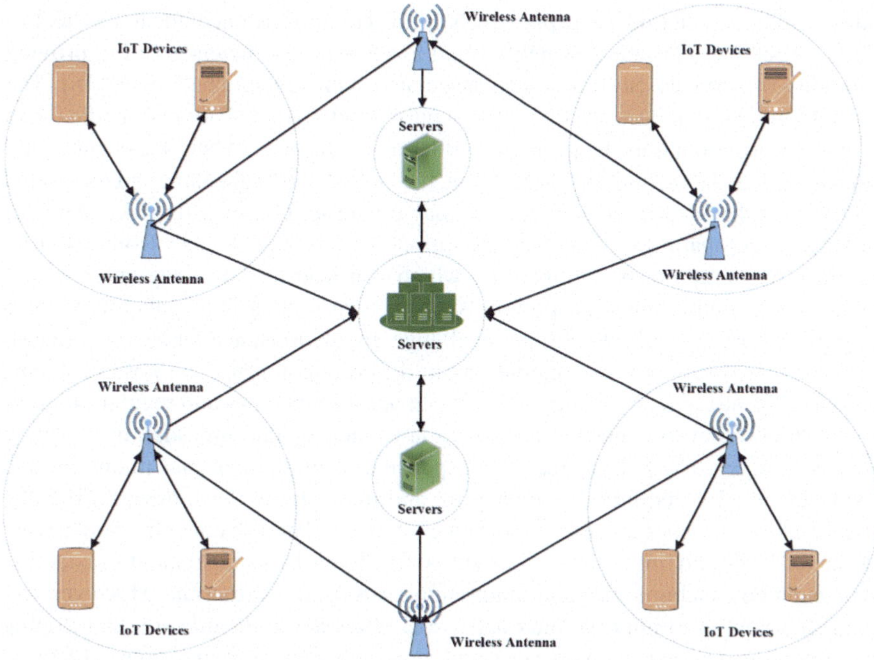

Fig. 3 IoT implementation architecture

promoting community health can be improved. Through these technologies, healthcare professionals and physicians have access to powerful tools, facilitating global improvements in the health system. The following will discuss the impact of each of these technologies on healthcare [17].

3.1 *IoT in Healthcare*

With the rapid growth of IoT technology, its applications in healthcare are also on the rise. This field has many applications for IoT, including monitoring physical activity and health, remote management of chronic diseases, access to online medical services, and even communication between medical devices and doctors through IoT. A major benefit of IoT in healthcare is that it increases the accuracy and efficiency of diagnosing and predicting diseases. Due to the automatic collection of data from sensors and devices connected to the Internet, this information can be analyzed more accurately and promptly for diagnosing and predicting diseases and health conditions. Additionally, IoT can improve accessibility to medical services. In addition to consulting and conducting online treatment sessions, traditional medicine, and even prescribing drugs remotely, IoT can also be used to provide a wide range of services. People living in remote areas or who require constant follow-up due to health conditions will find this especially helpful. The application of the IoT in healthcare is associated with many benefits, such as increasing efficiency and improving the quality of services. Still, it is also associated with several challenges [18]. The most important challenges are related to security and privacy. Patients' medical and personal information must be protected with great care since cyberattacks can result in the theft of such information. In addition, the lack of comprehensive and integrated standards for the security of IoT devices may create serious security risks. Another challenge is compatibility and standardization. Various IoT devices from different manufacturers may not be compatible, which can lead to data integration issues. Differences in communication protocols and device standards can also make data integration more difficult. As another challenge, regulations, and legislation are also considered. Various laws and regulations in different countries can pose problems for collecting, storing, and using medical data because the laws and regulations vary widely. In addition to constant changes in laws and regulations, adapting to these changes is necessary. IoT devices are expensive to implement and maintain, and expert personnel is required for managing and maintaining these devices. Finally, ethical and social issues are also important challenges. There is concern about excessive surveillance and limitations of individual freedoms. Ensuring patient satisfaction when using new technologies and maintaining trust is also important. However, IoT has great potential to improve the quality and efficiency of healthcare. Developing and improving relevant technologies and standards can overcome many of these challenges [19].

3.2 ML in Healthcare

ML technology has wide and diverse applications in the healthcare field that can significantly improve the quality and efficiency of services. A primary function of ML in the healthcare field concerns disease diagnosis. ML algorithms can identify hidden patterns in large medical data and highly accurately detect diseases. This ability is beneficial when diagnosing diseases such as cancer, heart disease, and diabetes, where early detection can make a significant difference in treatment success. In addition to diagnosing, ML is also used to predict treatment outcomes. Prediction algorithms can evaluate the probability of success of different treatments using past patient data and suggest the most appropriate treatment options [20]. With these predictions, doctors can make more informed decisions, and patients can better understand their treatment path. Additionally, this technology can be effective in reducing costs and optimizing resources. Another important application of ML in healthcare is the personalization of treatment. ML algorithms can design specific patient treatment plans based on their genetic data, lifestyle, and medical history. This approach is particularly effective in treating chronic diseases and long-term health management since it allows each patient to receive customized treatment. Despite the many benefits of ML technology in healthcare, this field also faces several challenges, most importantly regarding data security and privacy. Medical information is very sensitive, and using ML algorithms requires access to a large amount of this data, making it critical to protect it from unauthorized access and cyberattacks. Data quality and accuracy are also important challenges in this field. ML algorithms require accurate and complete data to obtain accurate results. Medical data may be incomplete or inaccurate, leading to inaccurate results or decisions. In addition, providing adequate and diverse data for the training of ML models can be challenging. Lastly, legal and ethical issues are also considered. Laws and regulations about medical data and ML algorithms vary from country to country, and complying with these laws can be difficult. Additionally, ethical issues regarding using artificial intelligence algorithms in medical decisions and their potential effects on patients' rights should be considered. Due to these challenges, using ML technology in health care requires careful planning, cooperation between specialists, and attention to security, legality, and ethical issues [21].

3.3 Blockchain in Healthcare

Blockchain technology can significantly increase healthcare systems' security, transparency, and efficiency. A significant application of blockchain technology in this area is medical records management. Hence, blockchain can store patients' medical records securely and immutably. When a new medical record is added to the system, it is added to a chain of blocks, each containing previous and new information. Using this feature makes it nearly impossible to change or delete past information, ensuring the integrity and security of medical records [22]. Blockchain can also be used to improve the supply chain of pharmaceutical products in healthcare. Drug fraud is a global problem that can have severe consequences for patients' health. Blockchain provides a transparent and reliable tracking system that can help alleviate this problem. Blockchain technology can track all medicine production and distribution phases, and any fraudulent activity or product changes can be quickly identified. With the use of this technology, the drug supply chain system can be improved, and counterfeit drugs can be prevented from entering the market. Also, blockchain can facilitate medical research and data sharing between researchers and health institutions. The greatest challenge in medical research is ensuring that reliable and high-quality data can be collected and shared. By creating a secure and transparent platform for sharing research data, blockchain can provide researchers with access to the necessary information without the fear of manipulation or accuracy. Hence, this can facilitate the research process, lead to the discovery of new treatments, and improve society's health [23]. Despite blockchain technology's many benefits in health care, several challenges exist. A significant challenge for blockchains is scalability. Public blockchains, including Bitcoin and Ethereum, may not be able to handle large volumes of transactions promptly. As a result, using blockchain in healthcare systems with a large amount of medical data and transactions can be challenging. This technology also needs to improve in terms of standards and compatibility. The absence of unified standards for implementing blockchain in health systems can create problems in terms of compatibility and cooperation between different organizations and systems. Creating an integrated and coordinated system can be difficult if each organization uses different protocols and standards. Additionally, the cost of implementing and maintaining a blockchain system is a significant challenge. Establishing and maintaining an efficient and secure blockchain system can be costly and require substantial financial and technical resources. Providing timed human resources to manage and develop these systems can also be challenging. Despite these challenges, the development and implementation of blockchain technology in healthcare, combined with careful planning, collaboration among specialists, and careful consideration of security, legal, and ethical considerations, will solve many of these problems and significantly improve this field [24].

4 Integrating Technologies in Healthcare: Prospects and Challenges

By integrating IoT, ML, and blockchain technologies into healthcare, healthcare services can be significantly improved in quality, security, and efficiency. In addition to solving challenges inherent in the independent use of these technologies, this combination can create new opportunities in the field. A significant benefit of integrating these technologies is improved accuracy and efficiency in diagnosing and treating diseases. Patients' heart rates, blood sugar levels, and other vital signs can be continuously collected in real-time using IoT devices. Data can be analyzed, and patterns can be identified using ML, leading to a faster and more accurate diagnosis of diseases [25]. Furthermore, the blockchain ensures that patient information remains accurate and unaltered by storing it securely and immutably. It is also important to note that this combination increases the security and privacy of medical data, which is extremely sensitive and requires high protection. The data collected by IoT devices is encrypted and stored in immutable blocks using blockchain, increasing information security. Furthermore, ML can help protect data from cyberattacks and security threats. Additionally, the integration of these technologies can improve the management of drug supply chains. Every step from production to distribution can be accurately recorded using blockchain technology, making the supply chain transparent and traceable. As a result, fraud and authenticity are reduced, and IoT devices can monitor the storage conditions of drugs throughout the supply chain, ensuring they are transported and stored correctly [26]. Using ML, it is also possible to optimize processes and reduce costs by analyzing data related to supply chains. On the other hand, one of the benefits of combining these technologies is improving drug supply chains. Blockchain technology can create a transparent and traceable supply chain by accurately recording every process step, from drug production to distribution. In addition to reducing fraud, IoT devices can assist in monitoring the storage conditions of drugs throughout the supply chain, ensuring they are transported and stored appropriately. Supply chain data can also be analyzed using ML to optimize processes and reduce costs. Combining these technologies also provides the opportunity for personalized treatment. Detailed information on each patient's health status can be obtained by collecting data from IoT devices. ML can analyze this data and suggest an individual treatment plan based on that data. This information can be stored securely on the blockchain and only accessed by authorized individuals. Therefore, these technologies can increase collaboration and data sharing between medical centers and researchers. Thus, medical data can be shared securely and transparently using blockchain technology, with every change and access being tracked [27]. By analyzing this data, ML can uncover new patterns and assist in developing new treatments. IoT devices can also be used to provide additional data for medical research. Despite the significant benefits of combining the IoT, ML, and blockchain technologies in healthcare, these technologies also present several challenges that must be addressed. Following is a review of some of the most important challenges associated with this area.

- **Security**: This combination of technologies may provide an easier means of accessing sensitive medical data, which may pose security risks. Strong cryptographic methods, stringent authentication mechanisms, and physical security measures, such as limiting access to IoT devices, must address this challenge.
- **Technical Complexity**: Integrating and coordinating these three technologies may present technical challenges. To overcome this challenge, standards and protocols must be developed that allow seamless communication between them [28].
- **Regulations and Legal**: Compliance with privacy laws, security regulations, and issues relevant to the use of medical data are possible challenges. The solution to this challenge is to develop rules and regulations related to these technologies.
- **Trust and Acceptance**: The ability to trust these procedures and accept these systems may require more training and awareness. Doctors and patients must receive the necessary information and training to solve this challenge [29].
- **Financial Problems**: Implementing and maintaining such a system may be costly, particularly if it is required to measure and address technical and security challenges. To resolve this challenge, appropriate budgets and financial resource optimization are necessary.
- **Complexity in Data Management**: Managing and maintaining sensitive medical data and ensuring its accuracy and quality can be challenging in an environment where three different technologies are used to collect data. Hence, advanced data management systems and standard processes must be used to resolve this challenge [30].

Aside from the many benefits this combination of technologies brings to the healthcare field, some challenges can be managed and controlled by appropriate management approaches and attention to technical, security, and legal issues.

5 Intelligent Early Detection Mechanism by Integrating Technologies

LDH (lactate dehydrogenase) is an important enzyme in the body's metabolic processes. In addition to converting lactate into pyruvate, it also contributes to cellular energy production. Many tissues and organs of the body contain this enzyme, and its levels are usually elevated in response to cell damage or tumor growth. In Wilms tumor disease, which is a rare and cancerous type of kidney tumor in children, LDH level can be used as an important biological marker for diagnosis and monitoring of disease recurrence. Detecting the recurrence of Wilms tumors early after initial treatment can pose a significant challenge for patients whose prognosis and treatment results can be adversely affected by the recurrence of the disease. Hence, regular monitoring of LDH levels can assist doctors in detecting the recurrence of disease earlier and initiating appropriate treatment measures as soon as the level of LDH in the blood increases [31]. By integrating the IoT, ML, and blockchain technologies, Wilms tumor recurrence can be detected early using the biological marker Lactate Dehydrogenase (LDH). With these technologies, medical data can be collected accurately and timely, analyzed to predict tumor recurrence, and ensured for security and accuracy.

- **Data Collection Using IoT**: IoT devices, including sensors and wearables, can continuously and noninvasively monitor the level of LDH in patients' blood. These devices collect real-time data and send it to a central server. Physicians and specialists can use this continuous data to detect changes in LDH levels and early signs of Wilms tumor recurrences. By using IoT in data collection, monitoring can be more accurate and efficient, as data is collected continuously and in real-time.
- **Analyzing Data Using ML**: ML algorithms can analyze data collected by IoT devices, identifying hidden patterns and trends and predicting the likelihood of tumor recurrence. ML models can provide early warnings about tumor recurrence by using changes in LDH levels and other health-related information. Doctors can then make better treatment decisions and initiate early treatment and prevention measures.
- **Ensuring Data Security and Integrity with Blockchain**: Blockchain technology can assist in maintaining the security and accuracy of medical records. Blockchain technology facilitates data storage in a distributed, immutable manner, preventing tampering and unauthorized access to sensitive data. Blockchain technology ensures that data collected from IoT devices and analyzed by ML algorithms is secure and reliable. Moreover, it contributes to the creation of greater transparency in the diagnostic and treatment processes, as well as an increase in patients' trust in digital health systems.

This section illustrates the proposed mechanism using a scenario whose architecture and implementation structure are shown in Fig. 4. The blockchain mechanism used in this scenario is HLF. Users can send blood samples through the proposed intelligent mechanism utilizing IoT machines that measure LDH. The machines can connect to a blockchain network and send a request to connect. Hence, only authorized IoT machines can access the network after the authentication mechanism has authenticated users. Secondly, IoT machines act as applications to access the blockchain network. After the Endorser nodes approve IoT device requests, they can request their information inserted into the Orderer Service (OS). OS queues and executes requests based on the order in which they were submitted. OS then sends the available requests to Chaincode (CC) for execution after placing the requests in the queue, which is communicated through a request. A request for implementation as a query is sent to the Ledger by CC when it receives a request from the OS. CC tries to provide the conditions for implementing the request as a query. In response to the request for data to be inserted in the form of a query, the Ledger attempts to store the data, and after it has successfully stored the data, it sends CC a response message. Once CC receives a response from the Ledger confirming that the data has been successfully inserted, it sends a response message to the authentication mechanism. Upon receiving a response from the authentication mechanism, it attempts to send a message to the IoT machines. As soon as the IoT machines receive the message, the machines display a response indicating that the storage process has been completed. Multiple users can utilize the proposed intelligent mechanism simultaneously. Thus, when users enter information regarding their LDH levels, the intelligence mechanism predicts the patient's state. Consequently, it attempts to obtain data from the OS. Upon receiving the request and reading the requested data, CC makes a Query and asks the Ledger to reply. As soon as the Ledger has received the request and read the data, the Ledger tries to notify CC through a response. When CC gets a response to its request, it tries to inform the intelligent mechanism by creating a response that the requested data is available. Upon receiving the preferred data, the intelligent mechanism uses ML methods to predict the trend of increased LDH levels in patients based on the data analysis. As a result, the desired data is sent to the ML system, which implements it by sending a request. The desired data and information of the intelligent mechanism are used in this step to implement ML techniques to predict the trend of increased LDH levels in patients and send the resulting results as a response. After receiving the results, the intelligent mechanism requests the assessing condition mechanism to determine the patient's condition. This stage involves the evaluation of their condition based on a prediction of the increasing trend in their blood levels of LDH. The next step is to send the IoT machines a request informing the patients of their condition and the need to contact the doctors as soon as possible. The IoT machines then display a warning message indicating the patient's condition.

The integration of these three technologies can lead to a significant improvement in the early detection of Wilms tumor recurrence. Hence, using received data and advanced analysis, this integrated system improves the accuracy and speed of diagnosis and can also help in the early detection of tumor recurrence. By ensuring the accuracy and security of data through blockchain, patients and healthcare professionals will have a greater trust in these systems and will be more inclined to accept them. Despite the many advantages, implementing this combination of technologies also comes with challenges. These challenges include high costs, the requirement for

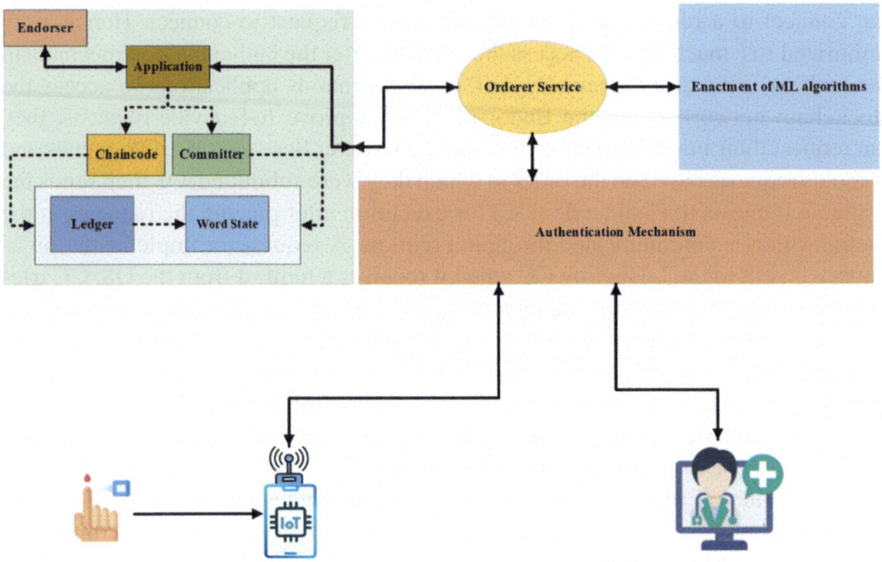

Fig. 4 Intelligent early detection mechanism architecture

advanced infrastructure, technical complexity, and privacy concerns. Hence, appropriate investments must be made to resolve these concerns, a culture of care must be developed among health workers and patients, and legislation and regulations must be drafted to address these challenges.

6 Conclusion

Since healthcare is directly related to human health, it is always considered a sensitive area. The lack of equipment and medical staff in hospitals always challenges this field. After the completion of the treatment period for a severe disease such as cancer, the patient is expected to remain under continuous supervision and care. A type of cancer commonly found in children is known as WT. When people get this disease, they must be monitored and controlled by their doctor if they recover, as the disease may recur. Considering the shortages, health concerns, and many patients, doctors and medical centers can only control and monitor a limited number. A delay in diagnosis can result in irreparable costs for patients who could experience disease recurrence. Due to this, emerging technologies are always considered as a means of reducing existing challenges and concerns. The solution includes the ability for doctors to control and monitor patients remotely and the ability for ML to predict disease recurrence and early diagnosis as part of the solution. This chapter presents an intelligent mechanism utilizing new technologies such as the IoT, ML, and Blockchain. This mechanism aims to address the challenges and concerns patients face in medicine. Patients can communicate with doctors and medical centers remotely using the proposed intelligent mechanism and share their health information. The proposed intelligent mechanism attempts to secure shared data and protect patient privacy and identity. A permission-based blockchain, such as HLF, protects all patient information on the Blockchain. The proposed intelligent mechanism analyzes data that patients have stored in the Blockchain to predict the risk of recurrence of WT cancer by using ML. As a result of the application of ML technology, the proposed intelligent mechanism attempts to categorize the patient's condition by examining the results. Therefore, the patient's condition may be divided into three categories: acute, requiring further investigation, and normal. Hence, the proposed intelligent mechanism attempts to predict the possibility of recurrence of WT cancer and transmit the results to physicians to facilitate early diagnosis of its recurrence. Therefore, patients can access physicians and specialists more quickly and without delay through the intelligent mechanism considered. This proposed intelligent mechanism allows people whose disease relapses are severe and acute to contact their doctor immediately and under their direct supervision and control.

In the future, the proposed approach will focus on utilizing cloud computing to improve accessibility for a broader audience. In addition, the availability of these machines can be increased if the implementation process is simplified and novel approaches, such as serverless computing, are considered to decrease the amount of energy consumed by IoT machines.

References

1. Spreafico, F., Fernandez, C.V., Brok, J., Nakata, K., Vujanic, G., Geller, J.I., Gessler, M., Maschietto, M., Behjati, S., Polanco, A., Paintsil, V.: Wilms tumour. Nat. Rev. Dis. Primers. **7**(1), 75 (2021)
2. de Carvalho, L.G., Kobayashi, T., Cypriano, M.D.S., Caran, E.M.M., Lederman, H.M., Alves, M.T.D.S., Abib, S.D.C.V.: Diagnostic errors in Wilms' tumors: learning from our mistakes. Front. Pediatr. **9**, 757377 (2021)
3. de Queiroz, D.A., da Costa, C.A., de Queiroz, E.A.I.F., da Silveira, E.F., da Rosa Righi, R.: Internet of things in active cancer treatment: a systematic review. J. Biomed. Inform. **118**, 103814 (2021)
4. Swanson, K., Wu, E., Zhang, A., Alizadeh, A.A., Zou, J.: From patterns to patients: advances in clinical machine learning for cancer diagnosis, prognosis, and treatment. Cell (2023)
5. Egala, B.S., Pradhan, A.K., Badarla, V., Mohanty, S.P.: Fortified-chain: a blockchain-based framework for security and privacy-assured internet of medical things with effective access control. IEEE Internet Things J. **8**(14), 11717–11731 (2021)
6. Singh, S., Rathore, S., Alfarraj, O., Tolba, A., Yoon, B.: A framework for privacy-preservation of IoT healthcare data using federated learning and blockchain technology. Futur. Gener. Comput. Syst. **129**, 380–388 (2022)
7. Karim, A., Shaikhyzada, K., Abulkhanova, N., Altyn, A., Ibraimov, B., Nurgaliyev, D., Poddighe, D.: Pediatric extra-renal nephroblastoma (Wilms' tumor): a systematic case-based review. Cancers **15**(9), 2563 (2023)
8. Zhu, Y., Fu, W., Huang, Y., Sun, N., Peng, Y.: Imaging features and differences among the three primary malignant non-Wilms tumors in children. BMC Med. Imaging **21**, 1–8 (2021)

9. Rankhambe, B.P., Khanuja, H.K.: A comparative analysis of blockchain platforms–Bitcoin and Ethereum. In: 2019 5th International Conference on Computing, Communication, Control and Automation (ICCUBEA), pp. 1–7. IEEE (2019)
10. Ghorbian, M., Ghobaei-Arani, M.: A blockchain-enabled serverless security mechanism for IoT-based drones. In: Building Cybersecurity Applications with Blockchain and Smart Contracts, pp. 55–82. Springer Nature, Cham, Switzerland (2024)
11. Shi, J., Li, R., Hou, W.: A mechanism to resolve the unauthorized access vulnerability caused by permission delegation in blockchain-based access control. IEEE Access **8**, 156027–156042 (2020)
12. Sarker, I.H.: Machine learning: algorithms, real-world applications and research directions. SN Comput. Sci. **2**(3), 160 (2021)
13. Arora, N., Singh, A., Al-Dabagh, M.Z.N., Maitra, S.K.: A novel architecture for diabetes patients' prediction using K-means clustering and SVM. Math. Probl. Eng. (2022)
14. Laghari, A.A., Wu, K., Laghari, R.A., Ali, M., Khan, A.A.: A review and state of art of Internet of Things (IoT). Arch. Comput. Methods Eng., pp 1–19 (2021)
15. Bhuiyan, M.N., Rahman, M.M., Billah, M.M., Saha, D.: Internet of things (IoT): a review of its enabling technologies in healthcare applications, standards protocols, security, and market opportunities. IEEE Internet Things J. **8**(13), 10474–10498 (2021)
16. Qian, Y., Wu, D., Bao, W., Lorenz, P.: The internet of things for smart cities: technologies and applications. IEEE Network **33**(2), 4–5 (2019)
17. Ghorbian, M., Ghobaei-Arani, M.: A Blockchain-enabled serverless approach for IoT healthcare applications. In: Serverless Computing: Principles and Paradigms, pp. 193–218. Springer International Publishing, Cham (2023)
18. Mathew, P.S., Pillai, A.S., Palade, V.: Applications of IoT in healthcare. In: Cognitive Computing for Big Data Systems Over IoT: Frameworks, Tools and Applications, pp. 263–288 (2018)
19. Gupta, N., Saeed, H., Jha, S., Chahande, M., Pandey, S.: Implementation of an IOT framework for smart healthcare. In: 2017 International Conference of Electronics, Communication and Aerospace Technology (ICECA), vol. 1, pp. 622–627. IEEE (2017)
20. Ghorbian, M., Ghorbian, S.: Usefulness of machine learning and deep learning approaches in screening and early detection of breast cancer. Heliyon (2023)
21. Abdullah, T.A., Zahid, M.S.M., Ali, W.: A review of interpretable ML in healthcare: taxonomy, applications, challenges, and future directions. Symmetry **13**(12), 2439 (2021)
22. Hölbl, M., Kompara, M., Kamišalić, A., Nemec Zlatolas, L.: A systematic review of the use of blockchain in healthcare. Symmetry **10**(10), 470 (2018)
23. McGhin, T., Choo, K.K.R., Liu, C.Z., He, D.: Blockchain in healthcare applications: research challenges and opportunities. J. Netw. Comput. Appl. **135**, 62–75 (2019)
24. Chen, H.S., Jarrell, J.T., Carpenter, K.A., Cohen, D.S., Huang, X.: Blockchain in healthcare: a patient-centered model. Biomed. J. Sci. Tech. Res. **20**(3), 15017 (2019)
25. Calarany, C., Indumathy, M., Senthilraja, P., Suganya, D., Vardhan, V.G., Rahul, Y.S.: Successful integration of IoT and blockchain technologies using several machine learning algorithms. In 2024 IEEE International Conference on Computing, Power and Communication Technologies (IC2PCT), Vol. 5, pp. 1980–1984. IEEE (2024)
26. Shahbazi, Z., Byun, Y.C.: Integration of blockchain, IoT and machine learning for multistage quality control and enhancing security in smart manufacturing. Sensors **21**(4), 1467 (2021)
27. Ogundokun, R.O., Arowolo, M.O., Misra, S., Awotunde, J.B.: Machine learning, IoT, and blockchain integration for improving process management application security. In: Blockchain Applications in the Smart Era, pp. 237–252. Springer International Publishing, Cham (2022)
28. Unal, D., Hammoudeh, M., Khan, M.A., Abuarqoub, A., Epiphaniou, G., Hamila, R.: Integration of federated machine learning and blockchain for the provision of secure big data analytics for Internet of Things. Comput. Secur. **109**, 102393 (2021)
29. Liu, Y., Yu, F.R., Li, X., Ji, H., Leung, V.C.: Blockchain and machine learning for communications and networking systems. IEEE Commun. Surv. Tutor. **22**(2), 1392–1431 (2020)

30. Thomas, G.A.S., Robinson, Y.H.: IoT, big data, blockchain and machine learning besides its transmutation with modern technological applications. In: Internet of Things and Big Data Applications—Part of the Intelligent Systems Reference Library book series, vol. 180, pp. 47–63. Springer (2020)
31. Chen, X., Liu, L., Kang, S., Gnanaprakasam, J.R., Wang, R.: The lactate dehydrogenase (LDH) isoenzyme spectrum enables optimally controlling T cell glycolysis and differentiation. Sci. Adv. **9**(12), eadd9554 (2023)

Leveraging Generative AI for Enhanced Predictive Maintenance and Anomaly Detection in Manufacturing

Vamshi Mugala

Abstract This chapter explores the potential of Generative AI in enhancing predictive maintenance and fault diagnosis within the context of Industry 4.0. As manufacturing companies strive to improve productivity and reduce downtime, this innovative approach goes beyond traditional predictive maintenance models by utilizing historical failure data, machine learning-based control limits, and optimal sensor thresholds to predict and mitigate issues. The chapter also integrates Standard Operating Procedures (SOPs) and historical maintenance records into a comprehensive diagnostic system. By implementing a Retrieval-Augmented Generation (RAG) system combined with Large Language Models (LLMs), this chapter demonstrates how this approach analyzes sensor data, SOPs, and maintenance logs to generate detailed, context-aware maintenance responses leading to more effective and timely decision-making. The study is illustrated through a simulated pump-related scenario, showcasing the successful application of the proposed methods. The findings reveal significant improvements in identifying and diagnosing equipment anomalies, offering a proactive maintenance strategy that enhances operational reliability and efficiency. By incorporating AI-driven techniques like Skope-Rules and RAG, this chapter highlights the critical role of AI in modernizing manufacturing processes and sets the stage for future research focused on real-time processing and broader equipment monitoring.

Keywords Predictive maintenance · Anomaly detection · Machine learning · Generative AI · Standard Operating Procedures (SOPs) · Maintenance records · Retrieval-Augmented Generation (RAG) · Large Language Models (LLMs) · Manufacturing · Industry 4.0 · Control limits · Sensor data · Diagnostic system · Smart manufacturing · Operational efficiency · Prescriptive analytics

V. Mugala (✉)
INVISTA, Wichita, KS 67220, USA
e-mail: vamshims128@gmail.com

N. R. Vajjhala et al. (eds.), *Generative Artificial Intelligence (AI) Approaches for Industrial Applications*, Information Systems Engineering and Management 24,
https://doi.org/10.1007/978-3-031-76710-4_13

1 Introduction

1.1 The Overview of Generative AI

Generative AI, a specialized sector within artificial intelligence, excels in creating new content like images, texts, or music by learning from existing data patterns. This capability is revolutionizing manufacturing processes, applying advanced algorithms and machine learning techniques to improve operations, enhance productivity, and drive innovation [1]. One notable application is customized manufacturing. AI enables systems to adapt to consumer demands and environmental conditions while assimilating vast process knowledge. This results in smarter production strategies, collaborative networks, and broader service offerings, which are crucial as consumer preferences shift towards personalized products [1]. Generative AI, particularly through machine learning, is transforming manufacturing by providing powerful tools for analyzing extensive datasets, often referred to as Big Data. This analysis allows manufacturers to extract key insights, streamline production lines, and make strategic decisions that enhance efficiency and product quality [2]. In smart manufacturing, Generative AI boosts automation and efficiency. The integration of machine learning and AI technologies enables firms to implement intelligent automation systems, enhancing operational smoothness and productivity. This integration supports real-time operational adjustments, predictive maintenance, and optimized production schedules [3]. Generative AI also improves resource efficiency by facilitating predictive maintenance, efficient production planning, fault detection, and proactive quality control. These AI-driven practices boost operational efficiency and promote sustainability by reducing energy consumption and waste [4]. As Generative AI becomes more embedded in manufacturing, it transforms the workforce. There is a growing need for advanced digital skills, while the demand for less specialized labor decreases. This shift reflects the profound impact of AI on the skillsets required in modern manufacturing [5].

Generative AI is crucial in developing the next generation of intelligent manufacturing systems. These systems combine traditional industrial methods with modern information and communication technologies, marking a new industrial revolution. They not only automate tasks but also significantly enhance communication, collaboration, and decision-making processes within manufacturing environments, boosting efficiency and market competitiveness [6]. Generative AI is not just a technological tool but a transformative force in modern manufacturing. It facilitates tailored products, optimizes procedures, and improves resource management. It also influences workforce dynamics and drives the creation of sophisticated manufacturing systems. As the sector continues to integrate these AI-driven technologies, it is poised to achieve unparalleled advances in productivity, innovation, and sustainability.

1.2 Large Language Models (LLMs) and Retrieval Augmented Generation (RAG) in Manufacturing

The manufacturing industry is witnessing transformative enhancements through the integration of Large Language Models (LLMs) and Retrieval-Augmented Generation (RAG) techniques. These large language models, like GPT-3, are built on AI algorithms trained extensively on vast corpora of textual data, enabling them to produce language that is not only coherent but contextually apt from general language inputs [7]. Their adeptness at mimicking human-like text makes them invaluable in a variety of applications within the manufacturing sector. Retrieval-Augmented Generation, or RAG, elevates the process of language generation by incorporating information fetched from external databases or knowledge repositories, significantly enriching the quality and applicability of the output [8]. When combined with the computational prowess of LLMs, RAG methods have set new benchmarks in Natural Language Processing (NLP) tasks, proving especially useful for knowledge-heavy tasks in manufacturing.

In the realm of quality control, these models offer a robust means to parse and analyze extensive amounts of textual data related to product specifications, defect reports, and customer feedback. They then generate detailed quality reports and suggestions for improvements [9]. This approach enables manufacturers to leverage a wide range of data, ensuring that products meet rigorous quality standards and comply with industry regulations. Predictive maintenance is another area where the synergy of LLMs and RAG methods is invaluable. By analyzing past maintenance records, equipment sensor readings, and maintenance manuals, these models can forecast maintenance needs and schedule timely alerts. Proactive maintenance helps in averting equipment failures, reducing downtime, and optimizing maintenance tasks, which in turn enhances efficiency and reduces costs.

In supply chain management, LLMs and RAG techniques can synthesize detailed reports, projections, and insights from data pulled from various sources such as inventory databases, communications with suppliers, and market analyses [10]. This capability supports manufacturers in making informed decisions about inventory management, procurement strategies, and demand forecasting, thereby streamlining supply chains and boosting operational efficiency. Additionally, for process optimization, these models analyze complex manufacturing processes to identify inefficiencies and suggest corrective actions [11]. Utilizing historical data, industry best practices, and standards, they provide manufacturers with insights that help optimize production workflows, minimize waste, and elevate productivity. Incorporating Large Language Models with Retrieval-Augmented Generation techniques marks a significant advancement in the manufacturing industry, impacting key areas like quality control, predictive maintenance, supply chain management, and process optimization. This integration not only fosters innovation but also improves operational efficiencies, ensuring that manufacturers remain competitive in a rapidly evolving industrial landscape.

2 Relevant Generative AI Technologies and Techniques

2.1 Large Language Models (LLMs) and Retrieval Augmented Generation Techniques

Large Language Models (LLMs), such as GPTs, Llama's, and more open-sourced Mistral, etc., are at the forefront of artificial intelligence advancements, gaining widespread recognition for their ability to generate text that mirrors human conversation. These models are trained on colossal datasets, allowing them to recognize and replicate the intricate patterns of language, which makes them invaluable tools across various sectors [12]. LLMs mark a significant leap in AI development by enabling natural language interactions between humans and computers, thus improving the interface where they intersect [13]. The operational essence of LLMs lies in their training process, which involves absorbing vast amounts of textual data. This extensive training equips them to produce contextually appropriate responses based on the cues they receive [14]. These models have excelled in numerous language-based tasks, such as generating text, translating languages, and retrieving information. Their success demonstrates a robust capability to process and articulate language in ways that are strikingly human-like [15].

In parallel, Retrieval-Augmented Generation (RAG) techniques are refining the capabilities of LLMs by integrating them with sophisticated retrieval systems. This enhancement not only boosts the quality of the generated content but also ensures its relevance. RAG models utilize a dual approach by combining pre-trained language models with expansive, retrievable external memories. This allows them to pull in supplementary data during the generation process, thus enriching the content's accuracy and depth [8]. RAG operates by fine-tuning a blend of pre-trained models with innovative memory mechanisms. This setup not only aids in effective knowledge assimilation but also enhances the scalability and cost-efficiency of language processing tasks. The practical benefits of RAG include streamlined knowledge integration and increased scalability, making these models particularly effective for tasks that demand extensive knowledge. By merging retrieval processes with generative capabilities, RAG models can tap into a wider array of information, producing outputs that are not only relevant but also deeply informed [26] (see Fig. 1).

RAG process for question answering: (1) Indexing: Documents are chunked, encoded into vectors, and stored in a vector database. (2) Retrieval: Top-k relevant chunks are retrieved based on semantic similarity. (3) Generation: The original question and retrieved chunks are input into an LLM to generate the final answer. Large Language Models (LLMs) operate by analyzing extensive datasets to comprehend and produce text that closely mimics human language. Concurrently, Retrieval-Augmented Generation (RAG) techniques elevate this text generation by integrating information from external knowledge sources, thereby enhancing both the quality and relevance of the output. These sophisticated AI models have demonstrated considerable potential across diverse applications, proving their prowess in processing, generating, and retrieving text data with remarkable precision and context sensitivity.

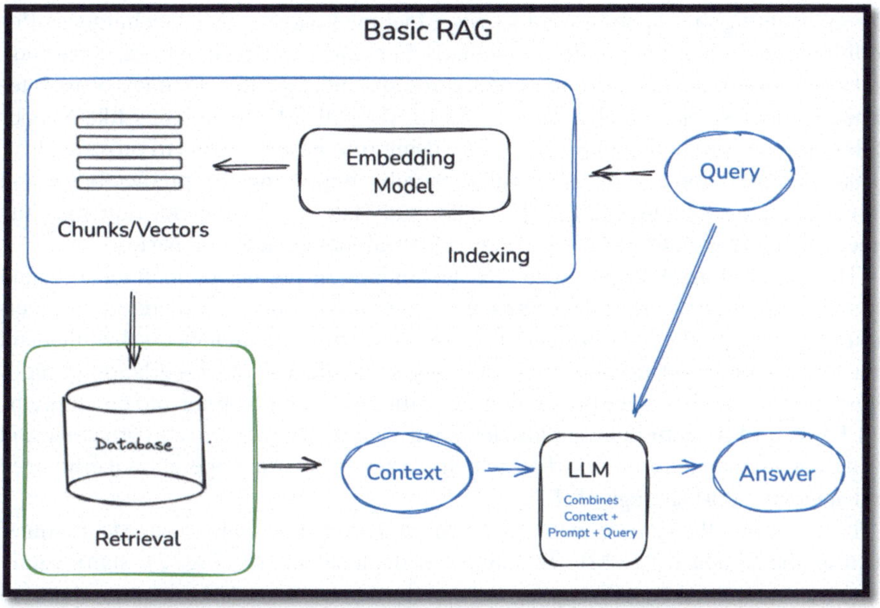

Fig. 1 Structure of basic RAG model

2.2 Leveraging Both LLMs and Machine Learning Techniques

Large Language Models (LLMs) have emerged as invaluable assets in industrial settings, particularly in the realms of predictive maintenance and anomaly detection. By leveraging LLMs, industries can enhance their maintenance approaches, boost operational efficiency, and minimize downtime through the early prediction of equipment failures and timely detection of anomalies [16]. In predictive maintenance, LLMs are instrumental in analyzing historical maintenance records, sensor data from equipment, and operational parameters to anticipate potential equipment malfunctions [17]. These models adeptly handle large datasets, identifying patterns that predict failures, thereby allowing industries to plan maintenance work proactively before any actual breakdown occurs. Integrating LLMs into maintenance strategies helps industries optimize maintenance intervals, cut costs related to maintenance, and improve the reliability of their equipment [18].

Moreover, LLMs are crucial in anomaly detection, where they assess real-time sensor data and performance metrics to identify deviations from normal operational conditions [19]. Detecting these anomalies early enables industries to execute corrective measures promptly, thus preventing potential equipment failures, reducing downtime, and ensuring uninterrupted operation of critical systems. LLMs enable industries to adopt proactive strategies for anomaly detection, thereby bolstering operational resilience and reducing the likelihood of unexpected disruptions [20].

Complementing the capabilities of LLMs, Machine Learning (ML) techniques offer additional analytical and predictive methods. For example, decision trees, clustering, and regression models can refine predictions and enhance the accuracy of maintenance strategies. Specifically, combining LLMs with ML techniques like Skope-Rules enriches predictive maintenance by using rule-based models to provide clear, visual interpretations of decision rules, thereby improving the predictability and transparency of outcomes [21, 25]. A case study later in this chapter illustrates the practical application of Skope-Rules in industrial maintenance scenarios.

The integration of LLMs with ML techniques in predictive maintenance and anomaly detection offers multiple benefits. These models not only increase the precision of failure predictions but also help in optimizing maintenance schedules and enhancing equipment reliability by utilizing sophisticated machine learning algorithms and natural language processing capabilities [21]. By tapping into the strengths of LLMs and ML techniques, industries can transition from reactive to proactive and predictive maintenance approaches, which are key to boosting operational efficiency and achieving cost savings [22].

Furthermore, the synergy between LLMs and predictive maintenance techniques, such as the Skope-Rules ML Technique as depicted in (see Fig. 2), significantly enhances maintenance efficacy in industrial environments [23]. These advanced models enable precise predictions of equipment failures, optimize the use of maintenance resources, and improve overall equipment effectiveness (OEE) through strategic, timely maintenance actions and proactive planning [24].

The utilization of Large Language Models along with ML Techniques in industrial applications, specifically in predictive maintenance and anomaly detection, offers significant advantages in enhancing maintenance strategies, improving equipment reliability, and minimizing operational disruptions. By leveraging the capabilities of LLMs, industries can transition towards proactive maintenance approaches, optimize maintenance schedules, and ensure the continuous and efficient operation of critical systems in today's dynamic industrial landscape.

3 Recent Machine Learning and GenAI Advancements in Manufacturing

3.1 Predictive Maintenance

Recent advancements in machine learning and Large Language Models (LLMs) have significantly impacted the field of predictive maintenance within manufacturing, reshaping how industries manage maintenance strategies and improve equipment reliability. By integrating cutting-edge technologies such as machine learning algorithms and LLMs, manufacturers are enhancing predictive maintenance practices, boosting operational efficiency, and reducing downtime through timely interventions, as noted by Carvalho et al. [27]. A study by Ayvaz and Alpay illustrates the

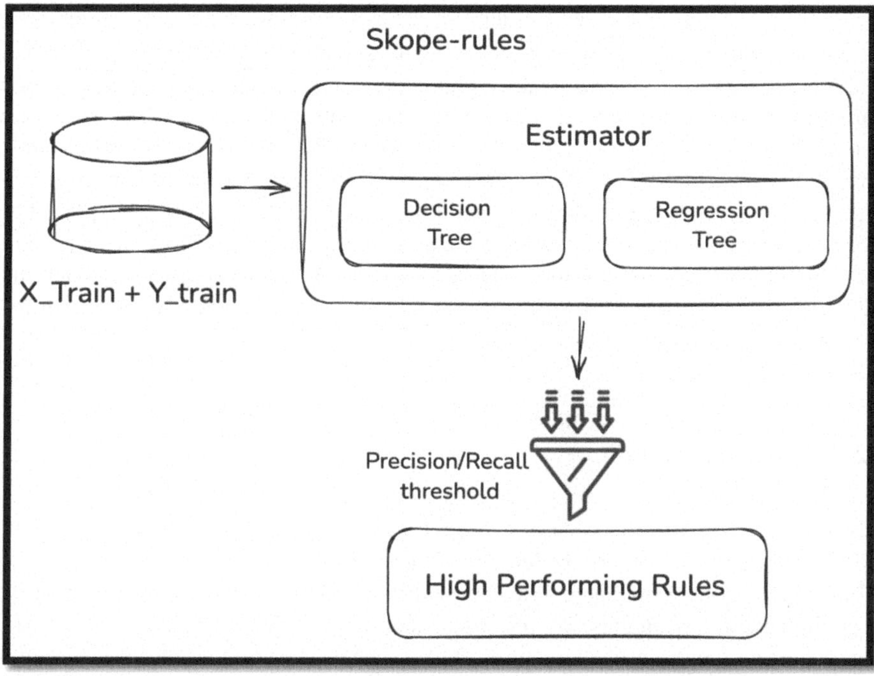

Fig. 2 As shown in Fig. 1, the Skope-Rules library provides a clear visualization of the decision rules applied in the model

development of a predictive maintenance system tailored for manufacturing production lines, which leverages real-time Internet of Things (IoT) data. This system skillfully combines machine learning techniques with IoT sensors to foster predictive maintenance in manufacturing settings, showcasing the robust potential of these technologies in boosting equipment reliability and operational performance [28].

Additionally, Dalzochio et al. explore the use of machine learning and reasoning for predictive maintenance in the context of Industry 4.0. Their research highlights the present challenges and developments in implementing sophisticated maintenance strategies, emphasizing the need to merge machine learning algorithms with reasoning capabilities to effectively navigate the complexities associated with predictive maintenance in this new industrial revolution [29]. Purnomo's work delves into utilizing deep learning data analytics to refine capacitated planned maintenance strategies. This approach underscores how advanced machine learning methods can significantly enhance maintenance planning accuracy, reduce equipment downtime, and optimize maintenance resources for more sustainable operational management [30].

Furthermore, Nacchia et al. conduct a systematic mapping study to examine the growing application of machine learning techniques for predictive maintenance in manufacturing. Their findings highlight the critical role of data-driven approaches in improving reliability engineering practices, pointing out how machine learning innovations can help optimize maintenance strategies and minimize downtime [31]. Abidi et al. focus on predictive maintenance planning within Industry 4.0, using machine learning to foster sustainable manufacturing. Their research stresses the pivotal role of predictive maintenance in reducing operational costs and achieving sustainable management practices, highlighting the benefits of leveraging machine learning techniques to enhance efficiency and sustainability in manufacturing processes [32]. The recent developments in machine learning and Large Language Models for predictive maintenance have revolutionized traditional maintenance approaches, enabling industries to implement more proactive strategies, optimize equipment reliability, and minimize operational disruptions.

3.2 Anomaly Detection

Advancements in machine learning and Large Language Models (LLMs) have substantially enhanced anomaly detection in industrial applications, enabling sectors to boost operational efficiency, reduce downtime, and improve equipment reliability through proactive strategies. By utilizing machine learning techniques and LLMs, manufacturers can identify deviations from normal operating conditions, pinpoint potential issues early, and initiate corrective measures to avert equipment failures before they become critical, as noted by Borghesi et al. [33]. Recent studies, including research highlighted by [34], demonstrate the effective implementation of machine learning, particularly deep learning, in anomaly detection at both the network and host levels. These advanced machine learning algorithms help industries achieve precise anomaly detection, maintaining operational resilience and preventing disruptions in manufacturing processes [34].

Quatrini et al. propose a two-step methodology for anomaly detection in industrial settings, using machine learning classification algorithms to spot anomalies and bolster safety and maintenance operations. The integration of machine learning models into anomaly detection systems allows industries to enhance safety measures, fine-tune maintenance practices, and ensure continuous industrial operations [35]. Pittino et al. discuss the benefits of automated anomaly detection systems in manufacturing, emphasizing how machine learning methods can reduce downtime from machine malfunctions and detect failures before they result in severe consequences. By implementing automated systems, industries can increase operational efficiency, minimize disruptions, and streamline maintenance tasks, all while reducing the dependence on expensive human expertise [36].

Additionally, Kammerer et al. provide insights into the application of anomaly detection in manufacturing based on sensor data. Their findings illustrate the effectiveness of such systems in monitoring machines and components in industrial environments. Through efficient anomaly detection, industries can proactively tackle equipment issues, optimize production processes, and enhance overall operational performance in manufacturing settings [37].

4 Bridging the Gap: From Theory to Practice

In the preceding sections, we examined the myriad ways machine learning and Large Language Models (LLMs) are being applied in manufacturing, particularly focusing on predictive maintenance and anomaly detection. The literature reviewed also highlights the transformative effects these advanced technologies can have on enhancing operational efficiency—primarily by optimizing maintenance schedules, predicting potential equipment failures, and detecting anomalies at an early stage. This ability to anticipate and mitigate issues before they escalate is crucial for minimizing operational disruptions and enhancing overall productivity. Studies have shown the successful deployment of these technologies at manufacturing plants, illustrating their practical benefits, and strengthening the need for ongoing innovation in this field.

Despite significant progress, several hurdles remain in the realm of predictive maintenance and anomaly detection. Traditional models often face integration issues and struggle to offer comprehensive diagnostic insights. They tend to rely excessively on historical data, which limits their effectiveness, as they do not capitalize on real-time information available from ongoing maintenance records and Standard Operating Procedures (SOPs). Moreover, there is a pressing need for these systems to produce outputs that are not only interpretable but also actionable—qualities essential for enabling maintenance teams to make informed, effective decisions. Addressing these challenges, this chapter proposes a novel approach by taking advantage of historical data on pump failures to establish control limits through sophisticated machine-learning techniques. Our project enhances traditional predictive maintenance by integrating it with Standard Operating Procedures (SOPs) and historical maintenance records, thereby crafting a more comprehensive diagnostic system. We employ a Retrieval-Augmented Generation (RAG) system that fuses a vector database with LLMs, enabling a thorough analysis of sensor data, SOPs, and maintenance logs. This method allows for the generation of precise, context-aware responses to maintenance inquiries, thereby aiding in more effective and timely decision-making. Our integration of machine learning with RAG systems showcases the practical application of these advanced AI techniques within the manufacturing sector. By leveraging the full capabilities of LLMs and real-time data retrieval, our approach provides actionable insights that significantly optimize maintenance schedules, diminish downtime, and boost equipment reliability. This initiative not only bridges existing gaps identified in the literature but also presents a scalable solution that can be adapted to various industrial scenarios, setting the stage for the development of more resilient and intelligent manufacturing systems.

Table 1 Showcasing the variables considered and data types

Variable name	Description	Units	Type
UDI	Unique identifier for each data point	–	Index
Product ID	Each product identifier	–	Categorical
Type	Type of product	–	Categorical
Air temperature (AT)	Ambient air temperature in the shop floor	Kelvin	Numerical
Process temperature (PT)	Operational temperature of the equipment	Kelvin	Numerical
Rotational speed (RS)	Rotational speed of the equipment	RPM	Numerical
Torque (T)	Torque applied to the equipment	Nm	Numerical
Tool wear (TW)	Wear and tear of the tool	Hours	Numerical
Target	Indicated whether a failure occurred	Binary	Categorical
Failure type	Type of failure	–	Categorical

5 Empirical Methods and Data

5.1 Dataset Description

This study utilized the AI4I 2020 Predictive Maintenance Dataset, sourced from the UCI Machine Learning Repository [38]. This dataset provides comprehensive sensor data from industrial machines, including various operational parameters. It is designed to support predictive maintenance and anomaly detection efforts by offering rich, labeled data that can be used to train and validate machine learning models. The dataset includes the following variables as shown in Table 1.

The failure types included in the dataset are:

- Tool Wear Failure (TWF)
- Heat Dissipation Failure (HDF)
- Power Failure (PWF)
- Overstrain Failure (OSF)
- Random Failure (RNF).

5.2 Tools Tested and Selected

Machine Learning Algorithm: Several algorithms, including decision trees, random forests, and gradient boosting machines. Each of these algorithms has its strengths, but we needed a model that not only provided accurate predictions but was also interpretable. This led us to Skope-Rules, a tool specifically designed for creating interpretable and actionable rule-based models.

Skope-Rules [25]: An important tool in our analysis was Skope-Rules, which is a machine learning algorithm designed for interpretable and accurate prediction models. Skope-Rules creates rule-based models that are easy to interpret, making them particularly useful for identifying critical thresholds and decision rules in maintenance data.

Large Language Model (LLM): We used Gemini LLM to analyze maintenance logs and SOPs. The RAG system was implemented using a vector database to store and retrieve relevant documents efficiently.

- **LLM**: Gemini 1.5 API
- **Vector Database**: Chroma (Langchain) [39]
- **Embeddings**: Google Generative AI Embeddings [40]

These tools helped us build a system capable of analyzing both structured sensor data and unstructured textual data, providing comprehensive diagnostic insights.

Skope-Rules and Its Role in the Usecase

Skope-Rules works by generating a set of if–then rules from the data, which can be easily understood and applied. For example, a rule might state: "If the temperature exceeds 300 K and the rotational speed is above 1500 RPM, then there is a high likelihood of heat dissipation failure." These rules are derived from the data using a combination of decision tree algorithms and association rule learning, ensuring both accuracy and interpretability.

Benefits of Using Skope-Rules

- Interpretability: The rules generated by Skope-Rules are easy to interpret, allowing maintenance teams to understand the reasoning behind each prediction and make informed decisions.
- Actionability: The clear if–then rules provide actionable insights that can be directly applied to maintenance schedules and protocols.
- Integration: Skope-Rules can be seamlessly integrated with other machine learning models and systems, enhancing the overall predictive maintenance framework.

This algorithm was used to establish control limits and identify key decision rules for predictive maintenance. By analyzing historical sensor data, Skope-Rules generated rules that helped in detecting anomalies and predicting failures. These rules were then integrated with the RAG system to provide comprehensive diagnostic insights.

LLMs, particularly the Gemini 1.5 API, played a crucial role in analyzing vast amounts of textual data (Considering it 1.5Million context length) from maintenance logs, SOPs, and sensor readings. This capability allowed us to generate detailed, context-aware insights, significantly enhancing our maintenance strategies and operational efficiency. We used Google Generative AI Embeddings to convert textual data into vectors, which were then indexed in the Chroma (Langchain) vector database.

5.3 Real World Use Case (Implementation Workflow)

Our implementation workflow is divided into two primary phases:

- Training phase
- Live-Production phase.

Training Phase: Establishing Control Limits and Building the Vector Database

In this phase, the focus is on establishing control limits and constructing a comprehensive vector database using historical data, SOPs, and maintenance logs. The steps involved are as follows:

Data Collection and Preprocessing

Sensor data was collected from various sensors installed on the manufacturing site. This data includes critical parameters such as air temperature, process temperature, rotational speed, torque, and tool wear. Simultaneously, SOPs and historical maintenance logs were gathered to provide insights into past maintenance activities and procedures.

Machine Learning for Control Limits

Machine learning algorithms were applied to analyze the historical sensor data and establish control limits for each parameter. These control limits define the normal operational ranges and thresholds that indicate potential equipment failures. Models like decision trees, random forests, and Skope-Rules were used to create interpretable and actionable rules that identify deviations from normal operations.

Figure 3 illustrates an example of control limits.

Embedding Model for Document Vectorization

The textual data from SOPs and maintenance logs was processed using an embedding model, specifically Google Generative AI Embeddings. This model converted the text into high-dimensional vector representations, capturing the semantic meaning of the content. The embedding model chunked the documents into manageable segments and encoded each segment into a vector.

Document Indexing

These embedded vectors were then stored in a vector database, Chroma (provided by Langchain). This database allows for efficient and accurate similarity searches, enabling quick retrieval of relevant documents based on semantic similarity (Fig. 4).

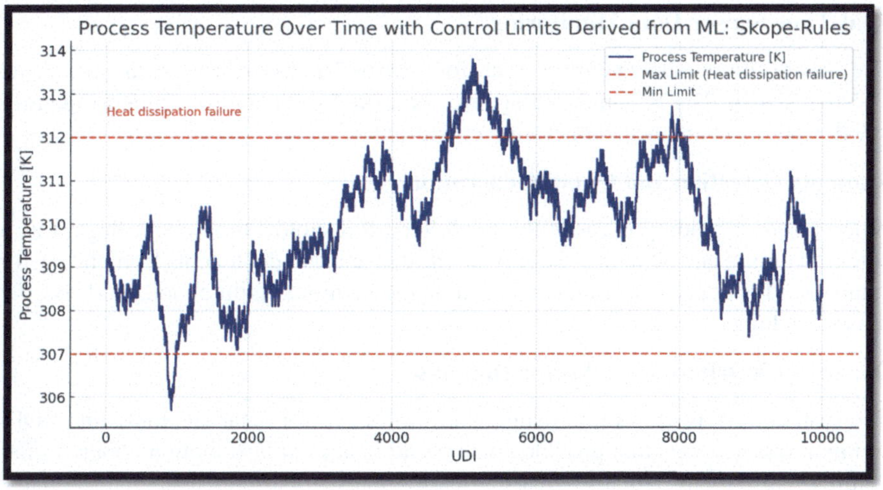

Fig. 3 Sample chart showcasing the control limits for process temperature variable

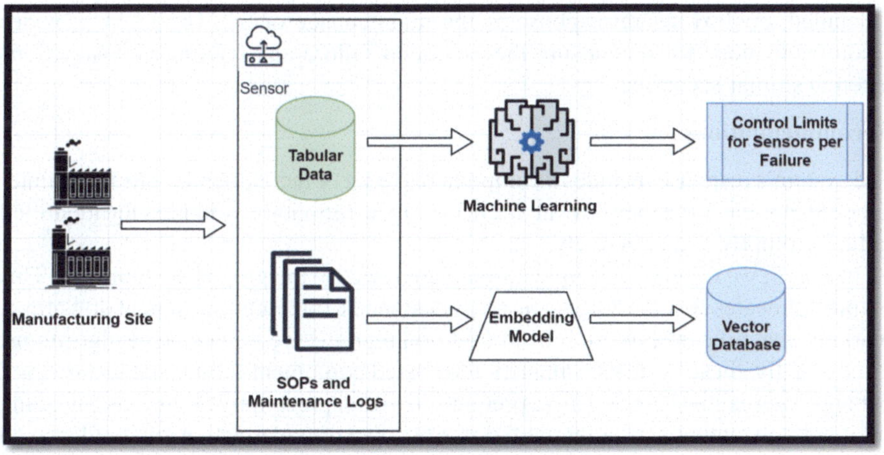

Fig. 4 Training phase workflow diagram

Live Production Phase: Real-Time Anomaly Detection and Diagnosis

In the live production phase, we use the established control limits and the vector database to monitor real-time sensor data, detect anomalies, and provide detailed diagnostic recommendations. The steps involved are:

Real-Time Sensor Data Monitoring

The system continuously monitors real-time sensor data from the manufacturing site. The live sensor data is compared against the established control limits to identify any deviations from normal operating conditions.

Anomaly Detection and Query Generation

When a sensor reading exceeds the predefined thresholds, the system flags it as a potential failure and generates a query to retrieve relevant documents from the vector database. This query is formulated based on the identified failure mode and specific sensor readings.

Document Retrieval from Vector Database

The system retrieves relevant documents from the vector database, including SOPs and maintenance logs that provide context and historical information about similar failures.

LLM for Detailed Diagnostics

The retrieved documents are then fed into an LLM (Gemini 1.5 API), which generates a detailed, context-aware response to the maintenance query. The LLM provides recommendations for diagnosing the identified failure and references past actions taken in similar scenarios.

Recommendations

The system's response includes a comprehensive set of recommendations for maintenance teams, ensuring they have all the necessary information to address the identified failure promptly and effectively.

In Fig. 5, we observe a sample output generated by the proposed algorithm. This output delineates specific diagnostic steps designed to preempt equipment downtime, marking a significant evolution from traditional predictive maintenance algorithms. Traditionally, these systems primarily forecasted equipment failure states; however, our approach extends beyond mere prediction to offer prescriptive solutions. By doing so, it not only anticipates potential failures but also provides actionable guidance on preventing them, thereby enhancing the reliability and efficiency of manufacturing operations.

6 Results

This case study, which integrates generative AI with traditional machine learning techniques, has achieved significant advancements in predictive maintenance and anomaly detection within the manufacturing sector. Key outcomes include:

Enhanced Predictive Maintenance: By leveraging sophisticated machine learning algorithms such as Skope-Rules, we established control limits that accurately predict

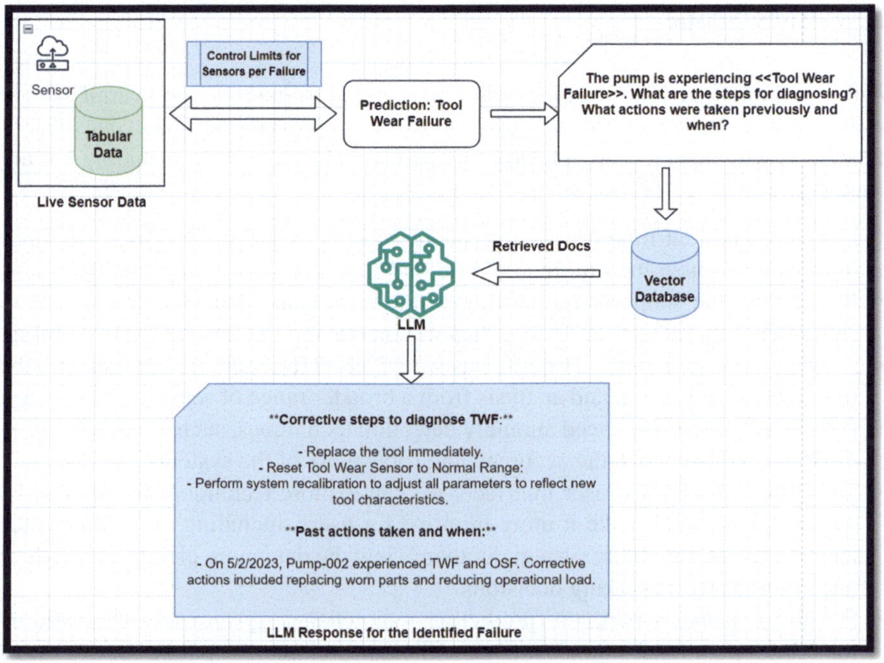

Fig. 5 Live production phase workflow with the results

equipment failures. These rule-based models provide clear and interpretable insights, aiding maintenance teams in making well-informed decisions.

Improved Anomaly Detection: Combining Large Language Models (LLMs) with Retrieval-Augmented Generation (RAG) techniques, we achieved precise anomaly detection. The system processes real-time sensor data and retrieves relevant documents from a vector database, offering detailed diagnostic recommendations.

Context-Aware Maintenance Recommendations: Utilizing the Gemini 1.5 API and Google Generative AI Embeddings, our system generates contextually relevant responses to maintenance inquiries, enhancing decision-making processes and minimizing downtime.

Operational Efficiency: Our approach has significantly boosted operational efficiency by preemptively addressing equipment issues, optimizing maintenance schedules as the next best step to diagnose, and reducing overall maintenance costs.

7 Future Scope

With traditional machine learning techniques added to improve the overall system in this project, we can speculate various possible improvements and expansions of the system with newer, more advanced generative AI capabilities, including but not limited to:

- For enhanced real-time processing, latency can be reduced, and response times improved by integrating more powerful real-time data processing and analytics.
- Broadening the scope of monitoring in factories can extend the application to equipment other than pumps, significantly increasing this system's applicability.
- Linking the system with IoT platforms would allow for easier integration with the IoT and data collection and analysis from a broader range of sensors and devices.
- Considering other advanced anomaly detection techniques, such as deep learning models, could enhance the accuracy and robustness of the system.
- To create a more ideal user interface, we can explore techniques to improve its design. This would make it more intuitive for users, including the maintenance team, to interact with the system, let them view the data more clearly, and reduce uncertainties while making decisions.
- For the expansion part, the predictive maintenance system can be explored for other operational domains, such as healthcare, energy, public services, and logistics, where machinery maintenance significantly contributes to operational efficiency.
- To continuously improve system performance, we can later create a feedback loop for consistent monitoring of the outcomes and existing models and allow for further updating of the models as new data and feedback come from the maintenance teams.

8 Conclusion

In this chapter, we examined how the synthesis of generative AI with machine learning technologies is transforming predictive maintenance and anomaly detection across the manufacturing landscape. Utilizing a blend of historical datasets, live sensor data, and sophisticated AI models, our methodology furnishes in-depth, actionable insights that significantly refine maintenance protocols and boost operational efficacy. The primary insights derived from our investigation highlight the effective implementation of control limits through machine learning techniques, such as Skope-Rules. These provide transparent, understandable insights that empower maintenance teams to make informed decisions. Moreover, the integration of Large Language Models (LLMs) with Retrieval-Augmented Generation (RAG) techniques has markedly enhanced anomaly detection capabilities and the generation of context-sensitive maintenance advice. The deployment of these technologies has substantially increased operational efficiency by proactively managing equipment issues, streamlining maintenance schedules, and curtailing expenses. These improvements emphasize the critical role of AI-driven approaches in modernizing manufacturing, setting the stage for the development of more robust and intelligent industrial systems.

References

1. Wan, J., Li, X., Dai, H., Kusiak, A., Martínez-García, M., Li, D.: Artificial-intelligence-driven customized manufacturing factory: key technologies, applications, and challenges. Proc. IEEE **109**(4), 377–398 (2021). https://doi.org/10.1109/jproc.2020.3034808
2. Arinez, J., Chang, Q., Gao, R., Xu, C., Zhang, J.: Artificial intelligence in advanced manufacturing: current status and future outlook. J. Manuf. Sci. Eng. **142**(11) (2020). https://doi.org/10.1115/1.4047855
3. Ghahramani, M., Qiao, Y., Zhou, M., Hagan, A., Sweeney, J.: AI-based modeling and data-driven evaluation for smart manufacturing processes. IEEE/CAA Journal of Automatica Sinica **7**(4), 1026–1037 (2020). https://doi.org/10.1109/jas.2020.1003114
4. Waltersmann, L., Kiemel, S., Stuhlsatz, J., Sauer, A., Miehe, R.: Artificial intelligence applications for increasing resource efficiency in manufacturing companies—a comprehensive review. Sustainability **13**(12), 6689 (2021). https://doi.org/10.3390/su13126689
5. Wei, W., Li, L.: The impact of artificial intelligence on the mental health of manufacturing workers: the mediating role of overtime work and the work environment. Front. Public Health **10**, 2022. https://doi.org/10.3389/fpubh.2022.862407
6. Li, B., Chai, X., Liu, Y., Li, T., Lin, T., Wei, D., et al.: Intelligent manufacturing enabled by information and communication technology in industrial environment. Chin. J. Eng. Sci. **24**(2), 75 (2022). https://doi.org/10.15302/j-sscae-2022.02.007
7. Habib, S.: Large language model performance on practice epilepsy board examinations. JAMA Neurol. (2024). https://doi.org/10.1001/jamaneurol.2024.0676
8. Yu, W.: Retrieval-augmented generation across heterogeneous knowledge (2022). https://doi.org/10.18653/v1/2022.naacl-srw.7
9. Reese, J., Daniš, D., Caufield, J., Casiraghi, E., Valentini, G., Mungall, C., et al.: On the limitations of large language models in clinical diagnosis (2023). https://doi.org/10.1101/2023.07.13.23292613
10. Rajasekharan, A., Zeng, Y., Padalkar, P., Gupta, G.: Reliable natural language understanding with large language models and answer set programming (2023). https://doi.org/10.48550/arxiv.2302.03780
11. Guu, K.: Realm: retrieval-augmented language model pre-training (2020). https://doi.org/10.48550/arxiv.2002.08909
12. Buehler, M.: Generative retrieval-augmented ontologic graph and multiagent strategies for interpretive large language model-based materials design. ACS Engineering Au **4**(2), 241–277 (2024). https://doi.org/10.1021/acsengineeringau.3c00058
13. Mensah, P.: All you need is context: clinician evaluations of various iterations of a large language model-based first aid decision support tool in Ghana (2024). https://doi.org/10.1101/2024.04.03.24305276
14. Liu, T., Xiong, Q., Zhang, S.: When to use large language model: upper bound analysis of bm25 algorithms in reading comprehension task (2023). https://doi.org/10.20944/preprints202301.0219.v1
15. Lewis, P.: Retrieval-augmented generation for knowledge-intensive NLP tasks (2020). https://doi.org/10.48550/arxiv.2005.11401
16. Çınar, Z., Nuhu, A., Zeeshan, Q., Korhan, O., Asmael, M., Safaei, B.: Machine learning in predictive maintenance towards sustainable smart manufacturing in industry 4.0. Sustainability **12**(19), 8211 (2020). https://doi.org/10.3390/su12198211

17. Görür, O., Yu, X., Sivrikaya, F.: Integrating predictive maintenance in adaptive process scheduling for a safe and efficient industrial process. Appl. Sci. **11**(11), 5042 (2021). https://doi.org/10.3390/app11115042
18. Mohan, R., Roselyn, J., Uthra, R.: Lstm based artificial intelligence predictive maintenance technique for availability rate and OEE improvement in a TPM implementing plant through industry 4.0 transformation. J. Qual. Maint. Eng. **29**(4), 763–798 (2023). https://doi.org/10.1108/jqme-07-2022-0041
19. Nordal, H., El-Thalji, I.: Assessing the technical specifications of predictive maintenance: a case study of centrifugal compressor. Appl. Sci. **11**(4), 1527 (2021). https://doi.org/10.3390/app11041527
20. Duan, X., Vasudevan, A., Bekar, E., Gandhi, K., Skoogh, A.: A data scientific approach towards predictive maintenance application in manufacturing industry (2022). https://doi.org/10.3233/atde220148
21. Tong, G., Xia, Q., Liu, Y.: Prognostics and predictive maintenance optimization based on combination BP-RBF-GRNN neural network model and proportional hazard model. J. Sens. **2022**, 1–17 (2022). https://doi.org/10.1155/2022/8655669
22. Bouabdallaoui, Y., Lafhaj, Z., Yim, P., Ducoulombier, L., Bennadji, B.: Predictive maintenance in building facilities: a machine learning-based approach. Sensors **21**(4), 1044 (2021). https://doi.org/10.3390/s21041044
23. García, F., Salgado, D.: Maintenance strategies for industrial multi-stage machines: the study of a thermoforming machine. Sensors **21**(20), 6809 (2021). https://doi.org/10.3390/s21206809
24. Canelón, R.: Design of a remote assistance model for truck maintenance in the mining industry. J. Qual. Maint. Eng. **30**(1), 175–201 (2023). https://doi.org/10.1108/jqme-02-2023-0024
25. Cohen, M.: Skope-Rules: Scikit-learn compatible rule-based models. GitHub repository (2024). Available at: https://github.com/scikit-learn-contrib/skope-rules
26. Lewis, P., et al.: RAG: Retrieval-augmented generation for knowledge-intensive NLP tasks. arXiv preprint arXiv:2312.10997 (2023). Available at: https://arxiv.org/pdf/2312.10997.
27. Carvalho, T., Soares, F., Vita, R., Francisco, R., Basto, J., Alcalá, S.: A systematic literature review of machine learning methods applied to predictive maintenance. Comput. Ind. Eng. **137**, 106024 (2019). https://doi.org/10.1016/j.cie.2019.106024
28. Ayvaz, S., Alpay, K.: Predictive maintenance system for production lines in manufacturing: a machine learning approach using IoT data in real-time. Expert Syst. Appl. **173**, 114598 (2021). https://doi.org/10.1016/j.eswa.2021.114598
29. Dalzochio, J., Kunst, R., Pignaton, E., Binotto, A., Sanyal, S., Favilla, J., et al.: Machine learning and reasoning for predictive maintenance in industry 4.0: current status and challenges. Comput. Ind. **123**, 103298 (2020). https://doi.org/10.1016/j.compind.2020.103298
30. Purnomo, M.: Incorporating deep learning data analytics techniques in the optimisation of capacitated planned maintenance. Jurnal Sistem Dan Manajemen Industri **6**(2), 167–175 (2022). https://doi.org/10.30656/jsmi.v6i2.5076
31. Nacchia, M., Fruggiero, F., Lambiase, A., Bruton, K.: A systematic mapping of the advancing use of machine learning techniques for predictive maintenance in the manufacturing sector. Appl. Sci. **11**(6), 2546 (2021). https://doi.org/10.3390/app11062546
32. Abidi, M., Mohammed, M., Alkhalefah, H.: Predictive maintenance planning for industry 4.0 using machine learning for sustainable manufacturing. Sustainability **14**(6), 3387 (2022). https://doi.org/10.3390/su14063387
33. Borghesi, A., Bartolini, A., Lombardi, M., Milano, M., Benini, L.: Anomaly detection using autoencoders in high performance computing systems. Proc. AAAI Conf. Artif. Intell. **33**(01), 9428–9433 (2019). https://doi.org/10.1609/aaai.v33i01.33019428
34. Dutta, V., Choraś, M., Pawlicki, M., Kozik, R.: A deep learning ensemble for network anomaly and cyber-attack detection. Sensors **20**(16), 4583 (2020). https://doi.org/10.3390/s20164583
35. Quatrini, E., Costantino, F., Gravio, G., Patriarca, R.: Machine learning for anomaly detection and process phase classification to improve safety and maintenance activities. J. Manuf. Syst. **56**, 117–132 (2020). https://doi.org/10.1016/j.jmsy.2020.05.013

36. Pittino, F., Puggl, M., Moldaschl, T., Hirschl, C.: Automatic anomaly detection on in-production manufacturing machines using statistical learning methods. Sensors **20**(8), 2344 (2020). https://doi.org/10.3390/s20082344
37. Kammerer, K., Hoppenstedt, B., Pryss, R., Stökler, S., Allgaier, J., Reichert, M.: Anomaly detections for manufacturing systems based on sensor data—insights into two challenging real-world production settings. Sensors **19**(24), 5370 (2019). https://doi.org/10.3390/s19245370
38. UCI Machine Learning Repository: AI4I 2020 predictive maintenance dataset [Online]. Available: https://archive.ics.uci.edu/dataset/601/ai4i+2020+predictive+maintenance+dataset
39. Langchain: Chroma: the Langchain vector database [Online]. Available: https://langchain.com/chroma
40. Google AI: Google generative AI embeddings [Online]. Available: https://ai.google.com/research/embeddings

The Transformative Role of Big Data Analytics and Generative AI in Redefining FinTech for New Business Models

Artor Nuhiu and Florin Aliu

Abstract This chapter elaborates on the role of Big Data Analytics and Generative AI's impact in redefining FinTech and transforming new business models in the financial services industry. This chapter addresses several issues regarding the integration of Generative AI and digital technologies in finance and describes how AI and machine learning systems are contributing in improved data analysis, process automation, and enhanced decision-making in the modern business environment. Handling and analyzing big data pose a considerable challenge primarily related to the quality of available data and ethical issues associated with using existing or acquired data. The chapter deals with the evolution of banking technologies, which have produced competitive customer-oriented financial services through digitization and AI's generative capabilities. The chapter provides a holistic perspective on the opportunities, challenges, benefits, and risks associated with the transformative potential of using generative AI and big data analytics in the FinTech industry.

Keywords FinTech · Big data analytics · Generative AI · Digital transformation · Blockchain technology · Machine learning

1 Introduction

In today's world oriented towards digitalization, data is considered a valuable (and vulnerable) resource, and for the FinTech sector, big data analytics plays a crucial role. The evolution of FinTech is closely related to the development of big data, which is defined by four main characteristics (the four V's): *Volume*—the size of

A. Nuhiu (✉)
Faculty of Law, Financial Law Department, University of Prishtina, 10000 Prishtina, Kosovo
e-mail: artor.nuhiu@uni-pr.edu

F. Aliu
School of Expertness and Valuation, Institute of Technology and Business, 37001 Ceské Budejovice, Czech Republic
e-mail: aliuf@mail.vstecb.cz

N. R. Vajjhala et al. (eds.), *Generative Artificial Intelligence (AI) Approaches for Industrial Applications*, Information Systems Engineering and Management 24,
https://doi.org/10.1007/978-3-031-76710-4_14

the data; *Velocity*—the speed at which data is generated and processed; *Variety*—the types and sources of data; and *Veracity*—the accuracy of the data [1, 2]. In the context of FinTech, big data analytics uses these characteristics to discover patterns, predict trends, and make more effective financial decisions [3]. The integration of Generative AI in digital finance has emerged as a game-changer for the FinTech sector, extending beyond traditional analytics to revolutionize how financial models and products are developed. Generative AI employs the extensive capabilities of big data to analyze, generate, and visualize large amounts of data quickly and efficiently through advanced algorithms and machine learning that simulate realistic financial scenarios and outcomes [4, 5]. This technology plays an essential role in areas such as automated financial advising, personalized banking experiences, and risk assessment, where it can predict and model financial behaviors with exceptional accuracy and swiftness [4, 6]. Particularly, Generative AI is transforming the FinTech sector through its ability to create complex financial models that can forecast market trends, enhance investment strategies, and deliver tailored financial advice. Generating synthetic financial datasets also aids in stress testing and scenario analysis without the need for historical data, thus providing a robust framework for decision-making under uncertainty [7]. This innovative approach is indispensable in developing resilient financial systems that adapt to dynamic market conditions, presenting a significant competitive advantage in the evolving digital economy.

It is challenging nowadays to imagine modern life without data. Technological devices are closely integrated into our daily lives, simplifying how we live and collecting data about our habits, preferences, and behaviors. According to a study by [1], the banking industry uses 700% more data every second. Meanwhile, the total quantity of data collected, copied, and used worldwide in 2021 reached a capacity of nearly 80 zettabytes or 80 trillion gigabytes and will continue to grow along with the rapid evolution of technology [8, 9]. Figure 1 below shows the exponential growth of data volume on the internet from 2019 to 2025, underscoring the rapid expansion of digital information.

Consequently, data has become necessary for managing a successful business, thus increasing the interest in data across numerous sectors, including FinTech. Big data in FinTech helps gain valuable insights and transform how companies build new

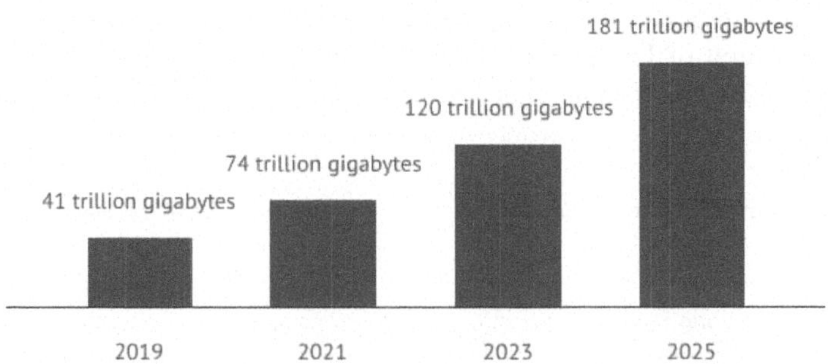

Fig. 1 The growth trend of global data volume on the internet

business models. Big data analytics reveals a pattern about the targeted audience, allowing businesses to provide better services and create products more tailored to customer needs [5, 10, 11].

Innovative technologies such as artificial intelligence and machine learning have become integral parts of FinTech. Artificial intelligence is instrumental in identifying patterns and fluctuations from large amounts of data, which can predict market behavior and personalize financial service offerings for clients [12]. Machine learning helps improve financial algorithms and automate decision-making processes [13]. Besides the benefits, legal and regulatory challenges are associated with using big data in FinTech. This includes data privacy and security issues and the need to comply with various national and international regulations [14]. In particular, the General Data Protection Regulation (GDPR) in the European Union significantly impacts how financial institutions manage and process personal data [15].

FinTech has paved the way for innovations such as virtual currencies, digital wallets, and financial intermediation platforms in the context of new financial products and services. Case studies, such as the use of Blockchain to manage groups of financial assets in financial markets, are significant examples showing how technology can solve complex problems and provide greater transparency [16, 17]. Another example is using algorithms to detect fraud in financial transactions, a significant challenge for banks in the digital era [18]. It is essential to anticipate and understand future developments in technological innovations and their potential to transform the financial industry. These developments include full digital banking, complete automation of financial services, and the increased use of artificial intelligence for financial decision-making [19, 20]. These developments affect how financial institutions operate and create new legal and regulatory challenges that must be addressed [17]. Figure 2 illustrates the key components of financial technology in a structured framework, detailing how each sector contributes to innovation and efficiency in financial services.

The use of big data in the FinTech sector is still in the early stages in most countries worldwide. The volume and diversity of the data and the importance of maintaining data quality are challenging for FinTech companies, primarily due to the need for proper information technology (IT) infrastructure in this sector. With the necessary tools, techniques, and infrastructure, the appropriate utilization and analysis of big data is a smooth process, not to mention integrating AI models and machine learning (ML) systems into existing organizational systems [5]. Additionally, most small FinTech companies need more capital to incorporate everything that involves big data into their businesses immediately. In contrast, other larger FinTech companies still need to operate with outdated systems [21, 22].

What matters is the quality of big data and what companies do with it. However, most FinTech companies, especially newer ones, must fully understand the importance of data quality. As a result, they risk making mistakes such as improperly marketed products, incorrect customer profiles, and inaccurate business risk assessments. These mistakes can cost companies the loss of customers, and they will need to re-spend large amounts of money to correct errors resulting from poor data quality.

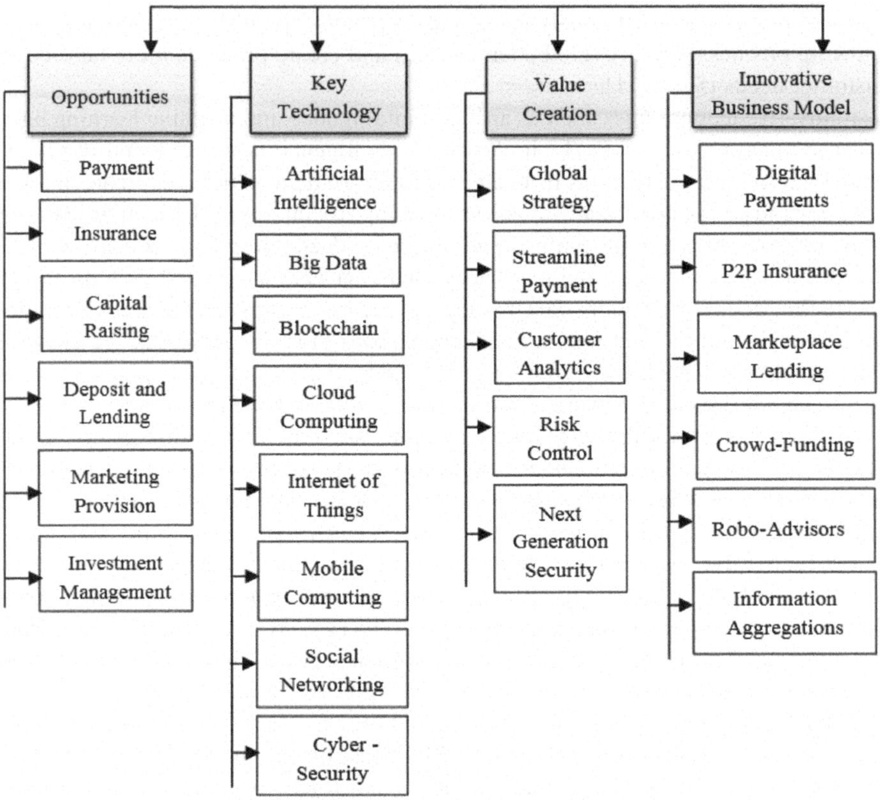

Fig. 2 Financial technology trends and opportunities

Moreover, poor data quality can also jeopardize regulatory compliance, data security, and the company's image [23].

1.1 The Role of Data in Traditional Decision-Making

Decision-making is widely regarded as one of the most essential characteristics of organizational activities. Many authors consider it the primary function of management. Managers and Executive Directors (CEOs) continually strive to produce the most optimal decisions and outcomes for their companies. However, making the right decision can be challenging. Personal biases, lack of information, uncertainty, and many external factors often challenge it. Many managers face biases that can disrupt their rational decision-making and must seek solutions to address or mitigate them [23, 24].

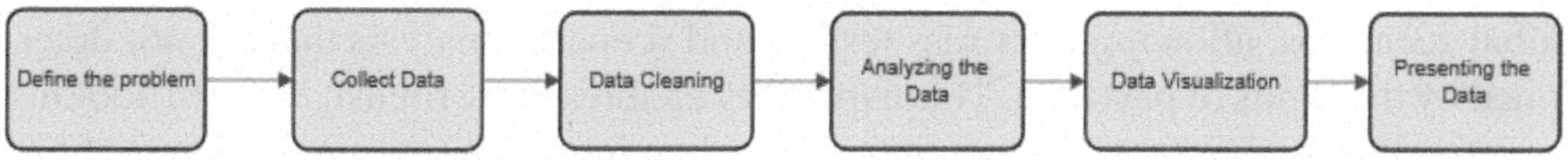

Fig. 3 Six steps of the data analysis process

Due to the increasing complexity of decision-making in business, the need for information becomes increasingly important to ensure effective decision-making. The growing amount of available data has made it more difficult for individuals to manage and process all the information. While the term 'decision' is often synonymous with 'choice,' viewing decisions as an institution is more beneficial. Individuals within an organization engage in what is called 'the decision-making process'. Rules govern these processes and involve multiple people; hence, they should be seen as institutions. The decision-making process typically involves various actors within a company, not just the upper management level [25].

In finance, traditional decision-making has been a process built on previous knowledge and experiences. This process includes using historical data and its analysis based on forecasts and expertise. For example, in investment portfolio management, traditional decision-making involves evaluating the performance of past investments to predict future outcomes. Historically, the use of data in financial decision-making has evolved from simply recording transactions on paper to complex statistical analysis. This evolution has been influenced by developments in information technology and the availability of large amounts of data [26]. Figure 3 illustrates a six-step process of data analysis, emphasizing a workflow of a structured approach to handling data, ensuring clarity, accuracy, and effectiveness from the data collected.

Qualitative and *quantitative* data are the primary data types used in traditional financial decision-making. Qualitative data includes information such as the company's reputation, while quantitative data includes information such as stock prices or interest rates. A significant challenge in using traditional data has been the need for more accuracy and timeliness. More than historical data is often needed to predict future events accurately, and decisions are made based on outdated data [23]. An example of traditional data use in finance could be the analysis of market fluctuations during the 1980s and 1990s. At that time, investors analyzed historical market price fluctuations to make predictions. With the advent of the digital era, financial decision-making has begun to incorporate more digital tools and advanced data analytics. This has led to a shift from traditional methods based on historical data to the use of algorithms and predictive analyses. Digital data may profoundly influence traditional financial decision-making practices. They enable faster and more accurate data analyses, improving decision accuracy and reducing risk levels [25].

Generative AI represents a considerable evolution in financial decision-making, delivering new solutions to traditional decision-making challenges. By creating synthetic data models and simulating financial scenarios, Generative AI provides financial managers and decision-makers with a richer, more diverse dataset than historical data alone could provide. For example, Generative AI can enhance investment strategy formulation by generating potential market conditions under various global scenarios, allowing for stress testing and scenario analysis that are not determined by the limits of past data. This capability enables more robust, forward-looking decisions that are important in today's rapidly changing economic environments [27]. Moreover, Generative AI is instrumental in identifying hidden patterns and correlations that human analysts might neglect. In credit risk assessment, for example, Generative AI can simulate the financial behavior of consumers under different economic conditions to predict defaults more accurately than traditional models. It enables better-informed lending decisions, optimizing financial institutions' risk and return profiles [28]. As we enter a new era of digital finance, the use of data is expected to become even more advanced. Artificial intelligence and machine learning are expected to be more significant in future financial data analysis.

New technologies have become one of the most important factors influencing the business environment of financial institutions, promoting new capabilities to support corporate activities and decision-making. Due to improved access to large data sets and the increasing data processing power, there is an evolution in the banking industry towards increasing investments in research and development (R&D) for technological improvements [29]. In the banking industry, new technologies are breaking down barriers to market entry and creating opportunities for new financial service providers. Competition from technologically advanced companies such as FinTech and increased regulations force banks to accelerate their digital innovations [30]. Banks must understand these innovations to compete in the new digital era. We are witnessing the global economy transforming into a digital economy in which artificial intelligence and digital transformation are becoming the new buzzwords in the business world. The International Data Corporation (IDC) offers some interesting perspectives on predictions for AI and digitization from 2021 onwards, where [30]:

- 65% of the world's economy is expected to be digitalized by 2022;
- 75% of all organizations will undergo a complete digital transformation by 2023; and
- 75% of businesses will leverage digital platforms and ecosystems by 2025.

Companies must strive to adapt to a more digitalized operating model if they want to survive in the era of the digital economy.

1.2 Applications of Generative AI in FinTech

Generative AI has begun transforming FinTech by introducing innovative applications that redefine traditional financial services. Several essential applications of Generative AI are shaping the future of FinTech, such as [6, 7, 31]:

Automated Financial Modeling: One of the primary applications of Generative AI in FinTech is automated financial modeling. This technology uses machine learning algorithms to generate models that can simulate financial outcomes based on various input variables and historical data patterns. It indicates the ability to quickly develop and adapt financial models that accurately predict investment risks and market movements for banks and investment companies. Automated financial modeling enables a faster reaction to market shifts, providing a competitive advantage by allowing companies to be first movers in response to financial insights.

Predictive Analytics: Generative AI significantly enhances predictive analytics in FinTech, delivering a more profound understanding of customer behavior, credit risk, and financial fraud detection. By analyzing vast datasets, Generative AI can forecast future behaviors and movements, such as predicting which customers may default on a loan or identifying unusual transaction patterns that may indicate fraudulent activity. These predictions help financial institutions mitigate risks and tailor products to meet their customers' expected demands.

Customized Financial Products: Through its advanced data processing capabilities, Generative AI facilitates the creation of customized financial products. By understanding individual customer patterns and preferences, financial institutions can use AI to design personalized banking and investment solutions to fulfill specific customer needs. For example, AI can suggest customized investment portfolios that align with a customer's risk tolerance and financial goals or recommend banking products that match their spending habits.

Operational Efficiency: Lastly, Generative AI significantly improves operational efficiency within FinTech companies. From automating routine customer inquiries via intelligent chatbots to processing claims and detecting anomalies in transaction data, AI systems reduce the need for manual intervention and allow human resources to focus on more strategic tasks.

By integrating Generative AI into their operations, FinTech companies improve their service offerings and enhance their innovation ability, thus remaining ahead in a highly competitive market.

1.3 Benefits of Big Data Analytics and Generative AI for Modern Business

The Financial Conduct Authority (FCA) defines 'big data' as using new or expanded data sets, including those from unconventional sources such as social networks [32]. Big data refers to adopting all necessary technologies to generate, collect, and store new forms of data; using advanced data processing techniques such as predictive analytics; and applying these data to improve decision-making in business activities [33]. Although the concept of 'big data' is relatively new, extensive data sets date back to the 1960s and 1970s when individuals and firms began using early data centers and developing databases [25].

Since the early 2010s, the term 'big data' has commonly been used to describe a new generation of technologies and approaches to data management. Key players in the internet industry developed this technology because traditional technologies could not adapt to an unpredictable number of users, rapidly increasing data volumes, and the growing need for fast data processing [34].

Big data plays a vital role in the FinTech revolution by catalyzing innovation in financial services. It is widely used in the financial sector in combination with advanced technologies such as machine learning (ML) and artificial intelligence (AI) applied to large data sets [35]. These technologies and big data are helping financial institutions use payment and mobile device data more effectively to understand credit risk better, often in partnership with Big Tech companies [36].

Big data is being utilized by both traditional financial institutions and FinTech companies, who are leveraging the power of 'big data' to predict customer behavior in real time and to provide advanced risk assessments. Additionally, big data contributes to developing new products and solutions tailored to specific clients. For example, big data is aiding financial sector companies in tracking customer preferences and demands for financial products in real-time, anticipating customer needs, and offering personalized products.

Customer segmentation has become more accessible with the help of big data, helping financial institutions better understand the choices and economic needs of different customers based on socioeconomic factors and other factors such as age and gender. By segmenting the market this way, financial institutions improve customer experiences by identifying and addressing specific market segments by providing customized solutions and products. Another way big data is helping to drive innovation is by developing reliable fraud detection systems and providing better risk assessments. For example, machine learning and natural language processing (NLP) technologies have enabled the development of new algorithms that can detect patterns in big data that can be useful in identifying potential fraud. One example is DataRobot, a U.S.-based FinTech company that offers an automated machine learning platform that analyzes traditional data and non-traditional data sets, such as social media data, to help banks and other financial institutions assess the credit risk of borrowers and detect fraud based on natural language processing (NLP) techniques [36]. Another example is MasterCard, which uses AI and machine learning to analyze detailed transaction data to evaluate bank card transactions in real-time to identify potentially fraudulent transactions. Another example is Feedzai, an American FinTech company that uses real-time machine learning to analyze big data and predict fraud with a claimed 95% accuracy [7].

Big data is transforming customer experiences by aiding in the development of next-generation virtual assistants such as Bank of America's chatbot 'Erica', launched in 2017, which utilizes predictive analytics and natural language processing built on big data, offering more advanced and personalized actions for the bank's customers, and can perform a wide range of banking services for clients [35]. Big data helps oil and gas companies identify potential resource locations and monitor pipeline operations in the energy industry. In the manufacturing industry, manufacturers and transportation companies rely on big data to manage their supply chains and optimize faster and cheaper transportation routes for goods.

Generative AI expands the capabilities of Big Data analytics by synthesizing new data sets and insights previously unreachable with traditional analytics alone. This advancement allows businesses to model complex scenarios and generate predictive data that improve decision-making processes across diverse sectors. One compelling case study is the use of Generative AI by JPMorgan Chase, a leader in the financial industry. The company utilizes Generative AI to develop advanced financial models that predict loan defaults more accurately than traditional models. By simulating numerous economic scenarios, Generative AI provides a broader range of outcomes based on diverse market conditions, thus improving the bank's ability to manage risk and tailor their financial products to customer demands effectively [37]. Another example is the collaboration between NVIDIA and American Express, where Generative AI is employed to detect fraudulent transaction patterns that deviate from the norm. By generating synthetic financial transactions that simulate legitimate and fraudulent behaviors, the model trains itself to identify unpretentious signs of fraud more accurately, thus improving security measures [6]. Moreover, Amazon has leveraged Generative AI to optimize its supply chain operations in the retail sector. By forecasting demand and supply variables under various scenarios, Amazon can adjust its inventory and logistic strategies in real time, minimizing costs and improving customer service. On the other hand, companies such as AIG utilize generative AI to create more accurate risk models for risk management. These models help predict potential losses from natural disasters or other significant events with greater precision, allowing for better strategic planning and resource allocation [6].

One of the key characteristics of big data is its speed. Data sets are collected and updated in real time, providing relevant information for more effective decision-making. This is important in digital finance, where every second can impact the outcome. The use of big data for rapid analysis and forecasting can provide competitive advantages in the market, such as [25]:

1. ***Risk Analysis***: Big data enables financial institutions to analyze business risks better. Through continuous market monitoring and analysis of big data, banks and other financial institutions can more quickly identify potential risks and intervene to minimize consequences.
2. ***Market Forecasting***: Big data analysis can be used to predict market movements. This is important for investors and investment portfolio managers who want to make informed investment decisions.
3. ***Fraud Prevention***: Big data is a powerful tool for identifying illegal activities and frauds. Analyzing transactions and behavior patterns can detect suspicious transactions and illicit actions.

Appropriate technological infrastructure and qualified human resources are necessary to use big data effectively. Therefore, companies must invest in specialized platforms for processing and analyzing big data. Additionally, expertise in data analysis and information security is essential.

2 Combining Big Data and Machine Learning to Improve Business Decision-Making

Data-driven decision-making is becoming increasingly common in today's digital world. Both large and small businesses are using this technology because data collection and analysis are more accessible than ever before. However, extracting essential insights can become more complex when data reaches 'big data' levels. It is a reasonable step for businesses to utilize big data and machine learning methods fully. Machine learning systems use data-driven algorithms and statistical models to examine and discover data patterns [38]. Big data provides the starting point from which machine learning systems can gain insights.

Obtaining and interpreting information from large quantities of data is at the core of big data. However, the volume of data is just one factor to consider when working with big data. Businesses must address several other essential characteristics of big data, including variety, accuracy, speed, validity, visualization, and data value. By gaining more insights from big data, machine learning, the foundation of modern artificial intelligence applications, adds significant value to big data applications.

Figure 4 presents the flow chart of the machine-driven process integrated with human-driven decisions during business decision-making. It starts with big data being processed through AI, which generates a menu of actionable options. After that, they are adjusted, if necessary, by non-digital information related to human judgment, contributing to informed business decisions.

Systems using artificial intelligence and machine learning for business can learn and change over time without specific instructions or following pre-programmed code. These machine-learning systems analyze data patterns and make judgments using statistical models [39]. In the past, businesses built complex rule-based systems for various reporting requirements. Still, they found that these solutions needed to be revised in order to be capable of adapting to ongoing changes. Now that artificial intelligence and machine learning have become powerful tools, businesses can use big data systems to make more effective decisions and perform more accurate business analysis and forecasting [40]. Big data has helped overcome the limitations of business intelligence (BI). Business analysis is now more accessible and more effective, thanks to big data analytics and artificial intelligence. For example, with the increase in various sources of big data, such as smart devices, the equipment

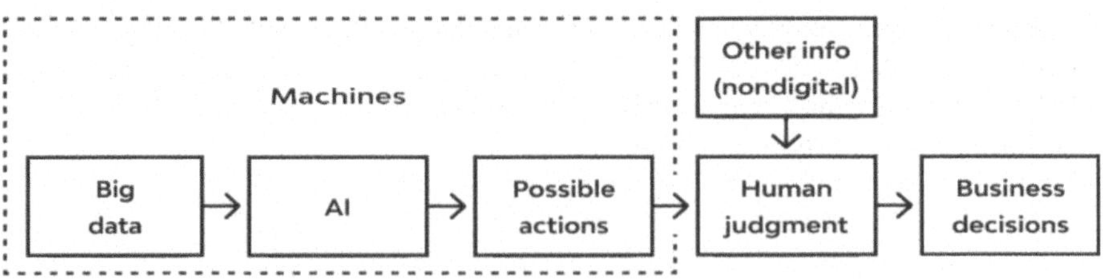

Fig. 4 The model of AI support in business decision-making

industry is no longer interested in static reports produced by business intelligence software. It is seeking new ways to use operational data in real-time. As a result, business intelligence software has evolved in three areas [41]:

(a) *Descriptive analysis*
(b) *Diagnostic analysis*
(c) *Predictive analysis.*

Business intelligence can contribute to more effective decision-making thanks to big data analytics and artificial intelligence. Big data transforms unstructured data and integrates it into AI systems, which then work on algorithms to turn these raw data into insights for decision-making [42]. This data processing method works wonderfully in predicting consumer behavior. Combining big data with AI can distinguish whether a platform or user is searching for customer relationship management (CRM) software. This data is beneficial for improving customer service management. This will encourage businesses to create more automated products, responding to customer demands. Today, with the application of big data analytics using artificial intelligence, businesses are focused on designing their products in a way that will allow them to meet customers' desires in the future [43]. Chatbot analytics, a standard part of any online business, is another exciting and important aspect of big data and AI. Data is collected from various sources and then analyzed through big data to identify potential questions and answers. These responses are processed in Chatbot programs using AI. Ultimately, Chatbots provide ongoing customer support while speeding up the sales process.

Enhancing the efficiency of industrial engineering without big data is currently almost impossible. By applying big data analytics, engineers can determine business constraints and how they relate to the business. As a result, they can immediately remove obstacles, significantly improving business performance [39]. Big data analytics enables businesses to predict future market fluctuations. Sales forecasting allows for a detailed examination of the best period to purchase inventory from the business. Likewise, operations in banking, healthcare, and many other industries have been improved by AI-based machine intelligence [43]. Robots are now used in healthcare procedures to measure outcomes more quickly and accurately. The use of AI systems in clinical settings has accelerated the development of new medical treatments, improving access to the healthcare system [41].

Generative AI introduces a layer to the interaction between Big Data and machine learning, delivering exceptional capabilities beyond traditional AI applications. Unlike conventional models that primarily analyze existing data to make predictions, Generative AI actively generates new data simulations that can model complex and unpredictable scenarios. This capability is specifically transformative in FinTech, where financial institutions can dynamically simulate various economic and market conditions to evaluate risk, compliance, and customer behavior. One of the Generative AI frameworks used in FinTech is Generative Adversarial Networks (GANs). It involves two neural networks, the generator and the discriminator, which work together to produce high-quality synthetic data. For instance, GANs can generate synthetic financial datasets that simulate real-world financial market conditions, enabling stress tests and risk assessments without catastrophic events [6, 37]. Another advanced model is the Variational Autoencoder (VAE), which compresses and generates data. Credit card companies employ VAEs to detect anomalies in spending behavior that could indicate fraud or identity theft by learning to reconstruct standard transaction patterns and flag deviations as potential fraud [7, 44]. The future of business is based on big data and Generative AI. Combining these new technologies is essential in improving business decision-making, especially in digital finance. These technologies offer new opportunities for deep and advanced analytics that can positively impact business performance and strategy.

3 Why Does a Business Need a Data Strategy?

A data strategy can bring innovation and revenue growth to a business. By accessing real-time data insights, companies can continuously discover new ways to improve their products or services. They can also identify market fluctuations and consumer preferences ahead of their competitors.

With a data strategy, a company can use data analysis to drive innovations. This may include developing new products or improving existing ones based on insights gathered from collected data. For example, the company might discover that a specific product is in high demand within a particular market segment, allowing it to focus on developing that product. Another important aspect is the identification of market fluctuations and customer preferences. Through data analysis, companies can anticipate changes in market demand and respond quickly to meet these demands [45]. This is particularly important in a business environment where staying one step ahead of the competition can be crucial. Figure 5 illustrates a framework for implementing a big data initiative, outlining three main components: prerequisites, process, and outcome. The prerequisites for a successful initiative include management support, infrastructure, a data-driven culture, and absorptive capacity. The process involves goal setting, team building, metrics selection, and plan implementation. The outcomes aimed for are improved human capital, innovation, and a new knowledge base, which together lead to a competitive advantage.

Moreover, access to real-time data enables companies to react quickly to changes in the market and consumer behavior. This speed of response helps them maintain a competitive edge and allows them to identify and exploit new opportunities whenever they arise. For example, by using data to track product performance in real-time, a company can make quick and effective changes to improve quality or address customer demands [46].

A *data strategy* is an action plan that describes how data will be used and analyzed to extract real-time, accurate, and actionable insights to address current and future business needs. With a robust data strategy, companies can maximize the utilization of their data resources through Business Intelligence (BI), resulting in more measurable return on investment and revenue growth. It assists company managers in setting the stage for a data-driven culture and aids in data-based decision-making.

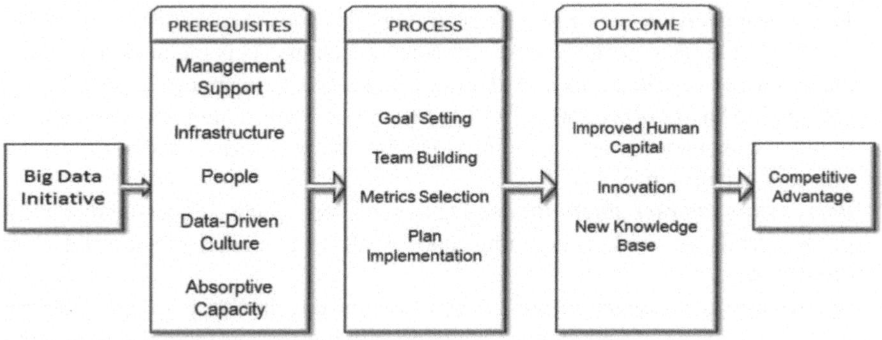

Fig. 5 Big data strategy framework

Integrating Generative AI into a company's data strategy can significantly enhance its capability to innovate and remain competitive. Generative AI, through its advanced algorithms, can create synthetic data sets that simulate various potential future scenarios, allowing companies to test responses and optimize strategies under different conditions. This ability is crucial for companies looking to innovate rapidly and with greater precision, as it provides a safe environment to explore the impact of changes before they are implemented in the real world. Generative AI can transform a company's approach to market analysis, product development, and customer service by providing deeper, actionable insights derived from predictive analytics and pattern recognition. For example, banks and investment firms in the financial sector can use Generative AI to model financial markets under various global economic scenarios, enhancing their investment strategies [6]. It improves decision-making and gives these companies a significant advantage over competitors who may still rely on traditional analytical methods [44]. A real-world application of this can be seen with companies like Spotify and Netflix, which use Generative AI to personalize content recommendations at a scale. Generative AI helps these platforms predict what content will keep users engaged by analyzing vast user preferences and viewing habits, increasing user satisfaction and retention rates. To effectively incorporate Generative AI into a data strategy, companies must consider several key elements [6]:

Data Quality and Diversity: Generative AI requires vast, varied, and high-quality data to avoid biases and inaccuracies in generated outputs.

Infrastructure and Expertise: Investing in necessary technological infrastructure and expertise to develop and maintain sophisticated Generative AI models.

Ethical Considerations and Compliance: Addressing ethical concerns and complying with data protection regulations, particularly when generating and using synthetic data.

Some of the main reasons for implementing a data strategy in an enterprise include [47]:

1. ***Improving decision-making***: A data strategy provides access to accurate information that helps managers make informed decisions. This includes improving the efficiency of operations and identifying new business opportunities.
2. ***Optimizing business processes***: With well-analyzed data, companies can identify inefficient or unnecessary steps in their business processes, improving overall business performance.
3. ***Forecasting market fluctuations***: Data helps companies understand current market trends and predict upcoming changes, allowing them to prepare and adapt proactively.
4. ***Identifying new opportunities***: A well-thought-out data strategy can uncover new revenue sources and identify new opportunities for business development.
5. ***Risk management***: Data analysis can help detect potential risks, allowing companies to take preventive measures and manage risk more effectively.
6. ***Improving customer relationships***: Companies can create and offer customers a better, more personalized experience and build stronger relationships by analyzing customer data.

A data strategy facilitates data-driven decision-making and paves the way for a data-oriented culture within the organization. This can result in increased efficiency, the discovery of new opportunities, and better risk management, ensuring a competitive advantage in the market and sustainable business growth.

3.1 Framework for Designing an Effective Data Strategy

The rapid growth of data can be a significant challenge for companies. Dealing with a large amount of unclassified and unanalyzed data can become a burden, making it difficult for companies to distinguish which data are essential for decision-making and business development [48]. Adding technological solutions that promise to improve data management and analysis can further add to the confusion and make choosing the right tools more challenging [49].

By implementing a data strategy, companies can set clear objectives aligned with the business's primary goals. This strategy will determine which data are most important for achieving these goals and how they should be handled. This clarification helps avoid the loss of focus on unnecessary or irrelevant data and allows companies to concentrate on efforts that will significantly impact business development.

Furthermore, a data strategy enables companies to set priorities effectively. Instead of trying to manage and analyze every piece of data, they can focus on the data that directly impacts effective decision-making and the development of key business strategies. This means that resources, such as time and budget, are used most efficiently, helping to achieve sustainable business results.

Data is becoming increasingly crucial for business success, and a data strategy facilitates data management and creates a competitive advantage. By effectively analyzing and interpreting data, companies can discover patterns, predict trends, and identify new opportunities for business growth. This is particularly important in a rapidly changing business environment, where the ability to respond to changes can be critical to the survival and prosperity of the business.

3.2 Implementation of the Data Strategy

Designing an effective data strategy is crucial for any organization that wants to stay competitive and adapt to changes. This framework includes the key steps and considerations for developing a data strategy that supports business objectives and provides a competitive advantage [48, 49]:

(1) *Defining business objectives and priorities*: The start of any data strategy is the precise definition of business objectives. What does the business want to achieve with the data? This may include increasing revenue, improving operational efficiency, or creating a better customer experience. Business objectives should be measurable and directly linked to the data strategy.
(2) *Assessing current data resources*: After defining the objectives, the next step is to determine the current data resources. This involves becoming familiar with the type and quality of data the business has and the tools and technologies used to collect and analyze this data. It is essential to recognize the limitations and opportunities of current data sources.
(3) *Developing an integrated plan*: An integrated plan that links the data strategy with business objectives should be created. This plan should define the specific steps to be taken and establish a timeframe for implementing these steps. It should also include how data will be collected, stored, processed, and analyzed.
(4) *Ensuring data quality and risk management*: Data quality is a critical issue. Inaccurate or unclear data can lead to poor business decision-making. Therefore, an efficient system must be established to ensure data quality by continuously verifying, improving, and cleaning data. Also, risk management is essential, especially concerning data privacy and security.
(5) *Involving and training the team*: It is essential to involve the team in developing and implementing the data strategy and ensuring the team has the necessary knowledge and skills to work with data. Ongoing training and development are essential to keeping up with the latest technologies and practices in data management. The success of a data strategy depends heavily on the level of engagement and knowledge of the team. All team members must be informed and appropriately trained to manage and use data effectively.

A data strategy should be dynamic and adaptable to the business environment and technology changes. Therefore, a continuous evaluation process should enable the review and improvement of the plan based on performance and market changes.

4 Leveraging Generative AI in Personalized Financial Products

Personalized financial products are revolutionizing the way customers interact with financial services. Unlike traditional financial products banks offer, these products are tailored to meet each customer's specific needs, preferences, and financial goals. Technological advancements and the growing variety of customer needs drive this evolution. Artificial intelligence, big data analytics, and machine learning are vital in personalizing financial products. They enable financial institutions to analyze large amounts of data to understand customers' preferences and design customized financial solutions [13, 50].

Personalized financial products encompass various financial services, including investment plans tailored to individual risk profiles, customized insurance policies based on personal needs, and credit options designed to suit specific financial situations [51].

For customers, personalized financial products offer many benefits. They provide solutions that are more aligned with individual financial goals, potentially generating better financial outcomes and higher consumer satisfaction. However, despite their benefits, personalized financial products also present challenges. Primary concerns include ensuring data privacy, meeting various regulatory requirements, and securing an advanced technological infrastructure to support personalization [52]. Examples of personalized financial products include robo-advisors in managing customized investment portfolios, which offer investment advice based on individual financial goals and the investor's risk preferences, adapting to changing customer behaviors in the market [53]. Another example could be when a bank offers personalized credit cards. These cards use a customer's past purchase data to provide customized rewards and discounts that match the customer's spending habits and trends [50].

Generative AI can contribute to the personalization of financial products by enabling the creation of highly tailored financial solutions that can adapt in real-time to changes in customer behavior and market conditions. Generative AI uses advanced algorithms to generate new data scenarios, enhancing predictive models that accurately foresee customer demands. One example is using generative AI by financial tech startups like Upstart, which offers personalized loan rates. Upstart's platform uses Generative AI to analyze thousands of data points from a potential borrower's financial history to predict creditworthiness more accurately than traditional credit score methods [31]. This approach improves the precision of loan offerings and expands access to credit for consumers whose conventional metrics need to be revised. Another innovative application is found in wealth management, where platforms such as Wealthfront use Generative AI to simulate various investment scenarios. These simulations help create customized investment strategies that dynamically adjust to market changes and personal financial milestones, providing customers with optimized investment paths tailored to their risk tolerance and financial goals [6, 7].

Personalized financial products represent a significant trend in digital finance. Their ongoing evolution and adoption are set to transform the financial sector, making financial services more customer-centric and responsive to individual needs. The future of personalized financial products foresees even deeper integration with emerging technologies such as Blockchain and the Internet of Things (IoT), providing efficient, personalized financial services [8]. Technology has paved the way for these developments, and the demand for personalized financial solutions will continue to grow as customers become more aware of the opportunities these products offer. As the financial industry adapts to these changes, we can expect further innovations to improve how we manage and interact with personal and business finances.

4.1 The Future of Smart Accounts with Generative AI

Smart accounts represent a significant evolution in the field of digital finance. They are not just digital versions of traditional bank accounts but are highly advanced accounts driven by technology that offer automation, advanced data analytics, and personalized financial products/services. The transformation from conventional bank accounts to smart accounts reflects a broad evolution towards digital finance. This evolution is driven by technological innovations and increasing demands for more efficient, secure, and beneficial financial services for customers. Smart accounts are characterized by their ability to seamlessly integrate with various digital finance tools, providing real-time analytics and forecasts. They utilize artificial intelligence and machine learning technologies to enhance users' services and increase financial management capabilities [54].

A blockchain enables smart accounts, which offer a new level of security and transparency by verifiably storing financial data. Artificial Intelligence and Machine Learning allow detailed analysis and personalization of financial services, giving each user a unique and personalized experience. These technologies ensure smart accounts can learn and adapt to clients' changing needs [7]. Individuals and businesses benefit from smart accounts in different ways. These accounts provide solutions in personal financial management with accurate and personalized analysis. For businesses, they provide advanced capabilities in analytics and forecasting, aiding in risk management and efficient decision-making [46].

Generative AI significantly advances the capabilities of smart accounts by introducing a deeper layer of intelligence and adaptability. By generating and analyzing synthetic data, Generative AI enriches smart accounts' analytics and forecasting abilities by generating and analyzing synthetic data, allowing for more personalized and foreseeing financial management solutions. It also enables smart accounts to perform sophisticated pattern recognition and predictive modeling beyond traditional data analysis. For instance, it can simulate various financial scenarios based on a user's spending habits and income changes to provide highly customized budgeting advice and saving tips. This type of predictive analysis helps users optimize their financial decisions in real-time. A practical application of this is seen in using AI-driven tools like Emma, a personal finance app. Emma uses Generative AI to analyze individual financial data, offering users personalized advice on avoiding overdrafts, finding unnecessary subscriptions, and predicting future cash flow. This allows users to make informed decisions that better align with their financial goals [6, 44].

Generative AI also plays an integral role in improving the security features of smart accounts. Generative AI systems can learn to detect unpretentious patterns of anomalous behavior that may indicate fraud by generating fraudulent and non-fraudulent transaction models. This capability significantly reduces the incidence of false positives and improves the accuracy of real-time fraud detection. For example, banks are increasingly integrating Generative AI into their smart account platforms to continuously monitor and analyze transaction data. One such innovation is implemented by Citibank, utilizing advanced machine learning algorithms to generate and explore thousands of transactional data points, enabling the detection of complex fraud patterns more effectively than ever [55]. Other examples of smart accounts include personalized investment recommendations in personal finance applications, automatic categorization of bank transactions, and real-time detection and prevention of fraud in financial transactions [54].

Despite smart accounts' advantages, they also present challenges, especially regarding data security and privacy. Data security is a significant challenge for smart accounts. Protecting personal and financial data from cyber-attacks and unauthorized use is critical. There is also the challenge of managing data privacy, aiming for high standards of security and accountability [42]. The future of smart accounts is expected to involve even deeper integration with emerging technologies, offering even more advanced financial management tools to transform the traditional banking model. Smart accounts are a vital part of the modern digital finance ecosystem. Their ongoing evolution will significantly shape the future of financial management, making it more efficient, secure, and customer-oriented.

5 Optimizing Risk Management and Fraud Prevention with Generative AI

Risk management and fraud prevention in digital finance ensure financial transactions' security and stability. They involve practices of identifying, assessing, and mitigating various types of risks and fraudulent activities that could impact financial systems. Effective risk management is essential in digital finance due to the high volume of transactions, the speed of operations, and the complexity of digital financial products. Proper risk management helps maintain customer trust and the integrity of financial systems. Risk management encompasses managing several types of risks, such as operational risk related to system failures, market risk due to fluctuations in financial markets, credit risk involving the possibility of borrowers failing to repay loans, and cyber risk related to digital transactions [56]. Figure 6 outlines the comprehensive application of big data analytics in risk management across four domains: Risk Assessment and Measurement, Risk Control and Monitoring, Front Office and Risk Operations, and Risk Reporting and Governance. Each domain includes specific activities such as risk modeling, fraud detection, front-office decision support, and regulatory reporting, emphasizing the integration of data analytics to enhance decision-making and operational efficiency in managing risks.

Fraud in digital finance can take various forms, from identity theft and phishing to complex fraudulent financial schemes. Understanding the nature of these fraudulent activities is essential for developing effective prevention strategies. Artificial intelligence and machine learning are increasingly used for risk analysis and fraud detection, providing advanced ways to monitor transactions and identify suspicious activities [57]. Blockchain technology also provides enhanced security features for digital transactions [56]. Figure 7 visually represents the integration of various domains in fraud analytics, including business intelligence (BI), data science, and artificial intelligence. The overlapping between management information, data warehousing, artificial intelligence (containing machine learning and deep learning), and big data technologies stresses how each contributes to enhancing fraud detection and prevention strategies in a business context.

Generative AI transforms risk management and fraud prevention by enabling more dynamic and adaptive strategies that improve security measures. This advanced form of AI goes beyond traditional data analysis by generating synthetic data and realistic simulations that can predict and mitigate potential risks before they materialize. Generative AI significantly enhances risk modeling by creating complex, predictive models that simulate possible future scenarios. For instance, in credit risk management, Generative AI can generate virtual profiles of potential borrowers based on existing data trends. These profiles can assess risk levels under various financial

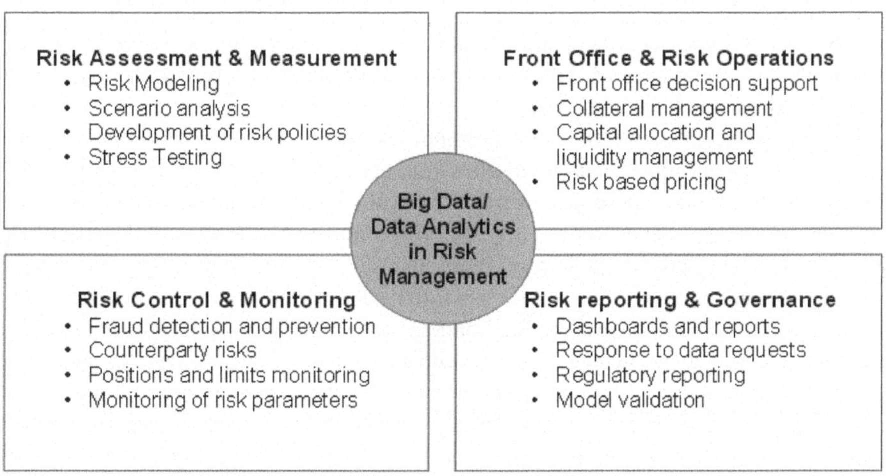

Fig. 6 Big data analytics in risk management activities

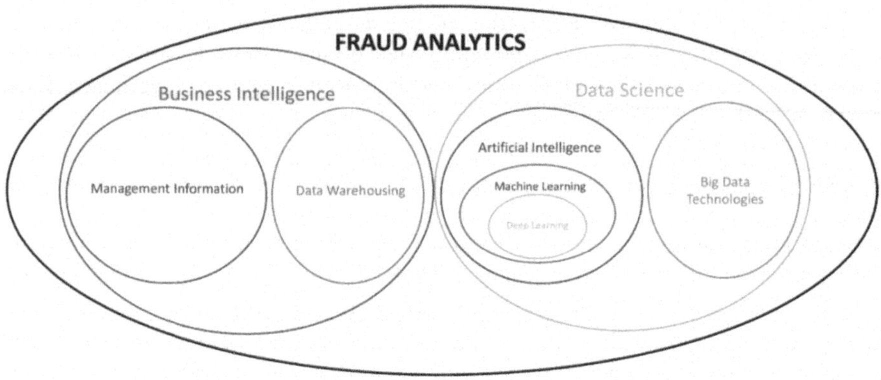

Fig. 7 The role of data science and BI in fraud detection and prevention

conditions without exposing the financial institution to actual risk. It allows for more complex risk assessments that are not solely dependent on historical data, reducing the likelihood of unpredictable losses [37, 44].

Generative AI is particularly effective in fraud detection because it produces large volumes of synthetic transaction data that simulate both normal and fraudulent behaviors. This capability is vital for designing fraud detection systems to identify unpretentious, sophisticated fraud patterns that traditional systems might overlook. One example is PayPal, which uses Generative AI to enhance its fraud detection algorithms. By analyzing and generating data on transaction behaviors, PayPal's systems accurately flag and investigate suspicious activities, drastically reducing false positives and improving customer trust [37]. Generative AI also supports personalized risk management solutions. For example, insurance companies are beginning to use this technology to offer customized insurance policies based on generative models that predict individual risk factors more accurately. These models consider various variables, including personal behavior and external conditions, to tailor insurance premiums and coverage to each client's specific demands, thereby optimizing risk pooling and pricing strategies [6].

Implementing effective risk management strategies involves continuous monitoring, developing robust financial models, and updating information and knowledge about potential risks. There are various examples where financial institutions and FinTech companies have successfully identified and prevented fraud, highlighting the effectiveness of advanced data analytics and real-time monitoring systems. These systems can analyze millions of transactions per second and identify suspicious transactions in real-time, preventing financial losses [22]. Despite advances, challenges in risk management and fraud prevention, such as adapting to rapidly evolving technologies, managing large volumes of data, and ensuring regulatory compliance.

Risk management and fraud prevention are essential for the security and reliability of digital information. As the financial sector continues to evolve, these practices will become increasingly important in ensuring the stability and integrity of financial systems. Future risk management and fraud prevention trends should include greater integration of artificial intelligence and machine learning, enhanced data analytics capabilities, and more robust cybersecurity measures.

6 Conclusion

This chapter emphasizes Big Data Analytics and Generative AI's evolving role in modern business decisions within the FinTech environment, providing valuable insights into the convergence between technological advancements and financial services. The main contribution of the chapter is to provide a thorough analysis of how Generative AI and Big Data Analytics catalyze innovation in FinTech, particularly influencing the increase in efficiency and personalization of financial products and services. Big data analytics is essential in modern financial ecosystems by enabling companies to use large amounts of data for efficient decision-making. By integrating Generative AI and machine learning technologies, companies can better predict consumer behaviors and improve risk assessments with unprecedented accuracy. Looking ahead, the potential of Generative AI in FinTech is vast, promising to unlock even more innovative approaches in data handling and decision-making processes. The chapter underlines the dual role of Generative AI and Big Data Analytics in enhancing existing financial institutions' capabilities by fostering an environment conducive to innovation. Furthermore, the chapter discusses the key challenges accompanying Big Data and Generative AI usage in FinTech, such as data privacy, security concerns, and the need for robust regulatory frameworks to match the pace of rapid technological advances.

References

1. He, W., Hung, J.-L., Liu, L.: Impact of big data analytics on banking: a case study. J. Enterp. Inf. Manag. **36**(2), 459–479 (2023)
2. Sun, Y., Shi, Y., Zhang, Z.: Finance big data: management, analysis, and applications. Int. J. Electron. Commer. **23**(1), 11–49 (2019)
3. Song, H., Li, M., Yu, K.: Big data analytics in digital platforms: how do financial service providers customise supply chain finance? Int. J. Oper. Prod. Manag. **41**(4), 410–435 (2021)
4. Kamel, M.A.: Big data analytics and market performance: the roles of customization and personalization strategies and competitive intensity. J. Enterp. Inf. Manag. **36**(6), 1727–1749 (2023)
5. Tian, X., He, J.S., Han, M.: Data-driven approaches in FinTech: a survey. Inf. Discov. Deliv. **49**(2), 123–135 (2021)
6. Marshall, A., Bieck, C., Dencik, J., Goehring, B.C., Warrick, R.: How generative AI will drive enterprise innovation. Strat. Leadersh. **52**(1), 23–28 (2024)
7. Yanting, Z., Ali, M.: Artificial intelligence, digital finance, and financial inclusion: a conceptual framework. In: Leong, C.-M., Ali, M., Raza, S.A., Puah, C.-H., Eksi, I.H. (eds.) Financial Inclusion Across Asia: Bringing Opportunities for Businesses, pp. 77–85 (2023)

8. Chauhan, S., Akhtar, A., Gupta, A.: Customer experience in digital banking: a review and future research directions. Int. J. Qual. Serv. Sci. **14**(2), 311–348 (2022)
9. Soltani Delgosha, M., Hajiheydari, N., Fahimi, S.M.: Elucidation of big data analytics in banking: a four-stage Delphi study. J. Enterp. Inf. Manag. **34**(6), 1577–1596 (2021)
10. Osei-Assibey Bonsu, M., Wang, Y., Guo, Y.: Does fintech lead to better accounting practices? Empirical evidence. Account. Res. J. **36**(2/3), 129–147 (2023)
11. Thomas, N.M., Mendiratta, P., Kashiramka, S.: FinTech credit: uncovering knowledge base, intellectual structure and research front. Int. J. Bank Mark. **41**(7), 1769–1802 (2023)
12. Belanche, D., Casaló, L.V., Flavián, C.: Artificial Intelligence in FinTech: understanding robo-advisors adoption among customers. Ind. Manag. Data Syst. **119**(7), 1411–1430 (2019)
13. Mohsen, S.E., Hamdan, A., Shoaib, H.M.: Digital transformation and integration of artificial intelligence in financial institutions. J. Fin. Rep. Account. (2024)
14. Singh, C.: Artificial intelligence and deep learning: considerations for financial institutions for compliance with the regulatory burden in the United Kingdom. J. Financ. Crime **31**(2), 259–266 (2024)
15. Deloitte: After the dust settles—how financial services are taking a sustainable approach to GDPR compliance in a new era for privacy, one year on (2018). Retrieve from: https://www2.deloitte.com/content/dam/Deloitte/uk/Documents/risk/deloitte-uk-the-impact-of-gdpr-on-the-financial-services.pdf. Last accessed 22 Feb 2024
16. Mora, H., López, F.A.P., Tello, J.C.M., Morales, M.R.: Virtual currencies in modern societies: challenges and opportunities. In: Visvizi, A., Lytras, M.D. (eds.) Politics and Technology in the Post-Truth Era, pp. 171–185 (2019)
17. Naheem, M.A.: Exploring the links between AML, digital currencies and blockchain technology. J. Money Launder. Control **22**(3), 515–526 (2019)
18. Kumar, J., Rani, V.: Journey of financial technology (FinTech): a systematic literature review and future research agenda. In: Rana, S., Sakshi, S., Singh, J. (eds.) Exploring the Latest Trends in Management Literature, vol. 1, pp. 89–108 (2022)
19. Arora, A., Gupta, S., Devi, C., Walia, N.: Customer experiences in the era of artificial intelligence (AI) in context to FinTech: a fuzzy AHP approach. Benchmark. Int. J. **30**(10), 4342–4369 (2023)
20. Ozili, P.K.: CBDC, Fintech and cryptocurrency for financial inclusion and financial stability. Digit. Policy Regul. Govern. **25**(1), 40–57 (2023)
21. Yan, J.: How the use of alternative information in risk management fintech platforms influences SME lending: a qualitative case study. Qual. Res. Fin. Mark. (2024)
22. Yang, J., Zhao, Y., Han, C., Liu, Y., Yang, M.: Big data, big challenges: risk management of financial market in the digital economy. J. Enterp. Inf. Manag. **35**(4/5), 1288–1304 (2022)
23. Stewart, H., Jürjens, J.: Data security and consumer trust in FinTech innovation in Germany. Inf. Comput. Secur. **26**(1), 109–128 (2018)
24. Amason, A.C., Schweiger, D.M.: Resolving the paradox of conflict, strategic decision-making, and organizational performance. Int. J. Confl. Manag. **5**(3), 239–253 (1994)
25. Dash, S., Shakyawar, S.K., Sharma, M., Kaushik, S.: Big data analysis for decision-making processes: challenges and opportunities for the management of health-care organizations. J. Enterp. Inf. Manag. **32**(6), 881–899 (2019)
26. Rasmussen, T.H., Ulrich, D.: People analytics in the era of big data: changing the way you attract, acquire, develop, and retain talent. J. Organ. Effectiveness People Perform. **2**(3), 304–306 (2015)
27. Hua, X., Huang, Y., Zheng, Y.: Current practices, new insights, and emerging trends of financial technologies. Ind. Manag. Data Syst. **119**(7), 1401–1410 (2019)
28. Gonçalves, A.R., Breda Meira, A., Shuqair, S., Costa Pinto, D.: Artificial intelligence (AI) in FinTech decisions: the role of congruity and rejection sensitivity. Int. J. Bank Mark. **41**(6), 1282–1307 (2023)
29. Zou, Y., Ali, M.: Artificial intelligence, digital finance, and financial inclusion: a conceptual framework. In: Financial Inclusion Across Asia: Bringing Opportunities for Businesses, pp. 77–85 (2023)

30. Sangwan, V., Harshita, Prakash, P., Singh, S.: Financial technology: a review of extant literature. Stud. Econ. Fin. **37**(1), 71–88 (2019)
31. Kumari, B., Kaur, J., Swami, S.: Adoption of artificial intelligence in financial services: a policy framework. J. Sci. Technol. Policy Manage. **15**(2), 396–417 (2024)
32. FCA: FCA publishes feedback statement on big data call for input, the financial conduct authority (2017). Retrieved from: https://www.fca.org.uk/news/press-releases/fca-publishes-feedback-statement-big-data-call-input. Last accessed 27 Feb 2024
33. De Mauro, A., Greco, M., Grimaldi, M.: A formal definition of big data based on its essential features. Libr. Rev. **65**(3), 122–135 (2016)
34. Saleh, I., Marei, Y., Ayoush, M., Abu Afifa, M.M.: Big Data analytics and financial reporting quality: qualitative evidence from Canada. J. Financ. Rep. Account. **21**(1), 83–104 (2023)
35. Mihet, R., Philippon, T.: The economics of big data and artificial intelligence. In Choi, J.J., Ozkan, B. (eds.) Disruptive Innovation in Business and Finance in the Digital World, vol. 20, pp. 29–43 (2019)
36. Elia, G., Stefanelli, V., Ferilli, G.B.: Investigating the role of Fintech in the banking industry: what do we know? Eur. J. Innov. Manag. **26**(5), 1365–1393 (2023)
37. Hirsch, P.B.: At the crossroads: generative AI and corporate risk management. J. Bus. Strateg. **44**(6), 426–429 (2023)
38. Özemre, M., Kabadurmus, O.: A big data analytics based methodology for strategic decision making. J. Enterp. Inf. Manag. **33**(6), 1467–1490 (2020)
39. Raza, S.A., Govindaluri, S.M., Bhutta, M.K.: Research themes in machine learning applications in supply chain management using bibliometric analysis tools. Benchmark. Int. J. 30(3), 834–867 (2023)
40. Singh, D., Raj, K.B., Choubey, S., Bhasin, N.K.K., Yadav, R., Gulati, K.: Role of machine learning in changing social and business eco-system—a qualitative study to explore the factors contributing to competitive advantage during COVID pandemic. World J. Eng. **19**(2), 238–243 (2022)
41. Banu, A.: Big data analytics—tools and techniques—application in the insurance sector. In: Sood, K., Dhanaraj, R.K., Balusamy, B., Grima, S. (eds.) Big Data: A Game Changer for Insurance Industry, pp. 191–212. Emerald Publishing Limited (2022)
42. Rashid, A., Baloch, N., Rasheed, R., Ngah, A.H.: Big data analytics-artificial intelligence and sustainable performance through green supply chain practices in manufacturing firms of a developing country. J. Sci. Technol. Policy Manage. (2024)
43. Shah, H.M., Gardas, B.B., Narwane, V.S., Mehta, H.S.: The contemporary state of big data analytics and artificial intelligence towards intelligent supply chain risk management: a comprehensive review. Kybernetes **52**(5), 1643–1697 (2023)
44. Dencik, J., Goehring, B., Marshall, A.: Managing the emerging role of generative AI in next-generation business. Strat. Leadersh. **51**(6), 30–36 (2023)
45. Berman, S.J., Hagan, J.: How technology-driven business strategy can spur innovation and growth. Strat. Leadersh. **34**(2), 28–34 (2006)
46. Giudici, G., et al.: Big data analytics in innovation processes: which forms of dynamic capabilities should be developed and how to embrace digitization? J. Bus. Ind. Mark. **33**(7), 1–23 (2018)
47. Zhang, Q., Sun, X., Zhang, M.: Data matters: a strategic action framework for data governance. Inf. Manage. **59**(4), 1–12 (2022)
48. Ciampi, F., Marzi, G., Demi, S., Faraoni, M.: The big data-business strategy interconnection: a grand challenge for knowledge management. A review and future perspectives. J. Knowl. Manage. **24**(5), 1157–1176 (2020)
49. Gomez-Trujillo, A.M., Gonzalez-Perez, M.A.: Digital transformation as a strategy to reach sustainability. Smart Sustain. Built Environ. **11**(4), 1137–1162 (2022)
50. Sheth, J.N., Jain, V., Roy, G., Chakraborty, A.: AI-driven banking services: the next frontier for a personalised experience in the emerging market. Int. J. Bank Mark. **40**(6), 1248–1271 (2022)

51. Gigante, G., Zago, A.: DARQ technologies in the financial sector: artificial intelligence applications in personalized banking. Qual. Res. Fin. Mark. **15**(1), 29–57 (2023)
52. Hentzen, J.K., Hoffmann, A., Dolan, R., Pala, E.: Artificial intelligence in customer-facing financial services: a systematic literature review and agenda for future research. Int. J. Bank Mark. **40**(6), 1299–1336 (2022)
53. Gao, Y., Liu, H.: Artificial intelligence-enabled personalization in interactive marketing: a customer journey perspective. J. Res. Interact. Mark. **17**(5), 663–680 (2023)
54. Brammertz, W., Mendelowitz, A.I.: From digital currencies to digital finance: the case for a smart financial contract standard. J. Risk Fin. **19**(1), 76–92 (2018)
55. Kommunuri, J.: Artificial intelligence and the changing landscape of accounting: a viewpoint. Pac. Account. Rev. **34**(4), 585–594 (2022)
56. Reepu, T., Taneja, S., Grima, S. (2023). The Risk Landscape in the Digital Transformation of Finance and Insurance. In: Sood, K., Balusamy, B., Grima, S. (eds) Digital Transformation, Strategic Resilience, Cyber Security and Risk Management, vol. 111C, pp. 163–175. Emerald Publishing
57. Mangala, D., Soni, L.: A systematic literature review on frauds in banking sector. J. Fin. Crime **30**(1), 285–301 (2023)

Enhancing Digital Security in Fintech Through Integration of Generative AI in Regulatory Practices

Artor Nuhiu

Abstract The chapter focuses on substantial aspects of enhancing FinTech's digital security by integrating Generative AI into regulatory frameworks and strategies. By examining the evolution and dynamic nature of cyber threats, the study attempts to explain the necessity of using AI technologies to develop defense mechanisms for data security and encryption to protect financial information. The chapter overviews the CIA Triad framework—confidentiality, integrity, and availability—and provides examples of how Generative AI can protect confidential digital information and promote data privacy trust. Moreover, it analyzes regulatory technology's (RegTech) role in adapting to a dynamic FinTech environment and providing regulatory compliance solutions. The chapter evaluates how Generative AI influences cybersecurity regulations and shapes security practices by maintaining the balance between Fintech innovation and cyber risk management. It considers the understanding of concepts needed to develop an advanced digital security strategy for FinTech related to data safety and the growth of trust and compliance in a financially integrated world.

Keywords Digital security · FinTech · Generative AI · Cybersecurity · Data encryption · RegTech · Data privacy

1 Introduction

With the rapid development of financial technology (FinTech), financial institutions and users of financial products and services have become increasingly aware of the importance of protecting data and digital transactions. The chapter addresses the challenges related to digital security in the financial sector, focusing mainly on the intricate aspects of information protection. In the early stages of the evolution of digital finance, security was primarily focused on protecting internal systems and customer data from cyber-attacks and internal misuse. Over time, the nature and

A. Nuhiu (✉)
Financial Law Department, University of Prishtina—Faculty of Law, Prishtina 10000, Kosovo
e-mail: artor.nuhiu@uni-pr.edu

N. R. Vajjhala et al. (eds.), *Generative Artificial Intelligence (AI) Approaches for Industrial Applications*, Information Systems Engineering and Management 24,
https://doi.org/10.1007/978-3-031-76710-4_15

complexity of security threats have evolved unexpectedly, increasing the need for advanced and tailored security strategies [1–3].

The application of Generative AI offers transformative potential for improving digital security and regulatory frameworks within the FinTech sector. By employing sophisticated algorithms, Generative AI technologies can learn and adapt from vast datasets and revolutionize how financial institutions mitigate risks and manage data [4]. These AI systems can generate realistic simulations and models that help predict and counter security threats before they impact the financial ecosystem. Furthermore, their capacity to automate compliance procedures and regulatory reporting by generating necessary documentation in real-time can significantly reduce administrative burdens and improve accuracy [5].

Cyber-attacks are considered one of the main threats in the FinTech sector. These attacks involve attempts to infiltrate the computer systems of a financial institution to steal sensitive data or cause financial damage. Standard cyber-attack methods include *Phishing, Malware, Ransomware, Initial Access Brokers (IAB), and DDoS Attacks* [6, 7]. In addition to these, financial institutions also face internal threats, such as fraud and employee misuse. To combat these threats, one of the critical components of digital security in finance is data encryption. Encryption is a process of coding information that ensures sensitive data, such as bank account details and clients' personal information, remain invisible to unauthorized individuals [8]. It is also essential that communication between clients and financial institutions is secure, which is achieved through secure communication channels and advanced security protocols [9]. Additionally, Generative AI integration can further enhance these security measures by facilitating dynamic encryption protocols and refined anomaly detection systems that learn from each interaction. Its ability to create predictive models based on emerging trends enables financial institutions to stay ahead of cybercriminals [10]. Generative AI can be instrumental in developing new forms of digital identity verification, thereby enhancing the integrity and authenticity of user transactions. Many FinTech companies employ Generative AI to create models that automatically learn from transaction histories to detect fraudulent activities. These systems continuously update their understanding of 'normal' versus 'anomalous' as new data comes in, allowing them to adapt to evolving fraud tactics [11].

The evolution of the FinTech sector now enables large amounts of money to be transacted online through financial transactions. The world relies on the systems of financial institutions that facilitate the making of payments and the circulation of money online. Digitalization has made financial transactions susceptible to privacy and security concerns. The impact of these transactions is significant, and privacy and security are vital for all of society. Therefore, today, countries invest billions of dollars annually in cybersecurity [12]. Digitalization has significantly affected the security and privacy of financial transactions, bringing both challenges and opportunities. First, it is vital to recognize the difference between privacy and security. Privacy relates to user data, where any exploitation, unauthorized access, or use of it is prohibited except with the consent and authorization of the users themselves. Thus, privacy means that individuals can choose how and how to interact with their surroundings or how much their data is appropriate for this inclusion [13]. Security is an older concept that deals with the protection of privacy. The main aspects concerning security are availability, authenticity, confidentiality, and integrity. Digitalizing financial transactions has influenced security and privacy through the emergence of financial technologies, which have provided innovative and adaptable financial services that lead to incredible speed and accessibility [14].

On the other hand, they have raised new concerns about security and privacy, one of the main concerns being the possibility of unauthorized access to sensitive information. Cybercriminals have become increasingly sophisticated in cyber-attacks, breaches of data privacy, and financial fraud. To address these concerns, financial institutions have taken measures such as [15]:

(a) improving security mechanisms in data storage and processing;
(b) conducting verifications and controlling access through cryptography or data encryption;
(c) using attribute-based access control for protecting client privacy; and
(d) developing a data storage mechanism with privacy ensured through cloud computing.

Despite these efforts, cyber-attack threats continue to exist, and financial fraud has emerged as a new challenge, forcing financial institutions to continually assess and improve their security and privacy strategies.

1.1 The CIA Triad's Role in Cybersecurity: Confidentiality, Integrity, and Availability for Data Security

One of the most significant challenges in the FinTech sector is ensuring data and services' confidentiality, integrity, and availability—a concept known as the *CIA Triad*. It represents the basic concept of any cybersecurity strategy and consists of three essential pillars for the secure and efficient functioning of any digital financial system. It is a model designed to guide information security policies within an organization. Each component of the CIA Triad addresses a distinct aspect of security. Generative AI technologies enhance the CIA Triad by introducing dynamic encryption keys for confidentiality, anomaly detection algorithms for data integrity, and predictive models for ensuring data availability [10]. AI-driven encryption duplicates keys in response to attack patterns, anomaly detection spots unauthorized data alterations, and predictive AI forecasts and mitigates system vulnerabilities, maintaining data accessibility and effectively balancing the Triad components [3]. Together, they provide a comprehensive framework for protecting sensitive information from cyber threats [16–18]:

Confidentiality: the first component of the CIA triad entails protecting information from unauthorized access. This means ensuring that only those who are authorized can access certain information. Confidentiality is often maintained through encryption, access control, and information verification procedures. Generative AI can play a crucial role by enhancing encryption methodologies. AI-driven encryption involves using machine learning algorithms to generate encryption keys dynamically, which can be more secure and less predictable than traditional methods [10]. A breach of privacy can lead to severe consequences, including loss of client trust, legal repercussions, and financial losses. Therefore, maintaining confidentiality is essential for the CIA Triad in cybersecurity.

Integrity: the second component of the CIA triad ensures that financial data and transactions are accurate, complete, and unaltered by unauthorized interventions. Integrity is often maintained through checksums and hashing, which verify data consistency. These methods can detect any modification to data, whether intentional or accidental, ensuring the integrity of the data. AI-powered anomaly detection algorithms can be invaluable for ensuring data integrity. These algorithms analyze historical and real-time data to establish a baseline of normal behavior. They can immediately detect deviations indicating a breach, such as unauthorized modifications to financial data or transactions [11].

Availability: the third component of the CIA triad ensures that data is accessible to authorized users when needed. This includes maintaining hardware, performing regular system updates, and creating backup copies to prevent data loss. DDoS attacks that aim to make a computer or network unavailable to its users are a common threat against availability in cybersecurity. Predictive AI models can forecast potential system downtimes or vulnerabilities that could lead to disruptions. These models can predict and mitigate future risks by analyzing vast amounts of data, including past incidents of system unavailability [3].

Figure 1 illustrates the CIA Triad model of Cybersecurity, which is centered on core concepts of various cybersecurity domains such as Network Security, Cloud Security, and Data Security, each playing a crucial role in ensuring the overall protection of information systems. The CIA is not just a theoretical model; it has practical implications in the real world. The principles of the CIA Triad in cybersecurity guide the development of security policies and procedures in organizations. By adhering to the principles of the CIA Triad, organizations can protect sensitive information from a wide range of cyber threats. This helps maintain the trust of clients and stakeholders and ensures the organization's smooth operation [19].

While the CIA Triad provides a robust framework for cybersecurity, its implementation is challenging. Balancing the three components of the CIA in cybersecurity can be a complex task. For example, increasing confidentiality through strict access control may unintentionally impact availability by making it harder for authorized users to access the necessary information. Similarly, ensuring data integrity may require additional resources, affecting the system's availability. Despite these challenges, the CIA Triad remains a fundamental concept in cybersecurity. With careful planning and execution, organizations can successfully implement the principles of the CIA in cybersecurity and protect their information from cyber threats.

Fig. 1 The CIA Triad components on cybersecurity

2 Cybersecurity Regulations for Financial Services

It is often said that regulations cannot keep pace with the speed at which technology changes—but that does not mean that regulatory agencies and other organizations that design and implement them are not trying their best. The social consequences of cyber-attacks and data breaches occur frequently and are increasingly damaging [20]. Today, in the business world, almost every company depends on software and technology, making cybersecurity a priority for organizations everywhere—this pressures cybersecurity leaders to improve their governance, risk, and compliance programs [21]. In 2019, cyber-attacks in the USA caused over $7.5 billion in damage to the public sector. However, the same year, no bank in the USA reported any cyber-attack incidents [20]. This shows a positive shift in the financial industry and indicates that cybersecurity regulations can work. Figure 2 shows the percentage of data breaches in various global regions for 2021 and 2022, based on the report data from [22].

North America experienced a significant decrease in data breaches from 52% in 2021 to 26% in 2022, indicating a possible improvement in security measures or a shift in cybercriminal focus. Conversely, Africa and Latin America show a marked increase in data breaches, particularly in Africa, where incidents rose from 4 to 22%, suggesting an escalating vulnerability in these regions' digital infrastructures.

In Europe, incidents from cyber-attacks are increasing exponentially year after year. Data breaches have become the "new normal phenomenon". Indeed, a 2019

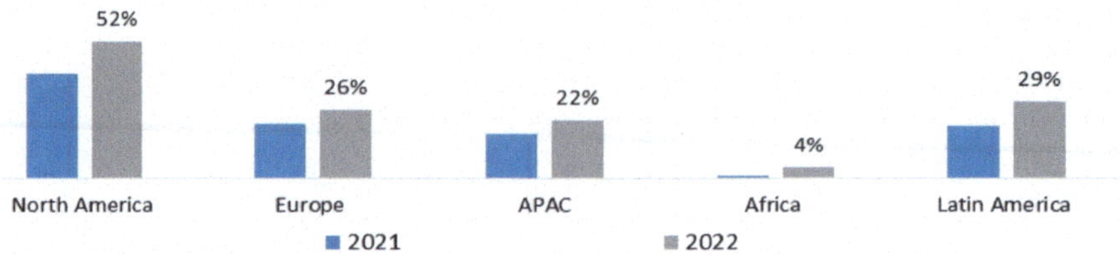

Fig. 2 Average weekly cyber-attacks per organization by region

report from *Carbon Black* reported that 88% of global businesses experienced one or more data breaches [21]. The response to these increasing cyber-attacks is the implementation of new cybersecurity regulations to protect organizations and their clients' data.

Figure 3 shows the trend in the number of data records breached globally from Q1 2020 to Q4 2023, based on data from [23]. The graph highlights significant fluctuations in data breaches over the period, with a notable peak in Q2 2021, suggesting a considerable cybersecurity event and subsequent declines, illustrating the dynamic nature of cybersecurity threats and the varying effectiveness of countermeasures over time. The overall trend underscores organizations' continuous challenge in protecting data amidst evolving security threats and the importance of implementing robust security measures to safeguard sensitive information.

As financial institutions encounter these evolving threats, Generative AI emerges as a vital technology in enhancing regulatory compliance and overall cybersecurity.

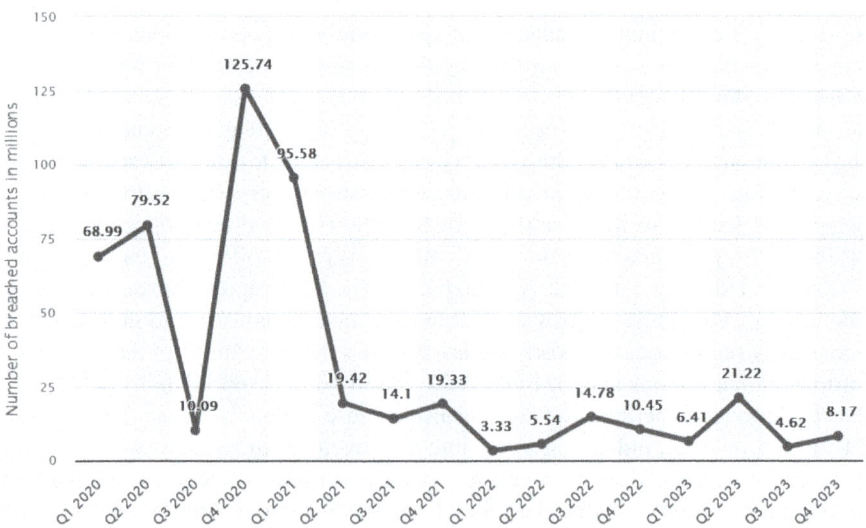

Fig. 3 Number of data records breached worldwide Q1 2020–Q4 2023

By integrating Generative AI, these institutions can automate the design and maintenance of security policies, adapt dynamically to new regulations, and generate compliance reports with high accuracy and minimal human intervention [4]. For example, AI-driven systems can analyze vast amounts of legislative documents and cybersecurity alerts to provide real-time updates on regulatory changes, ensuring that financial institutions remain compliant [5].

Cybersecurity regulations are crucial in shaping security practices and protecting data and financial assets. This regulation aims to provide a secure and trustworthy environment for digital financial transactions, protecting clients and businesses from cyber-attacks. Various standards and guidelines, such as *ISO/IEC* 27001, have been established to set a foundation for best cybersecurity practices [24]. These standards serve as a guide for financial institutions in managing security risks and protecting data. On a regional and national level, regulations such as the *General Data Protection Regulation* (GDPR) of the European Union have had a significant impact on the cybersecurity sector. GDPR has set high standards for the protection of personal data and has increased the responsibilities of financial organizations regarding the security of this data. Similarly, comparable legislation in various countries like the United States and China also influences the shaping of the cybersecurity component [21].

Financial institutions are increasingly turning to Generative AI solutions to manage and adhere to these rigorous standards effectively. These innovative technologies offer several advantages in tackling the compliance complexities of regional and international cybersecurity regulations [25]. One innovative application of Generative AI in this sector is the development of customized risk assessment tools. These AI-powered systems generate risk profiles based on a financial institution's specific operations and digital footprint, allowing for more precise and effective mitigation strategies [4]. Such tools can simulate potential attack scenarios and recommend the best preventative measures tailored to the institution's needs. Moreover, Generative AI can assist in the training and development of cybersecurity personnel. By creating simulated cyber-attack scenarios, AI systems can provide hands-on training experiences that are both realistic and up to date with the latest threat landscapes [25]. This helps build a more skilled workforce capable of effectively responding to and managing evolving cybersecurity challenges.

Although cybersecurity regulations are essential in raising the security standard, they also present significant compliance challenges for financial institutions. These challenges include managing compliance with multiple and sometimes conflicting rules. Financial institutions must develop effective compliance management strategies to align with these regulations [24]. Regulations have also shaped cybersecurity measures in digital finance, increasing client trust and security. Consequently, financial institutions must continue adapting and complying with evolving regulations to maintain a secure and trustworthy environment for their clients.

3 Regulating Digital Payment Services

The rapid technological advancement and shifting consumer expectations have led to the development of new payment system models. These models differ from traditional ones by offering various speed, cost, security, and usability advantages. New payment system models include the widespread use of digital and mobile payments. These payments allow transactions to be carried out anytime and anywhere using smartphones and tablets. Mobile banking apps and payment services like Apple Pay and Google Wallet have become commonplace. *Digital wallets* enable users to digitally store their credit and debit card data digitally, facilitating online and in-store payments [20, 21]. These wallets are linked to a bank account or credit card and provide a fast and secure way to make payments.

Blockchain technology and *cryptocurrencies* have introduced a new decentralized payment system model that operates without needing a third party as an intermediary. These technologies offer high security and transparency by reducing the need for intermediaries and are seriously considered as potential alternatives to the traditional payment system [26].

Instant payments are another new payment system model that allows the transfer of funds in real-time, 24 h a day, seven days a week. This model could replace the need for checks and bank transfers, offering a faster and more efficient solution [27].

Contactless payments and QR scans, such as chip cards and NFC (Near Field Communication), allow users to make payments simply by bringing their card or device close to a reader [28]. These technologies offer a fast and secure way to perform transactions and are becoming increasingly popular.

The open platform model allows the integration of various financial services into a single platform by directly interfacing with bank accounts and payment systems. This enables users to access and manage all their financial services from a single platform [29].

Artificial intelligence and automation are significantly transforming the payment system. They are being used to enhance its security, efficiency, and usability. These technologies help analyze financial data to detect fraud, optimize and automate processes, and manage risk. Generative AI is crucial in developing advanced fraud detection systems that generate and adapt real-time fraud indicators based on evolving transaction patterns. This capability significantly improves the security of digital payment systems, aligning well with regulatory frameworks aimed at protecting consumers and ensuring transaction integrity [4, 25]. Furthermore, Generative AI contributes to the efficiency of digital payments by generating predictive models that enhance the speed and accuracy of transaction processing. These models can predict and manage peak load times, optimize transaction pathways, and reduce processing delays, thus improving user satisfaction and compliance with service standards set by regulatory bodies [30].

Digital payment system models are helping to create a faster, more secure, and more efficient financial ecosystem. These innovations are changing the way individuals and businesses conduct transactions. They are built on new technologies and offer advantages over traditional models, including high speed, low cost, and improved usability. FinTech companies are developing innovative solutions to enhance and modernize the payment system. These include new platforms for managing and analyzing financial data and new solutions for providing loans, investments, and other financial services.

3.1 Security and Privacy Concerns in Digital Payment Systems

While digital payment systems offer many benefits, including improved efficiency and reduced transaction costs, they also bring new security challenges. These challenges must be carefully addressed to protect consumers' financial and personal data and maintain trust in the financial system [31–33]:

1. *Risk of Fraud and Theft:* New payment systems, especially those utilizing digital and online technologies, are susceptible to fraud and identity theft risks. Cybercrimes are becoming increasingly sophisticated, continually seeking ways to exploit security vulnerabilities to steal users' personal and financial data.
2. *Data Security:* Securing users' personal and financial data is a primary challenge for new payment systems. This includes protecting data during transmission and storage and ensuring it is accessible only to authorized individuals. New payment systems must protect data from unauthorized access and encryption to prevent theft.
3. *Risk of Technical Difficulties:* New payment systems are reliant on technology, and as such, they are susceptible to technical difficulties and defects. These issues can lead to service interruptions and significantly impact users and businesses.
4. *Compliance and Regulatory Issues:* New payment systems must comply with various laws and regulations to protect consumers and ensure financial stability. This requires a deep understanding of the legal framework and ongoing commitment to ensure the payment system complies with all legal requirements. Financial institutions and payment service providers must have clear and practical plans to respond to security incidents. This includes quickly identifying the incident, stopping damage spread, and transparent communication with affected users.
5. *User Education:* User education is critical to securing new payment systems. Users should be informed about potential risks and ways to protect their personal and financial data.

Security issues in new payment systems are complex and require ongoing attention and effort from all participants, including fintech companies, banks, regulatory authorities, and users. This involves data protection, security risk management, and educating users, who need to be aware of these issues and understand their role in helping to ensure a secure and reliable environment for digital payments.

3.2 PSD2 and SEPA Regulations for Digital Payment Systems

The need for updated regulatory frameworks has become more apparent with the evolution and innovation in payment systems. This has led to a regulatory response from the European Union through *PSD2* (*Payment Services Directive 2*) and *SEPA* (*Single Euro Payments Area*), which aim to foster competition, increase transparency, and protect consumers. *PSD2* is a directive from the European Union designed to create a more open and competitive market for payment services, improve the security of online transactions, and promote innovation. It achieves this by facilitating market entry for new payment service providers and ensuring consumer transparency [34]. PSD2 also enhances consumer protection by setting strict requirements for user identity verification for electronic payment transactions and providing better protection against fraud [35].

SEPA is another European Union initiative aimed at simplifying and streamlining Euro payment transactions within the European Union and other European countries. This is achieved by standardizing payment formats and eliminating differences between domestic and international payments. SEPA aims to make Euro payments as simple, secure, and efficient as domestic payments, removing barriers and facilitating trade and capital movement [36]. SEPA encompasses card transactions, bank transfers, and direct debits. This payment method has brought several significant benefits [37]:

- SEPA has enabled transactions in Euros across the European Union and other European countries under the same conditions, rights, and obligations, regardless of geographic location.
- SEPA has brought many benefits for consumers and businesses, including lower fees, faster payment processing times, and greater transparency.
- SEPA has driven banks and financial institutions to improve and modernize their payment infrastructure, leading to faster and more efficient services.

Regulatory authorities play a significant role in implementing PSD2 and SEPA. They oversee the payment services market, ensure all participants comply with regulatory requirements, and protect consumer interests [36]. The regulatory response through PSD2 and SEPA presents challenges and opportunities for banks and new payment service providers. Banks need to adapt their systems and processes to meet new requirements. In contrast, new payment service providers have a significant opportunity to enter the market and offer innovative solutions to consumers. PSD2 and SEPA have profoundly impacted the financial industry, pushing banks and other financial companies to adopt new technologies and improve their services [37]. However, these changes have also brought the need for significant investments in information technology (IT) infrastructure and enhanced security measures.

PSD2 and SEPA are part of the critical regulatory initiatives in response to recent developments in payment systems. They aim to create a more open, competitive, and transparent market for payment services while protecting consumers and fostering innovation.

4 Regulation and Supervision of FinTech

FinTech has revolutionized how financial institutions operate and how consumers use financial services. Yet, this innovation has also introduced new challenges, particularly in the regulatory aspect. Regulation for FinTech is crucial to ensure a safe and fair environment for all participants. One of the primary challenges in regulating FinTech is that technology evolves rapidly, whereas current regulations are often built on older principles and practices. It creates a gap between what is technically possible and what is allowed in the regulatory context. To address this issue, regulators must be more flexible and adaptable, updating regulations in line with new technological developments [38].

Moreover, FinTech regulations must protect consumers by ensuring that all products and services are safe and transparent and that consumer rights are protected. It is essential when dealing with sensitive personal and financial data. Internationally, regulating FinTech requires global cooperation. As FinTech often operates across borders, regulations in one country can significantly impact international markets [39]. Thus, collaboration and coordination among international regulators are essential to ensure a level playing field for all participants.

Another critical element is the inclusion of new technological innovations in the regulatory framework, which allows experimentation with new business models in a controlled regulatory environment known as the "Regulatory Sandbox". FinTech companies can test their products in a safe environment before launching them in the market [40]. Generative AI significantly enhances RegTech solutions' capability to manage risks and ensure compliance within the FinTech sector. For example, AI-driven platforms are increasingly used to automate and refine compliance processes [25]. These platforms can dynamically adapt to new regulations, automatically updating compliance systems and procedures without human intervention. The AI system continuously scans the bank's operations to ensure all data handling meets stringent GDPR requirements, effectively minimizing non-compliance risk. Another example is anti-money laundering (AML), where AI technologies have been deployed to track and analyze vast real-time transaction data. These systems can generate alerts for suspicious activities much more rapidly than traditional methods, thereby enhancing the effectiveness of financial monitoring and reporting mechanisms [4].

Risk Management is another vital aspect of FinTech regulation. Technological innovations such as blockchain, artificial intelligence, and big data have changed how risks are managed in the financial sector. Regulators need to understand these new technologies to assess and manage the risks they pose, including cybersecurity risks, fraud risks, and financial stability risks. For example, blockchain technology can increase transparency and reduce fraud risk but may also introduce new data privacy and cybersecurity challenges [41].

Integrating Generative AI into RegTech addresses the complexity of managing various compliance demands and provides predictive insights that help institutions preempt regulatory issues before they arise. By analyzing patterns and trends from

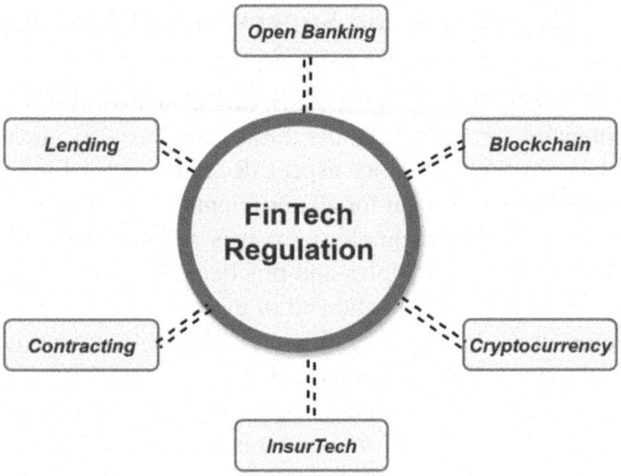

Fig. 4 Areas of interest for FinTech regulation

historical data, AI models can predict potential compliance failures, allowing institutions to implement corrective measures proactively [4, 25]. Figure 4 maps out the key areas affected by FinTech regulation, centralizing on the diverse segments within the financial technology sphere. The diagram emphasizes the extensive reach of regulations across different FinTech sectors, underscoring the importance of a regulatory framework in facilitating growth while managing risks.

Regulating cryptocurrencies, like Bitcoin, presents unique challenges for regulators, as digital currencies operate differently from traditional currencies. Questions about regulating their trade, ensuring they are not used for illegal activities, and protecting consumers are challenging to address. Therefore, regulators must develop clear guidelines for managing these risks, including cybersecurity measures and monitoring transactions to prevent illegal activities [42].

FinTech companies, through inclusivity and accessibility, have the potential to provide financial services to individuals traditionally excluded from the financial system [43]. However, regulators must ensure that FinTech services are accessible and suitable for many users, including those in rural or low-income areas. Regulation for FinTech also needs to address market integrity and the challenges of fair competition. FinTech companies can impact the structure of financial markets, bringing challenges related to competition and market integrity. Regulators must ensure that new FinTech companies compete fairly with traditional financial institutions and avoid creating harmful monopolies [44].

FinTech regulation needs to be dynamic, responsible, and consumer oriented. It requires balancing allowing innovation and ensuring that technological advancements do not introduce unacceptable risks. Regulators, FinTech companies, and users must work together to create a healthy and sustainable financial environment.

4.1 Foundations of Regulatory Technology (RegTech)

Regulatory Technology, known as RegTech, is a branch of financial technology focused on using technology to facilitate the implementation of regulatory requirements and compliance guidelines in the financial sector. RegTech has developed in response to the need for more efficient management of regulatory obligations in a continually changing and increasingly complex financial environment [45]. This technology involves using software, applications, and platforms that help financial institutions monitor, report, and comply with regulatory requirements. For example, RegTech may include systems for risk management, implementation of Anti-Money Laundering (AML) policies, and transaction monitoring. A key element of RegTech is using big data and data analytics to identify patterns and potential risks. It can help prevent fraud and detect illegal activities. Additionally, artificial intelligence (AI) and machine learning play a significant role in RegTech, offering ways to automate and optimize processes, easing the administrative burden and reducing the costs associated with implementing regulatory requirements [46].

The high regulatory compliance costs primarily drive the adoption of RegTech in the financial industry. In 2016, it was reported that American banks spent over 100 billion dollars on compliance, a figure that continues to rise. According to Bain & Co., spending on governance, risk, and compliance (GRC) constitutes the most considerable part of operational costs in banks. For instance, implementing specific regulations like Dodd-Frank and MiFID II has cost American banks billions of dollars [47]. By 2022, regulatory compliance costs are expected to account for about 10% of total expenses in financial institutions [48]. The complexity of financial institutions in terms of business models, legal structures, processes, and market services adds to the challenge of monitoring and compliance with regulations. This complexity is particularly pronounced for smaller companies. As a result, the financial industry is gradually shifting to RegTech-based solutions, with significant investments in consulting, professional services, and IT.

The application of RegTech is not limited only to the financial industry; it can also benefit other sectors. The significant expenditures of the financial industry on IT and big data, exceeding 360 billion dollars annually, underscore the need for efficient RegTech solutions, especially in the face of regulatory compliance challenges and data management issues [49]. Figure 5 illustrates the intersection of Regulatory Technology (RegTech) with Financial Technology (FinTech) and compliance within the financial sector. The Venn diagram highlights the overlapping areas of Regulation, Compliance, and FinTech, underscoring the critical role of RegTech in bridging these domains. Essential functions such as Supervision, Monitoring, and Reporting are pinpointed at the overlaps, demonstrating how RegTech facilitates the efficient management of regulatory processes, enhances transparency, and supports financial institutions in adhering to legal frameworks.

Digitization has significantly changed the financial regulatory environment, introducing new privacy and information security risks. This evolution has accelerated the growth of the RegTech industry, focusing on monitoring, reporting, and regulatory

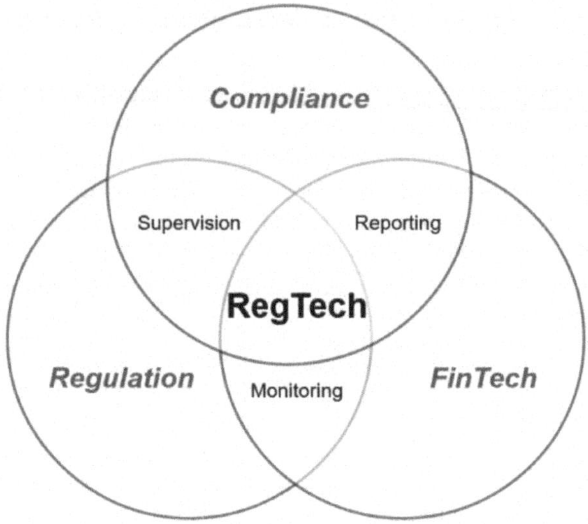

Fig. 5 The broad scope of RegTech

compliance. The shift from the traditional KYC (Know Your Customer) approach to a data-driven KYD (Know Your Data) approach requires a new regulatory perspective that addresses digital identity, data management, and algorithm supervision. Regulators face the challenge of fostering innovation while protecting consumers and ensuring financial stability. The use of regulatory "sandboxes" to test financial products and services allows for the design of new regulatory frameworks in this changing environment [50].

The European Union has built a RegTech ecosystem to support digital financial transformation in the European market. Its strategy is based on four pillars: extensive reporting requirements, strict data protection rules, open banking, and a legal framework for digital identity [51]. These pillars aim to transform the European market into a data-driven financial sector. RegTech offers considerable potential for regulators and institutions in the financial industry to leverage digitization for regulatory compliance and efficient risk management. However, achieving these benefits requires a coordinated approach among all stakeholders.

4.2 Preparing for a RegTech

Implementing the Regulatory Technology (RegTech) in the FinTech sector is a complex process that requires a comprehensive and strategic approach. Financial institutions must thoroughly assess their specific regulatory needs and challenges. This assessment will help identify areas where RegTech can provide better and more efficient solutions. For example, if a financial institution faces challenges in managing customer data for Anti-Money Laundering (AML) purposes, a RegTech solution focused on data analysis and identifying suspicious patterns would be appropriate. Another critical step is training and developing employees. Implementing RegTech requires a clear understanding of the technology and regulatory processes. Employees need to be trained not only in the use of technology but also in understanding its significance in the context of regulatory rules and guidelines [50]. It ensures a smooth and effective transition from traditional methods to new technologies.

It is essential to understand that RegTech is not just a technological solution but a shift in the organization's culture. Financial institutions must adopt an open approach to innovation and be willing to embrace the changes that come with using new technologies. It includes building an environment where staff feel encouraged to explore and experiment with new technological solutions [39]. FinTech is rapidly transforming the traditional environment of financial institutions. This evolution, although innovative, presents unique challenges such as data security, consumer protection, and financial stability. For this reason, regulators worldwide are developing regulatory frameworks to ensure that these challenges are effectively managed, promoting innovation while ensuring consumer protection and financial stability.

The traditional regulatory environment initially was variable and evolving, with countries adopting different regulatory approaches. Some countries chose a more direct strategy, implementing a "regulatory sandbox" that allows FinTech companies to experiment with new products without the total regulation weight [40]. Meanwhile, other countries adopted specific laws targeting online lending, payment systems, and digital currencies.

A critical focus has been on Anti-Money Laundering (AML) in FinTech. With FinTech companies handling large volumes of financial transactions and sensitive data, robust AML measures are essential. These include customer due diligence, transaction monitoring, and reporting suspicious activities, striking a balance between fostering innovation and ensuring financial integrity.

In the USA, the regulatory framework for FinTech is complex, involving multiple federal and state agencies. Key bodies include the Office of the Comptroller of the Currency (OCC), which regulates banks and federal savings associations, playing a significant role in allowing banks to offer security services for digital assets. The Consumer Financial Protection Bureau (CFPB) focuses on consumer protection, enforcing actions against deceptive practices. The Securities and Exchange Commission (SEC) oversees the trading of securities in financial markets, including ICOs and digital securities, and the Federal Reserve System (FED), which supervises banks to promote financial stability, also explores central bank digital currencies (CBDCs). In addition to federal regulation, American FinTech companies must also adapt to state regulations, which can vary significantly, with some countries offering regulatory sandboxes and others enacting specific laws for FinTech companies [50].

In the European Union, several regulations impact the operations of FinTech companies. The General Data Protection Regulation (GDPR) requires consent for data collection and provides data access and deletion rights. The Payment Services Directive 2 (PSD2) regulates payment services, requiring banks to enable payment infrastructure for third-party financial service providers. The Anti-Money Laundering Directive (AMLD) mandates due diligence and monitoring of suspicious activities. The Markets in Financial Instruments Directive 2 (MiFID II) and the Electronic Money Directive regulate the trading of financial instruments and the issuance of electronic money. The European Union has also established a regulatory sandbox for the controlled testing of new FinTech products [51].

FinTech regulations continue to evolve with new trends such as digital identity, open banking, cryptocurrencies, regulatory sandboxes, and international cooperation. Digital identity systems, like the EU's eIDAS (Electronic Identification and Trust Services), are crucial for online transactions. Open banking, promoted in the EU through PSD2, encourages competition and innovation in the financial sector. Cryptocurrency regulations are being implemented in various countries to address consumer protection and financial stability issues. Regulatory sandboxes in different countries facilitate innovation in complex regulatory environments. Finally, international collaboration is recognized as essential to address the global nature of FinTech, with organizations like the Financial Stability Board (FSB) and the International Organization of Securities Commissions (IOSCO) working to develop international standards for FinTech regulation.

5 Conclusion

The volume and complexity of digital financial transactions have increased dramatically due to FinTech's rapid development, making the industry a prime target for cyber threats like ransomware, malware, and phishing. The chapter discusses how these threats have changed over time, compelling financial institutions to create increasingly sophisticated cybersecurity defense mechanisms. Data encryption, which keeps sensitive information safe from unauthorized access, is one of the main strategies covered. In addition, secure communication channels and protocols must protect the interaction between clients and financial institutions. Regulatory technologies (RegTech) are now essential for assisting institutions in improving cybersecurity compliance and adjusting to changing financial environments. RegTech guarantees institutions adhere to international standards and assist with risk management and mitigation.

The CIA Triad—confidentiality, integrity, and availability—is a crucial framework emphasized for creating strict information security policies inside institutions to safeguard sensitive and confidential digital data from cyberattacks. The chapter addresses the difficulties in striking a balance between these three components of the CIA Triad, noting that tightening access controls to enhance confidentiality could unintentionally have an adverse effect on availability. Cybersecurity laws influence security procedures and contribute to developing a reliable and safe environment for online financial transactions. The potential of Generative AI to transform these security and regulatory practices is profound. Financial institutions can anticipate and mitigate risks more effectively by utilizing AI-driven analytics and machine learning algorithms. Generative AI can automate complex regulatory compliance processes, reducing the likelihood of human error and enhancing the efficiency of compliance audits. Moreover, AI's ability to analyze large datasets rapidly enables real-time threat detection and response, which is critical in maintaining the integrity and availability of financial services.

The role of Generative AI in FinTech security and regulation is expected to expand significantly in the future. Research areas focusing on integrating AI with blockchain technologies for enhanced transparency and security, advanced predictive analytics for fraud detection, and AI-driven behavioral biometrics for identity verification are poised to redefine the standards of digital financial transactions. Institutions should actively engage in the development of upcoming regulatory frameworks and concentrate on implementing modern technologies and procedures. This proactive strategy will maintain data security and protect the integrity of online financial transactions and address new cyber threats.

References

1. Ruhland, P., Wiese, F.: FinTechs and the financial industry: partnerships for success. J. Bus. Strat. **44**(4), 228–237 (2023)
2. Sangwan, V., Harshita, Prakash, P., Singh, S.: Financial technology: a review of extant literature. Stud. Econ. Financ. **37**(1), 71–88 (2020)
3. Yanting, Z., Ali, M.: Artificial intelligence, digital finance, and financial inclusion: a conceptual framework. In: Leong, C.-M., Ali, M., Raza, S.A., Puah, C.-H., Eksi, I.H. (eds.) Financial Inclusion Across Asia: Bringing Opportunities for Businesses, pp. 77–85. Emerald Publishing Limited (2023)
4. Marshall, A., Bieck, C., Dencik, J., Goehring, B.C., Warrick, R.: How generative AI will drive enterprise innovation. Strategy Leadersh. **52**(1), 23–28 (2024)
5. Mohsen, S.E., Hamdan, A., Shoaib, H.M.: Digital transformation and integration of artificial intelligence in financial institutions. J. Financ. Report. Account. (2024)
6. Akinbowale, O.E., Klingelhöfer, H.E., Zerihun, M.F.: Analysis of cyber-crime effects on the banking sector using the balanced score card: a survey of literature. J. Financ. Crime **27**(3), 945–958 (2020)
7. Bajwa, I.A., Ahmad, S., Mahmud, M., Bajwa, F.A.: The impact of cyberattacks awareness on customers' trust and commitment: an empirical evidence from the Pakistani banking sector. Inf. Comput. Secur. **31**(5), 635–654 (2023)
8. Creado, Y., Ramteke, V.: Active cyber defence strategies and techniques for banks and financial institutions. J. Financ. Crime **27**(3), 771–780 (2020)
9. Smikle, L.: The impact of cybersecurity on the financial sector in Jamaica. J. Financ. Crime **30**(1), 86–96 (2023)
10. Liu, H., Tang, T., Luo, J., Zhao, M., Zheng, B., Wu, Y.: An anomaly detection method based on double encoder–decoder generative adversarial networks. Ind. Robot. **48**(5), 643–648 (2021)
11. Bansal, K., Paliwal, A.C., Singh, A.K.: Analysis of the benefits of artificial intelligence and human personality study on online fraud detection. Int. J. Law Manag. (2024)
12. Odei-Appiah, S., Wiredu, G., Adjei, J.K.: Fintech use, digital divide and financial inclusion. Digit. Policy Regul. Governance **24**(5), 435–448 (2022)
13. Omoge, A.P., Gala, P., Horky, A.: Disruptive technology and AI in the banking industry of an emerging market. Int. J. Bank Mark. **40**(6), 1217–1247 (2022)
14. Sandner, P., Gross, J.: The digital Euro from a geopolitical perspective: will Europe lag behind? In: Kim, S.-J. (ed.) Fintech, Pandemic, and the Financial System: Challenges and Opportunities (International Finance Review, Vol. 22), pp. 223–240. Emerald Publishing Limited, Leeds (2023)

15. Cole, T.: How are financial institutions enabling online fraud? A developmental online financial fraud policy review. J. Financ. Crime **30**(6), 1458–1473 (2023)
16. Kapoor, B., Pandya, P., Sherif, J.S.: Cryptography: a security pillar of privacy, integrity and authenticity of data communication. Kybernetes **40**(9/10), 1422–1439 (2011)
17. Singh, S., Sahni, M.M., Kovid, R.K.: What drives FinTech adoption? A multi-method evaluation using an adapted technology acceptance model. Manag. Decis. **58**(8), 1675–1697 (2020)
18. Tian, X., He, J.S., Han, M.: Data-driven approaches in FinTech: a survey. Inf. Discov. Deliv. **49**(2), 123–135 (2021)
19. Kosutic, D., Pigni, F.: Cybersecurity: investing for competitive outcomes. J. Bus. Strat. **43**(1), 28–36 (2021)
20. Bechara, F.R., Schuch, S.B.: Cybersecurity and global regulatory challenges. J. Financ. Crime **28**(2), 359–374 (2021)
21. Al-Hawamleh, A.M.: Investigating the multifaceted dynamics of cybersecurity practices and their impact on the quality of e-government services: evidence from the KSA. Digit. Policy Regul. Governance **26**(3), 317–336 (2024)
22. MMR.: Evolving cyber attacks on digital payments. Maximize Market Research (2023)
23. Statista.: Number of user accounts exposed worldwide from 1st quarter 2020 to 4th quarter 2023, Cyber Crime & Security (2023)
24. Khaw, T.Y., Amran, A., Teoh, A.P.: Building a thematic framework of cybersecurity: a systematic literature review approach. J. Syst. Inf. Technol. **26**(2), 234–256 (2024)
25. Singh, C.: Artificial intelligence and deep learning: considerations for financial institutions for compliance with the regulatory burden in the United Kingdom. J. Financ. Crime **31**(2), 259–266 (2024)
26. Crosby, M., et al.: Blockchain technology: beyond bitcoin. Appl. Innov. Rev. **2**, 71 (2016)
27. Toapanta, F., et al.: The effect of blockchain technology on the payment system: a systematic review. J. Paym. Syst. Strat. **13**(1), 45–58 (2019)
28. Schmitz, J., Leoni, G.: Accounting and auditing at the time of blockchain technology: a research agenda. Aust. Account. Rev. **29**(2), 331–342 (2019)
29. Yang, D., et al.: Smart contracts: architecture, applications, and future directions. J. Syst. Softw. **162**, 110761 (2020)
30. Dencik, J., Goehring, B., Marshall, A.: Managing the emerging role of generative AI in next-generation business. Strategy Leadersh. **51**(6), 30–36 (2023)
31. Lin, H., et al.: Regulatory compliance in digital payment systems: challenges and strategies. J. Financ. Regul. Compliance **27**(1), 76–89 (2019)
32. Nguyen, T., et al.: User education and cybersecurity: an empirical analysis. J. Financ. Crime **27**(2), 467–482 (2020)
33. Smith, A.: Technical challenges in digital payment systems: a review. J. Inf. Technol. Polit. **17**(4), 385–403 (2020)
34. European Commission.: 2015/2366 Payment services directive (PSD2). Official Journal of the European Union. Directorate-General for Financial Stability, Financial Services and Capital Markets Union (2015)
35. Kokkola, T.: The SEPA Project: Technical and Legal Aspects (2010)
36. ECB.: Single Euro Payments Area (SEPA). European Central Bank (2014)
37. EPC.: SEPA Implementation Guidelines. European Payments Council AISBL (2019)
38. Zetzsche, D.A., Buckley, R.P., Arner, D.W., Barberis, J.N.: From FinTech to TechFin: the regulatory challenges of data-driven finance. N. Y. Univ. J. Law Bus. **14**, 393–446 (2017)
39. Gomber, P., Koch, J.-A., Siering, M.: Digital finance and FinTech: current research and future research directions. J. Bus. Econ. **87**(5), 537–580 (2017)
40. FCA.: Regulatory sandbox. Financial Conduct Authority (2015)
41. Peters, G.W., Panayi, E.: Understanding modern banking ledgers through blockchain technologies: future of transaction processing and smart contracts on the internet of money. In: Tasca, P., Aste, T., Pelizzon, L., Perony, N. (eds.) Banking Beyond Banks and Money: A Guide to Banking Services in the Twenty-First Century, pp. 239–278. Springer (2016)

42. Foley, S., Karlsen, J.R., Putniņš, T.J.: Sex, drugs, and bitcoin: how much illegal activity is financed through cryptocurrencies? Rev. Financ. Stud. **32**(5), 1798–1853 (2019)
43. Demirguc-Kunt, A., et al.: The Global Findex Database 2017: Measuring Financial Inclusion and the Fintech Revolution. World Bank (2018)
44. Carney, M.: The Promise and Perils of FinTech. Bank of England (2017)
45. Arner, D.W., Barberis, J., Buckley, R.P.: The evolution of Fintech: a new post-crisis paradigm? Georg. J. Int. Law **47**, 1271–1319 (2016)
46. Anagnostopoulos, I.: Fintech and RegTech: impact on regulators and banks. J. Econ. Bus. **100**, 7–10 (2018)
47. Arnold, B.: The cost of compliance: regulatory challenges in the financial sector. J. Account. Public Policy **35**(2), 175–191 (2016)
48. Gandhi, P.: RegTech and the future of financial regulation. J. Oper. Inst. Compliance **11**(3), 213–226 (2020)
49. Philippon, T.: The FinTech opportunity. NBER Working Paper No. 22476 (2016)
50. Zavolokina, L., Dolata, M., Schwabe, G.: The fintech phenomenon: antecedents of financial innovation perceived by the popular press. Financ. Innov. **3**(26) (2017)
51. European Commission.: Overview of digital finance. Directorate-General for Financial Stability, Financial Services and Capital Markets Unions (2018)

Text Summarization: An Application of Generative AI

Tapan Kumar Das and Arati Mohapatro

Abstract This chapter explores the application of Generative AI in text summarization, a technique used to condense long pieces of text into concise, informative summaries. The primary goal is to create summaries that retain the key points of the original document while being coherent and contextually relevant. The chapter highlights two main approaches to text summarization: extractive and abstractive. Extractive summarization involves selecting key sentences directly from the source text, whereas abstractive summarization generates new sentences that convey the essential information. The chapter provides a comparative analysis of these two methods, evaluating their effectiveness in different contexts. With the advent of generative AI powered by deep learning models, such as recurrent neural networks (RNNs) and convolutional neural networks (CNNs), the field of text summarization has seen significant advancements. These AI-driven approaches offer enhanced capabilities for producing high-quality summaries that can significantly improve information processing workflows across various industries. The chapter concludes by discussing the potential of generative AI in transforming the landscape of automated text summarization and its implications for future research and practical applications.

Keywords Generative AI · Extractive text summarization · Abstractive text summarization · LSTM · NLP

T. K. Das (✉)
School of Computer Science Engineering and Information Systems, Vellore Institute of Technology, Vellore 632014, India
e-mail: tapan.das@vit.ac.in

A. Mohapatro
Department of Computer Science, Indian Academy Degree College (Autonomous), Bengaluru 560043, India

N. R. Vajjhala et al. (eds.), *Generative Artificial Intelligence (AI) Approaches for Industrial Applications*, Information Systems Engineering and Management 24,
https://doi.org/10.1007/978-3-031-76710-4_16

1 Introduction

The enormous amount of content on the internet and archives grows day-by-day. To have information about a specific event/news, it requires a lot of effort to read and comprehend. Reading vast amounts of text information is also time-consuming. Moreover, there are many repeated and redundant information in between. To avoid wasting effort, summaries are getting preferred in common daily [1]. However, manual summarization and briefings consume a lot of time and effort. Hence it does not serve the purpose as summarization result must be readily available. In recent times, generative AI, powered by deep learning models, offers a competitive solution for automatically summarizing the text. These models leverage neural network architectures, such as recurrent neural networks (RNNs), convolutional neural networks (CNNs), or transformer models, to learn intricate patterns and structures within textual data. Generative AI models are trained with a vast corpora of text so that they can understand language semantics, syntax, and context. Generative AI tools based on highly parameterized large language model (LLM) can be used to address the issue of automatic text summarization.

Automatic text summarization (ATS) is a fascinating technique within the field of natural language processing (NLP) comes into play. It's a method of generating a quick and succinct quantity of text from multiple text resources like books, news articles, blog posts, analysis papers, emails, etc. The ATS systems can be categorized into either (i) single-document summarization system where a single summary is generated for a single document or (ii) multi-document summarization system where a single summary is created for multiple documents. To automatically generate summaries, three main techniques are generally employed:

(i) Extractive technique: Important sentences from a document are chosen and combined to generate a final summary [2].
(ii) Abstractive technique: In this approach, rather than selecting meaningful sentences directly, it generates new sentences that convey the same information using NLP algorithms
(iii) Hybrid approach: This approach combines both the abstractive and the extractive approaches to generate the summary.

Abstractive summarization relies on deep learning [3], in this approach a trained model is used to predict the summary of the text provided. Extractive summarization evaluates important keywords in the text [4]. The summary is produced using the words present in the text itself. This chapter is intended to compare different aspects of these two methods and select the better approach for the application. The contributions of the chapter are listed below:

- Elucidating major ATS approaches and their advantages.
- Providing step-by-step implementation of extractive summarization and abstractive summarization techniques over a dataset.
- Evaluating the performances of each technique.
- Discussing future horizons and research directions.

The rest of the chapter is organized as follows: Sect. 2 throws light on literature pertaining to the text summarization. Extractive summarization approach and its implementation is explained in Sect. 3. Section 4 describes abstractive summarization approach. Result analysis of both the approaches is carried out in Sect. 5. The chapter is concluded in Sect. 6 with highlighting future improvements.

2 Literature Survey

In recent years, machine learning and deep learning techniques are extensively used for numerous NLP applications [5]. ATS is an important NLP task. Few of the deep learning techniques including Recurrent Neural Network (RNN), Gated Recurrent Unit (GRU), Long Short-Term Memory (LSTM), and pre-trained models (BERT) are quite efficient in condensing the text [6]. A study of automatic text summarization techniques for Indian and foreign languages is carried out [7], the authors employed extractive and abstractive summarization techniques. They have tested it for Indian languages including Hindi, Punjabi and Bengali. However, it is concluded that a single strategy for different language contents may not be efficient.

Another literature [8] about text summarization investigates the differences between different algorithms for summarization. This work also compares the results to human-made summaries and the accuracy achieved. They used the PSO method, Genetic probabilistic based summarization. They reveal extractive approaches faster, but abstractive methods give better results. The abstractive approaches perform detailed language-based analysis of the text data and generate a summary like a summary generated by humans. So, they outperform extractive approaches but are more computationally expensive. Text summarization using particle swarm optimization (PSO) was carried out [9], however, model produces summaries that are 43% similar to the human-generated summaries. Another summarization method based on the features selected from the text was studied by Abuobieda et al. [10]. Mokhale (2019) presented a study on multi-document summarization, they used Hidden Markov Model for this work [11].

Meena (2015) presented research on extractive summarization based on feature priorities and they noted that specific combinations of words or vectors need to be given priority depending on the type of document to make it more efficient and optimized [12]. Different evolutionary algorithms for extractive automatic text summarization employing genetic algorithm (GA), PSO are carried out and it is revealed that usage of GA increased the performance [13]. A work related to variations of the similarity function of text rank for automated summarization concluded that cosine similarity improved the performance [14]. Ercan et al. used lexical chains for keyword extraction and concluded that the lexical chain features improve the precision significantly [15].

In the vector space model, cosine is commonly helpful in calculating the similarity between two different vectors. Its calculation is done efficiently, to remove any irregularities especially for sparse vectors, as only the dimensions which have some value need to be considered. It has been applied in solving different kinds of complex text mining problems, including text differentiation, text summarization, information retrieval, chat bots. Li (2013) achieved a similarity score of 0.875 for the case problem discussed in their paper [16].

3 Extractive Summarization

The detailed methodology of extractive summarization is presented in Fig. 1.

The methodology includes following steps:

3.1 Web Scraping

Web scraping is the process of gathering information from the Internet [17]. The web application will be able to enter a website link into the input box. The backend recognizes the link using a variable, say regex.

- A regex is given for checking the format of a website link.
- The regex is compiled and stored into an object "p".
- The text stored in text_str is then checked with the regex compiled. An URL opening request is passed to the given link after it is verified. The request is stored in a scraped_data object.
- All the html code of the website is read and stored into a parsed_article object.
- All the paragraph tags are stored into an array named paragraphs.
- For each entry in array paragraphs, the data is appended to a text string articles_ text.
- Text cleaning is done using regex to get rid of unwanted numbers, tags etc.
- Cleaned text is printed and ready to be sent to the text_summarisation module.

3.2 Porter Stemming

Stemming is the method of producing completely different variants of the same original word [18], e.g. "wait", "waiting", "waited", "waits" reduced to root word "wait" Stemming is a crucial part of the pipelining method in natural language processing especially in the use of summarization and efficient morphological conversion. The input to the stemmer algorithm is tokenized words.

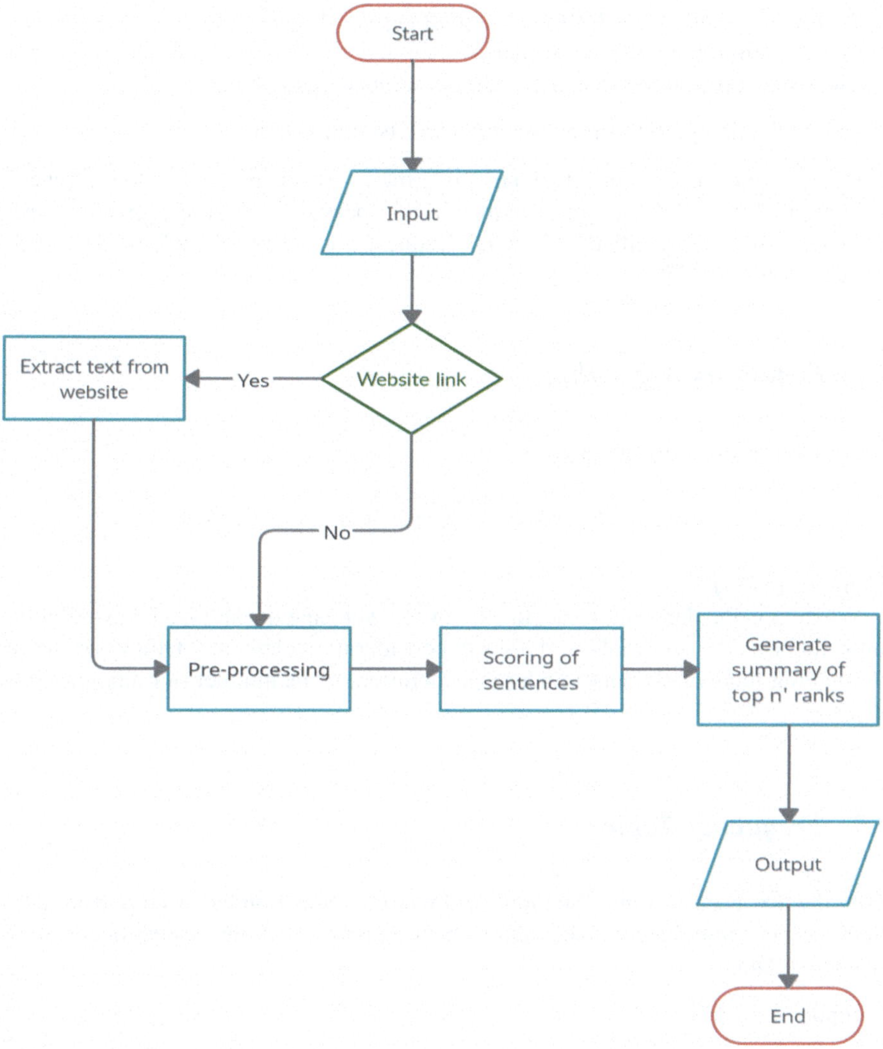

Fig. 1 Extractive summarization process flowchart

3.3 Tokenization

Tokenization refers to ripping a sentence/group of words into an array of words or smaller sentences [19]. This can be useful once an oversized size of text is given and that we wish to control text at a smaller scale. In this chapter, we tend to employ a word frequency table, thus tokenization is needed for enumeration every word singly. This can be explained by following example:"

Sentence = "'The text hold on in text_str is then checked with

the regex compiled. An URL gap request is passed
to the given link when it's verified.
The request is hold on in an exceedingly scraped_data object'"

Upon tokenizing, the following output will be generated:

['The', 'text', 'hold', 'in', 'text_str', 'is', 'then', 'checked', 'with', 'the', 'regex', 'compiled.An', 'URL', 'gap', 'request', 'is', 'passed', 'to', 'the', 'given', 'link', 'when', 'it', 'is', 'verified', '.', 'The', 'request', 'is', 'stored', 'in', 'a', 'scraped_data', 'object', '.']

3.4 PageRank Algorithm

The weight of ith webpage is given by

$$S(Vi) = (1 - d) + d * \sum j(1/|Out(Vi)|) * S(Vj)$$

where j $\in$ in (Vi)

Where d is the damping issue, in case of the webpage doesn't have any outgoing links. In(Vi) is the set of inbound links of consideration element *i*. Out(Vj) is the set of outgoing links of element j under consideration, is the number of outbound links of j.

3.5 Frequency Table

After the text is passed into the input, and word tokenization has been performed, a dictionary is maintained which contains the frequency of all the words/tokens of the given text data.

Input: text_str = ""

Text summarization refers to the technique of converting long pieces of text into shorter ones. The intention is to create a brief and fluent summary having the main points outlined in the text document. In this study, a comparison will be made between different text summarization techniques. The most efficient technique will be used finally. The project will be made in the form of a web app. The backend framework used will be Django. In today's world everything must be economical and quick, reading giant amounts of text information takes long time. To avoid wasting effort, summaries are getting common now-a-days. Text summarization is amongst the foremost difficult and attention-grabbing issues within the field of NLP. It's a method of generating a cryptic and outline of text from multiple text resources like books, news articles, blog posts, analysis papers, emails, and tweets. The demand for automatic text summarization systems is spiking of late because of the provision of enormous amounts of material information. The goal of this study is designing a web app that can produce summaries straight away.

The output is shown in the following Table 1.

Table 1 Word-frequency table depicting the number of occurrences of the word

Word	Frequency	Word	Frequency	Word	Frequency
text	9	summar	4	one	2
refer	1	techniqu	3		12
convert	1	long	1	intent	1
piec	1	shorter	1	creat	1
brief	1	fluent	1	summari	4
main	1	point	1	outlin	1
document	1	thi	2	project	4
,	8	comparison	1	made	2
differ	1	effici	3	use	2
final	1	form	1	web	2
app	2	backend	1	framework	1
django	1	today	1	‘	1
world	1	everyth	1	need	1
fast	2	read	1	larg	2
amount	2	data	2	time-consum	1
save	1	effort	1	becom	1
common	1	everi	1	day	2
automat	2	challeng	1	interest	1
problem	1	field	1	natur	1
langaug	1	process	2	(	1
nlp	1	)	1	gener	1
concis	1	meaning	1	multipl	1
resourc	1	book	1	news	1
articl	1	blog	1	post	1
research	1	paper	1	email	1
tweet	1	demand	1	system	1
spike	1	becaus	1	avail	1
textual	1	aim	1	make	1
produc	1	veri	1	–	–

3.6 Sentence Tokenization

Consider the following example

['\nText summarization refers to the technique of converting long pieces of text into shorter ones.', 'The intention is to create a brief and fluent summary having the main points outlined in the text document.', 'In this project, a comparison will be made between different text summarization techniques.', 'The most efficient technique will be used in the final project.', 'The project will be made in the form of a web app.', 'The backend framework used will be Django.', 'In today's world once everything must be economical and quick, reading giant amounts of text information are often long.',' to avoid wasting effort, summaries are getting additional common a day. '.Automatic Text account is one amongst the foremost difficult and attention-grabbing issues within the field of tongue process (NLP).' It's a method of generating a cryptic and meaty outline of text from multiple text resources like books, news articles, blog posts, analysis papers, emails, and tweets. 'The demand for automatic text account systems is spiking of late because of the provision of enormous amounts of matter information.', 'This project will aim at making a web app that can produce summaries very fast and efficiently.']

Table 2 shows the sentence tokens extracted from the above paragraph.

Table 2 Sentence tokens extracted from the input paragraph

Sentence token	Sentence token
Text summarization refers to the technique of converting long pieces of text into shorter ones	The intention is to create a brief and fluent summary having the main points outlined in the text document
In this project, a comparison will be made between different text summarization techniques	The most efficient technique will be used in the final project
The project will be made in the form of a web app	The backend framework used will be Django
In today's world once everything must be economical and quick, reading giant amounts of text information are often long	To save effort, summaries are becoming more common every day
Automatic Text account is one amongst the foremost difficult and attention-grabbing issues within the field of tongue process (NLP)	It's a method of generating a cryptic and meaty outline of text from multiple text resources like books, news articles, blog posts, analysis papers, emails, and tweets
The demand for automatic text account systems is spiking of late because of the provision of enormous amounts of matter information	This project will aim at making a web app that can produce summaries very fast and efficiently

Table 3 Calculated scores of the sentences

Sentence	Score
Text summ	3.545
The intent	3.0
In this pr	4.545
The most e	4.166
The projec	3.833
The backen	3.4
In today's	3.0
To save ef	3.777
Automatic	2.75
It is a pr	2.578
The demand	3.066
This proje	3.363

3.7 Graph Module

The words of the text are represented as nodes in the graph. The nodes are connected to nearby words found in the text provided. More number of edges to a node means higher frequency of the word implying a more important word. The nodes are ranked by their importance in the final step. Top N ranked words/nodes are used for the summary.

3.8 Sentence Scoring

The output is shown in Table 3.

3.9 Generating Summary

From Table 3, the average score is calculated.

Top sentences with respective scores are given below:

4.545: In this project, a comparison will be made between different text summarization techniques.
4.166: The most efficient technique will be used in the final project.

Hence the final summary is computed as below:

In this project, a comparison will be made between different text summarization techniques. The most efficient technique will be used in the final project.

4 Abstractive Summarization

The detailed experimental process of abstractive summarization is represented in the Fig. 2.

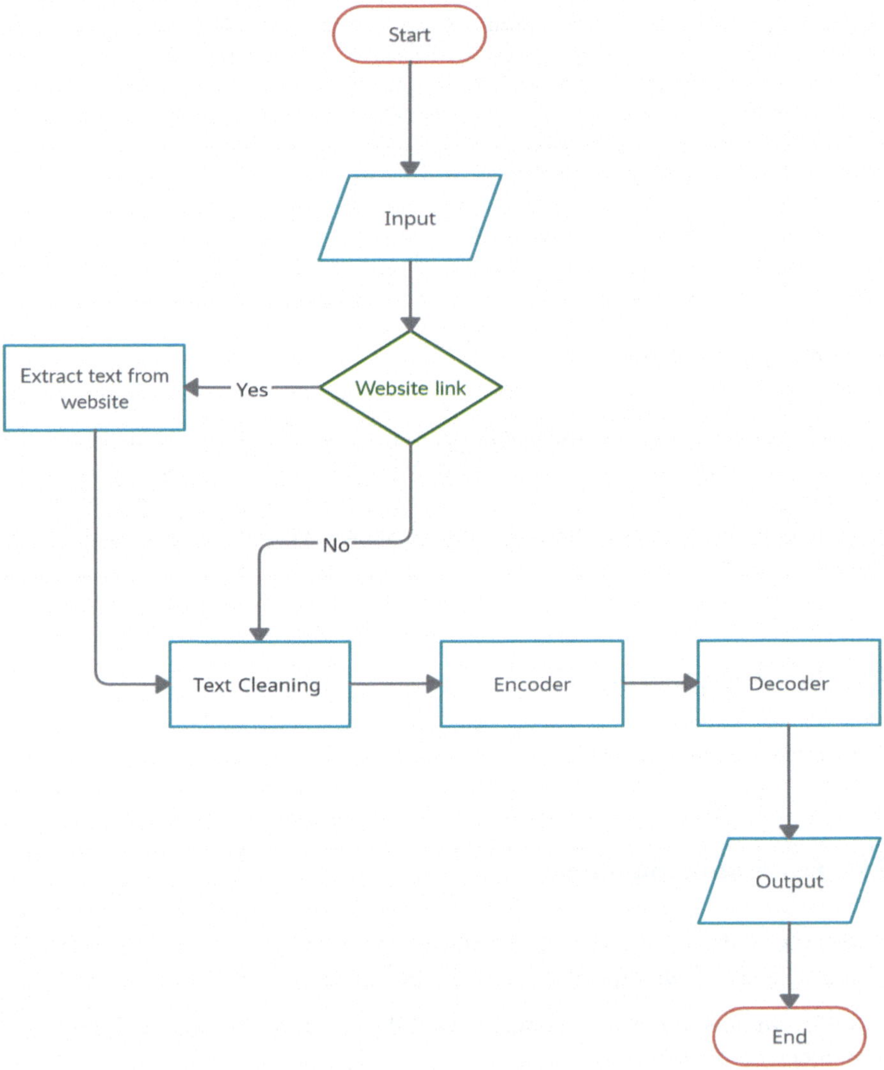

Fig. 2 Detailed abstractive summarization process

4.1 *Sequence-to-Sequence (Seq2Seq) Modeling*

Our objective is to design a text summarizer for an input of long sequence of words and the output may be a short outline. So, we are able to model this as a many-to-many Seq-2-Seq downside. The encoder-decoder design is especially accustomed for solving the sequence-to-sequence (Seq-2-Seq) design issues wherever the input and output sequences are of various lengths. Below we outlined the step- wise algorithm of the Sequence-to-Sequence (Seq-2-Seq) modeling.

Encoder

An encoder LSTM reads the complete input sequence. The data is then processed, at each step and record information within the input order and it is represented in Fig. 3.

Decoder

The decoder may be a Long Short-Term Memory (LSTM) model that reads the complete target sequence word-by-word or unit-by-unit and predicts a similar sequence offset by one step. The decoder is trained to predict ensuing word within the sequence given the previous word. A decoder architecture is shown in Fig. 4.

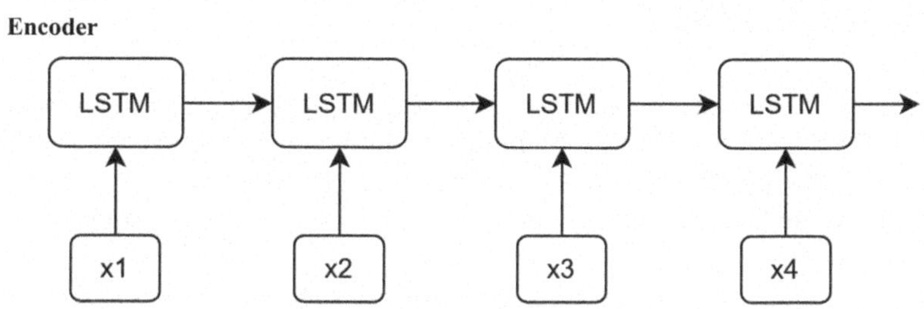

Fig. 3 A schematic encoder architecture

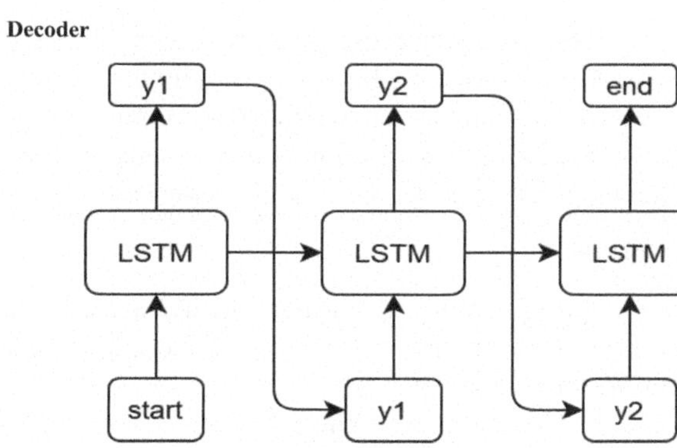

Fig. 4 The architecture of a decoder

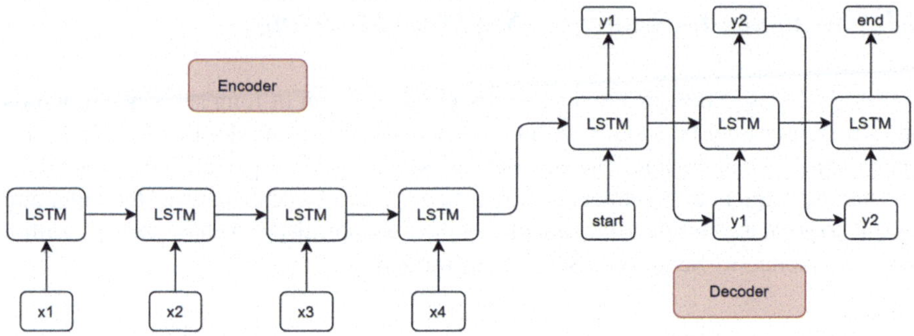

Fig. 5 Inference phase

Inference Phase

After training, the model is saved offline and then is tested on new source sequences for which the target sequence is not known to check if it is working properly. The trained model with combined encoder and decoder is represented in Fig. 5.

4.2 Model Building

For experiment, the data is collected from Kaggle, text are cleaned by removing HTML tags and ('s) from words (stemming), eliminating punctuations and special characters (including @, $, %, & etc.), removing stop words etc.

The LSTM model architecture with layer details is shown in Fig. 6 and it is visually represented in Fig. 7.

Model Layers

Embedding Layer

Keras offers an Embedding layer which is very helpful for working with neural networks on textual data. It uses various input parameters which have been discussed below. The parameter demands that the input data provided should be encoded in integer, so that each word (embedding) is matched to a unique integer. This makes data manipulation easier for neural networks.

Input layer

Input function is used for starting a neural network tensor. A neural network tensor represents a tensor-like object. It can take multiple parameters to run properly. Below mentioned are some of them.

Shape: This parameter known as the shape tuple. Elements of this input vector can be None; 'None' elements indicate that the shape is not known. It is dynamic in nature.

Layer (type)	Output Shape	Param #	Connected to
input_1 (InputLayer)	[(None, 80)]	0	
embedding (Embedding)	(None, 80, 500)	14533000	input_1[0][0]
lstm (LSTM)	[(None, 80, 500), (N	2002000	embedding[0][0]
input_2 (InputLayer)	[(None, None)]	0	
lstm_1 (LSTM)	[(None, 80, 500), (N	2002000	lstm[0][0]
embedding_1 (Embedding)	(None, None, 500)	3877000	input_2[0][0]
lstm_2 (LSTM)	[(None, 80, 500), (N	2002000	lstm_1[0][0]
lstm_3 (LSTM)	[(None, None, 500),	2002000	embedding_1[0][0] lstm_2[0][1] lstm_2[0][2]
attention_layer (AttentionLayer	((None, None, 500),	500500	lstm_2[0][0] lstm_3[0][0]
concat_layer (Concatenate)	(None, None, 1000)	0	lstm_3[0][0] attention_layer[0][0]
time_distributed (TimeDistribut	(None, None, 7754)	7761754	concat_layer[0][0]

Total params: 34,680,254
Trainable params: 34,680,254
Non-trainable params: 0

Fig. 6 LSTM model architecture

Batch size: static batch size is an integer. It is an optional parameter.

LSTM layer

We are using LSTM because of its longer memory property compared to normal recurrent neural networks. Our data is of textual nature due to which words coming earlier in the sequence of input text lose their value over time and may not play a major role in the final output. LSTM solves this problem by not losing the value/effect of earlier text inputs. Hence the LSTM layer is used in this project.

Attention layer

Attention mechanism gives weights to different words in a sentence based on their importance in the meaning. This helps produce better summaries as the converted text will have more important words in the final summary. The attention mechanism is depicted in Fig. 8.

Concatenation layer

The concatenation layer takes input as a list of keras tensors and returns a single tensor as output to the next layer. In this model architecture also, concate layer takes input from attention and LSTM layer provide the output to the next layer.

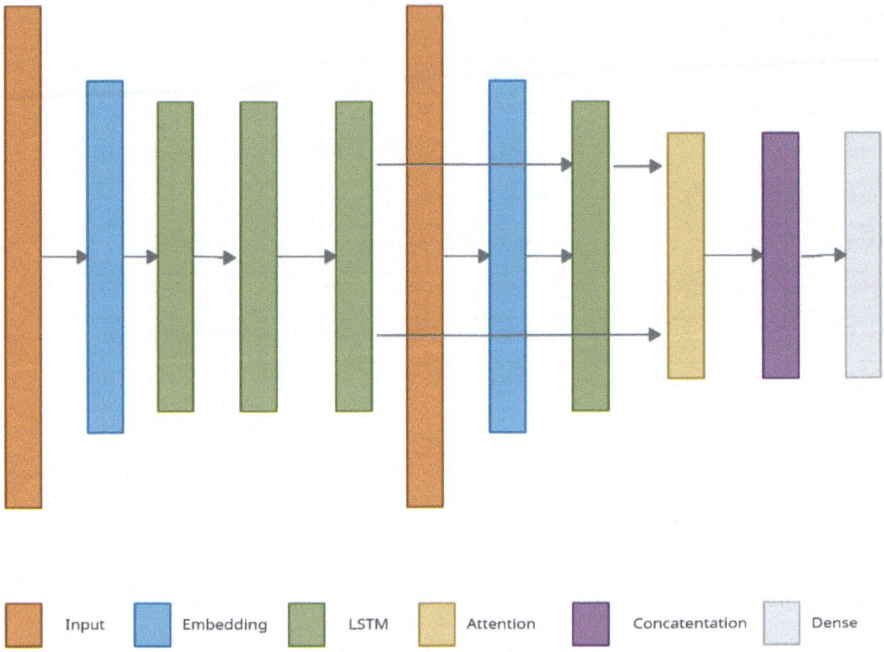

Fig. 7 Model architecture visualization

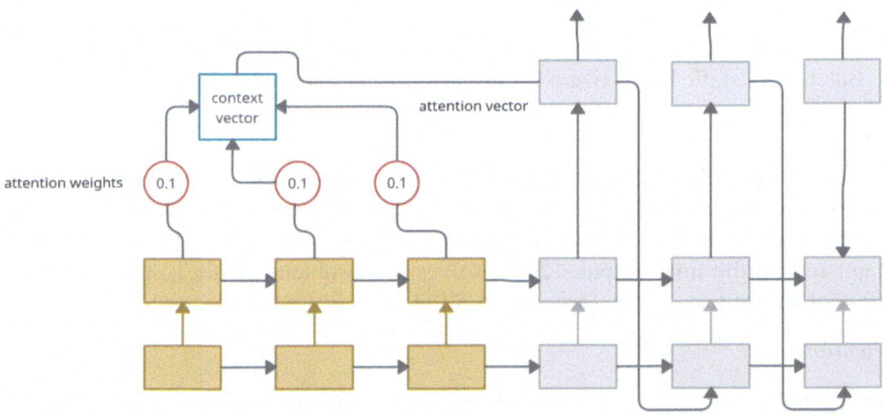

Fig. 8 Attention mechanism

4.3 Testing Various Configurations

A test was conducted for different number of batch size and epochs during the training phase. The model was trained for 10 K reviews to speed up the process and get an overall idea of the best configuration. After observing different graphs, it was concluded that 50 epochs performs the best in our case. Batch size was chosen as 512 because smaller size slows down the process considerably.

5 Results and Discussion

Generic evaluation criteria such as Precision, Recall and F-score are used for assessing the performance of the summarization approaches. Precision provides an information about how many sentences are common in references and candidate summary divided by number of sentences in candidate summary. However, recall is division of number of sentences common in references and candidate summary with number of sentences in references summary. The F-score is calculated from Eq. (1).

$$F-score = (2 * Precision * Recall)/(Precision + Recall) \quad (1)$$

ROUGE (Recall-Oriented Understudy for Gisting Evaluation) [20].

It is the most widely used evaluation metrics for text summarization. ROUGE is largely a recall-oriented evaluator that works by examination of variety of machine generated words/phrases that square measure an area of the reference sentence with relation to the overall number of words within the reference sentence. ROUGE has four types, i.e. ROUGE-N, ROUGE-L, ROUGE-W and ROUGE-S [21].

BLEU (Bilingual Evaluation Understudy)

The BLEU evaluation compares consecutive words of the automated translation with the consecutive words it finds within the reference conversion, and counts the number of matches, with weights associated. These matches aren't smitten by the position. The next match worth symbolizes the next degree of similarity with the reference translation, and better score. Comprehensibility and grammatical factors aren't taken under consideration for the analysis method.

5.1 *Evaluation for Abstractive Summary*

We have exhibited the results of abstractive summarization in Table 4. Recall and F-score of articles number 1 … 10 are provided in the table.

5.2 *Evaluation for Extractive Summary*

We have presented the results of extractive summarization in Table 5. Recall and F-score of articles 1 … 10 are provided in the table.

Table 4 Experimental results of abstractive summarization approach

Article number	Recall	F-score
1 (compiled)	0.12	0.05
2	0	0
3	0.5	0.3
4	0.25	0.15
5	0	0
6	0	0
7	0.25	0.15
8	0	0
9	0	0
10	0	0

Table 5 Experimental results of extractive summarization process

Article number	Recall	F-score
1	0.41	0.58
2	0.49	0.65
3	0.51	0.67
4	0.511	0.67
5	0.56	0.71
6	0.54	0.7
7	0.7	0.82
8	1	1
9	0.43	0.6
10	0.47	0.63

We compare the efficiency of both the approaches. The better approach can be used to implement in a web application. We used the ROUGE metric to evaluate the performance of both the approaches. Clearly, the average score of extractive approach is greater than the abstractive approach. Abstractive approach performs poorly due to the internal algorithm of ROUGE metrics. These evaluation metrics work on similarity principle in which same words help increase the score. Extractive method uses the same words from the text and gives out the summary of most important keywords. Due to this, the F- score is more than the abstractive approach which may use other words.

In spite of the wide acceptance of ROUGE as a standard for measuring the accuracy of a summarization model, it has the limitation that it only matches strings between the summaries without considering the meaning in single words or series of words. Therefore, in order to address the meaning issue, evaluation methods such as Basic Elements (BE) [22] and DEPEVAL (summ) [23] have been proposed.

6 Conclusion

The demand for automatic text summary is spiking due to the provision of huge amounts of information. To save time, complexity of the literature, and to highlight the main theme of the text, text summarization has become a necessity in the sphere of knowledge workers. In this chapter, a comparison is carried out between two different approaches of ATS, and it is found that extractive summarization is a good choice. In future, we plan to create an online app which will generate summaries readily and expeditiously.

Extractive summarization is a bit easier and according to the current trend, extractive summaries are mostly prevalent. Abstractive summarization is not mature enough and users are skeptical on using this due to its black box nature of functioning. Hence future research could be more focused towards streamlining abstractive summarization approach and even hybridizing extractive and abstractive approaches for more accurate summarization. Furthermore, the following topics are complex in nature and require more research:

- Semi-structured text summarization
- Multi-document text summarization
- Determining the most appropriate features pertaining to documents to be summarized.

Moreover, text summarization studies specific to applications such as extracting summaries from legal documents, summarizing tourist attractions, scientific papers summarization, tweet summarization, opinion summarization and news summarization will more interesting and appealing.

References

1. El-Kassas, W.S., Salama, C.R., Rafea, A.A., Mohamed, H.K.: Automatic text summarization: a comprehensive survey. Expert Syst. Appl. **165**, 113679 (2021)
2. Widyassari, A.P., Rustad, S., Shidik, G.F., Noersasongko, E., Syukur, A., Affandy, A.: Review of automatic text summarization techniques and methods. J. King Saud Univ. Comput. Inf. Sci. **34**(4), 1029–1046 (2022)
3. Zhang, M., Zhou, G., Yu, W., Huang, N., Liu, W.: A comprehensive survey of abstractive text summarization based on deep learning. Comput. Intell. Neurosci. **2022**(1), 7132226 (2022)
4. Abu Nada, A.M., Alajrami, E., Al-Saqqa, A.A., Abu-Naser, S.S.: Arabic text summarization using Aribert model using extractive text summarization approach (2020)
5. Roy, P.K., Tripathy, A.K., Das, T.K., Gao, X.Z.: A framework for hate speech detection using deep convolutional neural network. IEEE Access **8**, 204951–204962 (2020)
6. Reang, R., Dehalwar, V., Pateriya, R.K.: Deep learning techniques for automatic text summarization: a review. In: 2024 IEEE International Students' Conference on Electrical, Electronics and Computer Science (SCEECS), pp. 1–6. IEEE (2024)
7. Shah, P., Desai, N.P.: A survey of automatic text summarization techniques for Indian and foreign languages. In: 2016 International Conference on Electrical, Electronics, and Optimization Techniques (ICEEOT), pp. 4598–4601. IEEE (2016)

8. Dalal, V., Malik, L.: A survey of extractive and abstractive text summarization techniques. In: 2013 6th International Conference on Emerging Trends in Engineering and Technology, pp. 109–110. IEEE (2013)
9. Binwahlan, M.S., Salim, N., Suanmali, L.: Swarm based text summarization. In: 2009 International Association of Computer Science and Information Technology-Spring Conference, pp. 145–150. IEEE (2009)
10. Abuobieda, A., Salim, N., Albaham, A.T., Osman, A.H., Kumar, Y.J.: Text summarization features selection method using pseudo genetic-based model. In: 2012 International Conference on Information Retrieval & Knowledge Management, pp. 193–197. IEEE (2012)
11. Mokhale, S.V., Dhopawkar, G.M.: A study on different multi-document summarization techniques. In: 2019 3rd International Conference on Inventive Systems and Control (ICISC), pp. 710–713. IEEE (2019)
12. Meena, Y.K., Gopalani, D.: Feature priority based sentence filtering method for extractive automatic text summarization. Procedia Comput. Sci. **48**, 728–734 (2015)
13. Meena, Y.K., Gopalani, D.: Evolutionary algorithms for extractive automatic text summarization. Procedia Comput. Sci. **48**, 244–249 (2015)
14. Barrios, F., López, F., Argerich, L., Wachenchauzer, R.: Variations of the similarity function of textrank for automated summarization. arXiv preprint arXiv:1602.03606 (2016)
15. Ercan, G., Cicekli, I.: Using lexical chains for keyword extraction. Inf. Process. Manage. **43**(6), 1705–1714 (2007)
16. Li, B., Han, L.: Distance weighted cosine similarity measure for text classification. In: International Conference on Intelligent Data Engineering and Automated Learning, pp. 611–618. Springer, Berlin, Heidelberg (2013)
17. Singrodia, V., Mitra, A., Paul, S.: A review on web scrapping and its applications. In: 2019 International Conference on Computer Communication and Informatics (ICCCI), pp. 1–6. IEEE (2019)
18. Jivani, A.G.: A comparative study of stemming algorithms. Int. J. Comp. Tech. Appl **2**(6), 1930–1938 (2011)
19. Mohapatra, S.K., Prasad, S., Bebarta, D.K., Das, T.K., Srinivasan, K., Hu, Y.C.: Automatic hate speech detection in English-odia code mixed social media data using machine learning techniques. Appl. Sci. **11**(18), 8575 (2021)
20. Rennard, V., Shang, G., Hunter, J., Vazirgiannis, M.: Abstractive meeting summarization: a survey. Trans. Assoc. Comput. Linguist. **11**, 861–884 (2023)
21. Yadav, D., Katna, R., Yadav, A.K., Morato, J.: Feature based automatic text summarization methods: a comprehensive state-of-the-art survey. IEEE Access **10**, 133981–134003 (2022)
22. Hovy, E.H., Lin, C.Y., Zhou, L., Fukumoto, J.: Automated summarization evaluation with basic elements. In: LREC, vol. 6, pp. 604–611 (2006)
23. Owczarzak, K.: Depeval (summ): dependency-based evaluation for automatic summaries. In: Proceedings of the Joint Conference of the 47th Annual Meeting of the ACL and the 4th International Joint Conference on Natural Language Processing of the AFNLP, pp. 190–198 (2009)

The manufacturer's authorised representative in the EU is Springer Nature Customer Service Centre GmbH, Europaplatz 3, 69115 Heidelberg, Germany. If you have any concerns regarding our products, please contact ProductSafety@springernature.com

Printed and bound by CPI Group (UK) Ltd, Croydon, CR0 4YY

30/09/2026

02225600-0011